Exploring the Moon
The Apollo Expeditions (Second Edition)

David M. Harland

Exploring the Moon

The Apollo Expeditions
(Second Edition)

 Springer

Published in association with
Praxis Publishing
Chichester, UK

David M. Harland
Space Historian,
Kelvinbridge,
Glasgow, UK

SPRINGER-PRAXIS BOOKS IN SPACE EXPLORATION
SERIES EDITOR: John Mason B.Sc., M.Sc., Ph.D.

ISBN 978-0-387-74638-8 Springer-Verlag Berlin Heidelberg New York

Springer is a part of Springer Science + Business Media (*springer.com*)

Library of Congress Control Number: 2007939116

Cover design: Jim Wilkie
Typesetting: BookEns Ltd, Royston, Herts. UK

Printed in Germany on acid-free paper

"Man must explore"

Dave Scott
always inspirational

Contents

Illustrations

Tables

Acronyms

The following list has been compiled to assist readers who are unfamiliar with the NASA vernacular of the Apollo era.

AGS	Abort Guidance System		LCRU	Lunar Communications Relay Unit
ALS	Apollo Landing Site		LEAM	Lunar Ejecta and Meteorites Experiment
ALSCC	Apollo Lunar Surface Close-up Camera		LGA	Low-Gain Antenna
ALSEP	Apollo Lunar Surface Experiment Package		LM	Lunar Module
ANT	Anorthosite, Norite, Troctolite		LMP	LM Pilot
AOS	Acquisition Of Signal		LNPE	Lunar Neutron Probe Experiment
APS	Ascent Propulsion System		LOI	Lunar Orbit Insertion
APS	Alpha-Particle Spectrometer		LOS	Loss Of Signal
ASE	Active Seismic Experiment		LPD	Landing Point Designator
BSLSS	Buddy Secondary Life Support System		LRL	Lunar Receiving Laboratory
CapCom	Capsule Communicator		LPM	Lunar Portable Magnetometer
CCGE	Cold-Cathode Gauge Experiment		LRRR	Laser Ranging Retro-Reflector
CDR	Commander		LRV	Lunar Roving Vehicle
CM	Command Module		LSCRE	Lunar Surface Cosmic-Ray Experiment
CMP	CM Pilot		LSG	Lunar Surface Gravimeter
CPLEE	Charged-Particle Lunar Environment Experiment		LSM	Lunar Surface Magnetometer
CRDE	Cosmic-Ray Detector Experiment		LSPE	Lunar Surface Profiling Experiment
CSM	Command and Service Modules		LTGE	Lunar Traverse Gravimeter Experiment
DGE	Doppler Gravity Experiment		MCC	Mission Control Center
DPS	Descent Propulsion System		MESA	Modular Equipment Stowage Assembly
DSN	Deep Space Network		MET	Modular Equipment Transporter
EASEP	Early Apollo Surface Experiment Package		MIT	Massachusetts Institute of Technology
EMU	Extravehicular Mobility Unit		MOCR	Mission Operations Control Room
EVA	Extravehicular Activity		MSC	Manned Spacecraft Center
GCTA	Ground-Controlled Television Assembly		MSFN	Manned Space Flight Network
HFE	Heat-Flow Experiment		NASA	National Aeronautics and Space Administration
HGA	High-Gain Antenna		OPS	Oxygen Purge System
HTC	Hand Tool Carrier		PAO	Public Affairs Office
IAU	International Astronomical Union		PDI	Powered Descent Initiation
IFR	Instrument Flight Rules		PGNS	Primary Guidance and Navigation System
JPL	Jet Propulsion Laboratory		PI	Principal Investigator
KSC	Kennedy Space Center		PLSS	Portable Life Support System
LACE	Lunar Atmosphere Composition Experiment			

PSE   Passive Seismic Experiment
RTG   Radioisotope Thermal Generator
SCB   Sample Collection Bag
SEQ   Scientific Equipment
SEPE   Surface Electrical Properties
            Experiment
SESC   Sealed-Environment Sample
            Container
SIDE   Suprathermal Ion Detection
            Experiment
SIM   Scientific Instrument Module
S-IVB   The third stage of the Saturn V launch
            vehicle
SM   Service Module
SPS   Service Propulsion System
SRC   Sample Return Container
SWC   Solar Wind Collector
SWCE   Solar Wind Composition Experiment
SWP   Science Working Party
SWS   Solar Wind Spectrometer
TCU   Television Control Unit
TEI   Trans Earth Injection
TLI   Trans Lunar Injection
USGS   US Geological Survey
UVC   Far-Ultraviolet Ultraviolet Camera/
            Spectrograph
VHA   Very-High Alumina
VIP   Very Important Person

Foreword

Less than three months after the near-tragic loss of the Apollo 13 crew, NASA-of-Apollo made one of its boldest decisions. In the face of that near disaster, dwindling public support and a rapidly declining budget, NASA-of-Apollo decided to skip the final 'H'-type mission, press on with upgrading the 'system' (hardware, software, science and operations) to the 'J' configuration, and launch *three* full-up 'J'-missions to the most significant scientific sites on the Moon. This upgrade from 'H' to 'J' included the Lunar Roving Vehicle, double the lunar stay time, double the EVA surface excursion time, significantly more scientific equipment and experiments, and many other enhancements.

To go beyond the Apollo 11 'G'-mission demonstrated considerable courage and confidence, especially after achieving the political objective of 'landing a man on the Moon and returning him safely to Earth'. To advance beyond 'H' into even *one* 'J'-mission (much less *three*), fully seven months before the launch of the Apollo 14 'H'-mission, required a very bold and aggressive decision. But this commitment to extend the scientific exploration into 'J'-missions was surely one of the most rewarding decisions of the Apollo Program. It would have been a lot easier, safer and cheaper to finish the program with the final two 'H'-missions as scheduled (for if one of the final missions were to be a failure, the program would surely end, and 'Apollo' would forever have been considered a 'failure'). Fortunately for the overall results and success of Apollo, as well as for providing some great material for a book on exploration geology, the 'right' decision had been made!

The majority of this book describes the results of that bold decision, from both engineering and scientific aspects. However, the real significance of this narrative may well be its value to future planetary explorers, including robotic explorers, and even 'virtual' explorers. When these future explorers look back, they will surely ask: What did we learn from Apollo? How does it apply to the future? When will such an adventure happen again?

What did we learn from exploring the Moon? Perhaps two things are cogent: (1) years after the final three Apollo 'J'-missions, the scientific community reached a consensus on the most visible scientific goal of Apollo – how the Moon was formed; and (2) in preparing and conducting the Apollo missions, 'we' (scientists, engineers and astronauts) learned how to develop and practice the art and science of *planetary* field geology (lunar in this case).

How was the Moon formed? Before Apollo explored the Moon, there were three theories on the origin of the Moon, often called the 'daughter', 'sister' and 'spouse' concepts (see, e.g., Wilhelms, *To A Rocky Moon*, 1993, pp. 352–353). It was not until 15 years after the first lunar sample was returned to Earth, and after countless papers and analyses of the subsequent lode of lunar material and photos, that the scientific community finally reached a consensus on the manner in which the Moon was very likely formed. It now appears that our Moon is a daughter of not **one**, but **two** parents – during the early formation of the Earth, the collision of a Mars-sized object ejected part of the Earth's mantle which joined with much of the object's mantle in orbit around the Earth whereupon the disk of mixed substances accreted to form the Moon (Wilhelms, 1993, p. 353).

How do the lessons learned from Apollo apply to the future? From the Apollo expeditions emerged a new scientific discipline that one might define as '*planetary* field geology', which, as so clearly described in this book, is actually quite different from *terrestrial* field geology – it must be conducted in a hostile, unforgiving environment under extreme time limitations, and with no opportunity to ever return (at least within a contemporary lifetime!). During the geological exploration of the Moon, 'we' (the field geology team, on the Moon and on the Earth) had to make instant decisions on the scientific value of the objects of our search, and frequently instant analyses of the objects we selected – there was never sufficient time to dwell on our finds and relish their meaning (pick up a sample – 5 second look – 10 second description – in the bag – on to the next one!). Fortunately, the reader has the opportunity to dwell on the pages and sample the quick-step of our actions from station to station, from sample to sample, and from photo to photo. And even at the time, we knew that we could never, ever, return to the beautiful pristine sites of lunar geologic discovery – sites that are unparalleled on Earth. The terrestrial field geologist, on the other hand ...

When will we go again – to explore first-hand the Moon, Mars, or another planet? Or perhaps more specifically, when will *humans* ever again walk on another planet? One could argue that even the current status of 'robotic technology' precludes the need to risk the life and cost of humans *in situ* exploring the surface of an extraterrestrial body. The remarkable advances in computer science and robotics, including software that produces human-like capabilities, seem to indicate that it will not be long before many will say that artificially-intelligent robots of the future should replace the artificially-robotic humans of the past (picture those somewhat-intelligent Apollo beings of the mid-20th century in those old stiff, bulky, heavy pressure suits!). But will robots ever be able to experience the high adventure of exploring the unknown sights of a new frontier?

There are generally two schools of thought on why explorations of such magnitude occur: (1) technology has advanced to the point where it can be readily used to expand the horizons of exploration; and (2) certain individuals come together at unique periods of history to cause major advances in exploration even though the requisite technology has not yet arrived. Today, almost 30 years after the first lunar landing, with the remarkable technology now available, one might very well opt for the latter – Apollo happened primarily because of people, not technology. The technology available at the time, which was crude by today's standards, was used by 400,000 gifted and inspired individuals who joined in a unified goal stimulated by the aspirations, intellect and leadership of three unique individuals: John F. Kennedy, Wernher von Braun and Sergei Korolev. Read once again the stories in this book; and as you 'take yourself there' with us, ask yourself what is really happening at this moment in time – humans or technology?

In any event, it should be pointed out, for whatever meaning it may have today, that in historical equivalents the age of planetary exploration is barely post-Columbian in its development. Columbus has returned to Spain and perhaps the Earth is not flat after all; but Cabot and Magellan have yet to make their voyages – and Captain James Cook of the *Endeavour* has not even been born.

Before they go back to the Moon, or on to Mars and other planetary destinations, even in their visions, hopefully the Cabots, Magellans, Cooks and Armstrongs of the future will enjoy the adventure of 'Exploring the Moon' and learn from the lessons of 'The Apollo Expeditions' through the pages of this exceptional book.

David R. Scott
Commander, Apollo 15
December 1998

Author's preface

The first edition of this book was written to mark the 30th Anniversary of the Apollo 11 mission. In contrast to the plethora of earlier books about Apollo, it concentrated on what the astronauts did on the Moon's surface, which was to deploy geophysical instruments and perform field geology. Of necessity, therefore, the scientific theme was lunar geology. In relating this story, I used the mission transcripts to develop a travelogue of exploration, and illustrated the text with the pictures that were taken at the time. When astronauts arrived at a new sampling 'station', they took a series of photographs to form a panorama to enable their location to later be determined by triangulation of horizon features, and I assembled these for the book. The book was complementary to those devoted to the political, managerial and engineering aspects of the program, and also to those that focused on the astronauts as people caught up in momentous events. The dialogue attributed to the astronauts was derived from the transcripts in the *Apollo Lunar Surface Journal*, which is published by Eric Jones on the Internet. Although I edited quotations for clarity, for brevity and to eliminate intermingling of speech, I endeavoured to preserve the sense of the moment.

When NASA recently made high-resolution scans of the original Hasselblad film, I suggested to Praxis that I reassemble the panoramas for a completely re-illustrated edition of the book. On seeing my samples, they decided to enlarge the format. And in reworking the manuscript I decided to tighten up the text. So that is how this new edition came about. If you already have a copy of the first edition and liked it, I hope that you welcome the look and feel of this new one.

David M. Harland
Kelvinbridge, Glasgow
October 2007

Acknowledgements

This book has benefitted greatly in various ways from the assistance of Dave Scott, Eric Jones, Roger Launius, Mike Gentry, Patrick Moore, Paul Spudis, Brian Harvey, Philip Harris, David Schrunk, Neville Kidger, Frank O'Brien, Karl Dodenhoff, Ed Hengeveld, Mauro Freschi, Kipp Teague, Ron Wells, W.D. Woods, Keith Wilson, Marc Rayman, Markus Mehring, Harald Kucharek, Mark Gray, Ken MacTaggart and Ken Glover; and of course Clive Horwood of Praxis.

The Apollo challenge

On 25 May 1961, in a speech entitled 'Urgent National Needs', President John F. Kennedy said: "I believe that this nation should commit itself to achieving the goal, before this decade is out, of landing a man on the Moon, and returning him, safely, to the Earth."

Kennedy was not considering the scientific benefit of a landing on the Moon, although, for sure, this would be substantial. His motivation was national prestige.

Six weeks earlier, the Soviets had launched Yuri Gagarin into orbit. The most that America had been able to do, rather belatedly, was send Al Shepard on a suborbital arc over the Atlantic. Kennedy had selected the goal of landing a man on the Moon precisely because it represented a vast technical challenge. By literally 'shooting for the Moon' he was betting that America would not only be able to catch up with the Soviets in space, but would forge ahead. He had reached the conclusion that space was the arena of Superpower Politics, and was laying down the challenge to Nikita Khrushchev. He had imposed a deadline in order to ensure that reaching the Moon was perceived as a 'race'. To emphasise the scale of the task, Kennedy went on: "No single space project in this period will be more impressive to Mankind or more important for the long-range exploration of space; and none will be so difficult or expensive to accomplish."

Sending a man to the Moon was to be the modern form of the ancient practice of 'single combat', whereby opposing armies lined up and then sent a single soldier to decide the issue; the first man to put his boot on the lunar surface would decide this particular issue. Kennedy made

On 25 May 1961 John F. Kennedy challenged his nation to land a man on the Moon before the decade was out.

sure that everyone understood this analogy: "It will not be one man going to the Moon ... it will be an entire nation." Then, to emphasise what was at stake, he warned: "If we are to go only half way, or reduce our sights in the face of difficulty, in my judgement it would be better not to go at all."

As we all know, the race was won on 20 July 1969 when Neil Armstrong stepped onto the lunar surface. However, this was not the end of the Apollo program ...

1

The robots

At the dawn of the Space Age, the Soviet Union was able to exploit its R-7 rocket's 'heavy lift' to send probes far into space, and the immediate target was the Moon.*

Launched on 2 January 1959, the first Luna, affectionately dubbed Mechta ('Dream'), was intended to hit the Moon, but it missed by 5,000 km. Nevertheless, this marked a significant achievement. A follow-up probe in September hit near the crater Autolycus. In October a probe was put into an elongated Earth orbit which looped around the Moon. Unlike its inert predecessors, Luna 3 had a camera. After the film had been developed, the pictures showing its passage across the eastern limb were scanned and transmitted. The Moon's rotation is tidally locked to its orbit around the Earth, so its rear hemisphere was a mystery. If it resembled the visible side, it would be dominated by dark plains known as maria. Although the quality was poor, Luna 3's pictures gave a new perspective on the maria on the eastern limb, which are highly foreshortened when seen from Earth, and revealed that the other side is predominantly highland terrain. In the early days, the purpose of the space program was to be *first* to achieve each feat. As Luna 2 had shown, flights would be repeated until the goal was achieved, and then the program moved on. Luna 3 had succeeded, so rather than follow it up with a probe which would return better imagery the effort switched to landing a camera on the surface.

Because the new probe was a complex spacecraft with its own propulsion system, it was 1963 before it was ready. Unlike its predecessors, the new vehicle was to be put in a 'parking orbit' prior to being sent to the Moon. A retrorocket was to slow the 'bus' just above the lunar surface, so that

A 'raw' version of the first photograph of the far side of the Moon taken by Luna 3.

it could eject a capsule which would make a 'hard' landing. Unfortunately, perfecting this spacecraft proved difficult. In April 1963 Luna 4 missed the Moon by 8,000 km. Things progressed no further in 1964: Luna 5's retrorocket failed to fire in May; a stuck thruster made Luna 6 miss the Moon in June; an early firing left Luna 7 with a long fall to destruction in October; and in December Luna 8 smashed into the surface because its engine fired late. On 3 February 1966 Luna 9, reputedly the last of the batch, performed perfectly and its spherical capsule rolled to a halt in Oceanus Procellarum. After it had opened its 'petals' to extend its antennas and expose its solitary instrument, a TV camera, it broadcast a panoramic view of its landing site. As luck had it, this transmission was received by the radio telescope at Jodrell Bank in England, and a national newspaper fed it to a commercial wire-facsimile machine and produced an image. Unfortunately, they set the wrong aspect ratio, and the compressed panorama depicted a jagged 'moonscape'. Because nobody knew what the surface of the Moon looked like,

---

* A Spaceflight log is provided on page 363.

A part of the panoramic vista transmitted by Luna 9 from the surface of the Moon, as released by the Soviet Union.

Harold C. Urey, chemist.

Gerard P. Kuiper, astronomer.

the error went unnoticed until the official version was published in Pravda the next day.

Was the mare a splash of "impact melt" from the formation of a basin, as Harold Urey supposed? Was it the "frothy vacuum lava" that Gerard Kuiper believed? Was it the fine dust that Tommy Gold had inferred from radar reflections? The pirated image appeared so harsh that those who believed the mare to be lava characterised it as scoriaceous.* If correct, this would make the task of lunar explorers devilishly dangerous. Although the correctly processed picture was rather less intimidating, the rocks looked to be surprisingly fresh! Was the Moon still active?

Having achieved a lunar landing, the program moved on to the next objective, namely the placing of a probe into orbit. The Luna spacecraft had been designed as a modular system, with a bus and a payload. All that had to be done was to replace the landing capsule. After launch on 31 March 1966, Luna 10 used its retrorocket not to bring itself to a half, but to slow down just sufficiently to enter lunar orbit, which it did successfully, and thereby achieved another worthy 'first'. Its instrument capsule carried a gamma-ray spectrometer to make an initial survey of the chemical composition of the lunar surface. The data from the two lunations during which it was operative not only indicated that the maria were basaltic, but also that there were no significant extrusions of silicic lava. The basalt 'confirmed' Kuiper's assertion that the Moon had once been 'hot' and was thermally differentiated, and it effectively ruled out Urey's 'cold' Moon hypothesis. The surprise was the lack of silicic lava. In the mid-1960s, most lunar geologists were swinging behind the theory that the highlands were shaped by volcanism. It is a pity that the Luna 10 data was not more conclusive, because it was to sample the supposed rhyolite that Apollo 16 was sent to Descartes-Cayley six years later.

The objective now switched back to photography. In August and October 1966, Luna 11 and Luna 12 were placed in near-equatorial 150 × 1,200-km orbits for 20-metre-resolution mapping. Luna 13, in December, was a lander carrying penetrometers to measure the mechanical properties of the lunar surface. As had its predecessor, it landed in Procellarum. Luna 14 entered a 160 × 870-km lunar orbit in April 1968 and drew to a

---

* Technical terms are explained in the Glossary on page 373.

close the second phase of the Soviet investigation of the Moon. The versatile bus had facilitated a number of 'firsts', and its payloads had made significant discoveries, but there had been no real follow-through, and, for sure, the results had little impact on Apollo.

RANGER

The National Aeronautics and Space Administration (NASA) was formed in October 1958, at President Dwight Eisenhower's behest, to manage America's civilian space program.

The Ranger project was initiated in December 1959, as NASA's flagship for the exploration of the Moon. It was managed by the Jet Propulsion Laboratory (JPL) of the California Institute of Technology in Pasadena. In May 1961, however, John F. Kennedy committed the nation to "landing a man on the Moon", and Ranger was subordinated to the new Apollo program.

The first two flights, in August and November 1961, were intended to test the spacecraft's basic systems, particularly the innovative 3-axis stabilisation, but the Agena upper stages left them stranded in parking orbit. Despite these losses, it was decided to proceed with the second batch of spacecraft, which were to execute the full flight profile, ending with a lunar impact. This plunging dive was to be documented by a TV camera. A split second before the spacecraft hit, it was to

A model of the Ranger spacecraft.

eject a spherical shock-resistant 'hard landing' capsule that was to deploy a seismometer once it had settled. Unfortunately, Ranger 3's Agena over-performed, with the result that the spacecraft missed the Moon by 35,000 km. The next Agena was so accurate that Ranger 4 hit the Moon, but an electrical fault had crippled the spacecraft. A power failure disabled Ranger 5, which missed the Moon by 750 km.

In December 1962, with its best result being a dead spacecraft hitting the Moon, the Ranger project was in danger of cancellation. After a review of spacecraft assembly procedures, NASA redefined the project's goals: the next vehicles would have only the TV package, and their sole objective would be to provide close-up pictures of the lunar surface in order to assess whether it was likely to support a spacecraft.

The position of the target was constrained by flight dynamics considerations. The initial TV viewpoint would match the best telescopic pictures. Ideally, the spacecraft was to make a near-vertical dive so that successive images would overlap. Watching the final moments of the spacecraft's descent in 'real time' would be stunning, but the analysis would be painstaking. And not only would the film be replayed to follow the manner in which the character of the surface altered on different scales, 'zooming out' by running the film backward would put fine detail into context. Ranger 6's TV system was crippled by an electrical arc at launch, but this did not become evident until it failed to start as the spacecraft approached its target, which was in Mare Tranquillitatis, 100 km east of the crater Julius Caesar.

The project's luck changed spectacularly on 31 July 1964, when Ranger 7 struck a patch of mare in the general vicinity of Mare Nubium. The area was crossed by bright 'rays' from two craters: Copernicus 600 km to the north and Tycho 1,000 km to the southeast, so valuable information was forthcoming throughout the descent. The nature of the rays was disputed. The leading theories explained them either as secondary craters made by debris ejected by larger impacts, or as being pits created by endogenic gas venting from crustal fractures. The Ranger pictures not only showed them to be secondary craters, it was also possible to distinguish the craters in the Copernicus ray from those in the Tycho ray because the size of a secondary is inversely proportional to its distance from its primary.

Kuiper noted that the last picture showed detail half a metre across, which was a *one-thousand-fold*

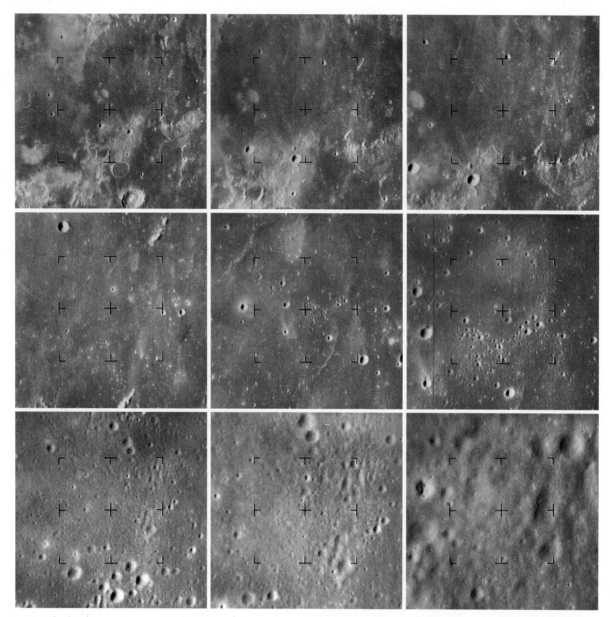

A sequence of TV pictures (left to right, top to bottom, starting from an altitude of 1,225 km) from Ranger 7 as it made its almost vertical dive onto the cratered plain of Mare Nubium.

improvement in resolution over the best telescope. Gene Shoemaker pointed out that the terrain was fairly soft and rolling; there were none of the jagged features common in science fiction stories. Although an automated spacecraft might well come to grief by setting down on a rock or in a crater, there were evidently wide open spaces as well, and an Apollo crew should be able to manoeuvre to a safe spot. And the presence of boulders implied that the surface was likely to support a spacecraft, so it was not the sea of dust that Gold had supposed. Kuiper said that a set of shallow ridges indicated that the mare was a lava flow, and that this supported his contention that

the mare was a "crunchy" layer of "frothy" lava. Urey insisted that the pictures did not positively identify the nature of the surface. Although Shoemaker agreed that the mare was a lava flow, he insisted that the profusion of craters of all sizes meant that the lunar surface is subjected to ongoing bombardment by space debris: large impacts dig up rocks, and these are progressively eroded by smaller impacts, until the fragmental debris is a lunar 'soil'. Urey promptly called this process 'gardening'. An analysis of the extent to which craters of various sizes had excavated rock provided a measure of the depth of the debris; it was several metres thick. Bill Hartmann used

Eugene M. Shoemaker, astrogeologist.

crater counts – the bane of the photo-geologist – to 'date' this particular patch of mare to 3.6 billion years old; this was to prove to be a remarkably shrewd assessment.

In his book *To A Rocky Moon*, Don Wilhelms says of Shoemaker's analysis of the lunar surface in the light of the Ranger 7 results:

[He] characteristically specified the properties of the gardened layer of shattered and pulverised rock so accurately that one might conclude that the Moon had [already] been explored:

(1) it rests on a cratered mare substrate with irregular relief, and varies in thickness up to a few tens of metres;

(2) about half of its fragments were ejected from craters less than a kilometre away, but some fragments could have come from anywhere on the Moon;

(3) the number of times fragments are re-ejected and overturned increases greatly toward its surface;

(4) its surface is pockmarked by craters of all sizes from submillimetre to tens of metres; and

(5) only its uppermost few millimetres are the fragile open network inferred by the astronomers; so

(6) its bearing strength increases rapidly with depth.

In fact, because the gardening process would produce a seriate distribution of particle sizes, the term 'regolith' was coined to describe the lunar soil.

On 31 August 1964 the International Astronomical Union recognised Ranger 7's success by renaming the impact site Mare Cognitum (Known Sea).

Having examined the ray-crossed Nubium, the Apollo site selectors wanted to inspect a patch of mare which was free of rays, in order to determine whether the surface was significantly smoother. Accordingly, in February 1965 Ranger 8 followed a shallow trajectory over the Central Highlands to Mare Tranquillitatis. On the way, it flew over Sabine and Ritter. Some workers believed the unusually shallow floors of these two 30-km craters to signify that they were calderas. Those who preferred an impact origin for *all* craters insisted that any volcanic activity must have been stimulated by the excavation of the crater. Whilst Ranger 8's increased surface coverage was welcomed by the mappers, it resulted in substantial smearing in the final frames. This patch of Tranquillitatis turned out to be just as heavily cratered and rock strewn as the ray-crossed Nubium. Indeed, given only the last few frames of each site, it was difficult to tell them apart. On a local scale, maybe the maria were basically similar? Shoemaker ventured that this realisation was "one of the most striking Ranger results".

After inspecting two mare sites, NASA released the final Ranger to the scientists. The selectors considered the craters Copernicus, Kepler and Aristarchus, then settled on Alphonsus, a 110-km crater east of Nubium. Not only did Alphonsus have a central peak and a flat floor full of interesting rilles and dark-halo craters that appeared to be volcanic, but it was high on the list of sites associated with 'glows' that were widely believed to be gas emissions. At the very least, it was hoped to confirm that the 1-km-high central peak was a volcano. On 24 March 1965, Ranger 9 fell into the crater and, for the first time, the TV was fed 'live' to the commercial networks.

The trajectory tracked northeast across the crater. Both the inner wall of the crater and its central peak proved to be surprisingly smooth. The absence of craters suggested that the slopes, which were fairly shallow (no more than 20 degrees), were thick with regolith that had masked surface relief by slumping. Contrary to expectation, there was no clear evidence to confirm the central peak as a volcano. Kuiper insisted that the dark-halo craters were volcanic. Urey reluctantly agreed, allowing that they were probably "some sort of plutonic activity". While the litter of boulders implied a surface with considerable bearing strength, Gold continued to say that any spacecraft would sink in dust. He observed aptly that the Ranger pictures "are like a mirror, and

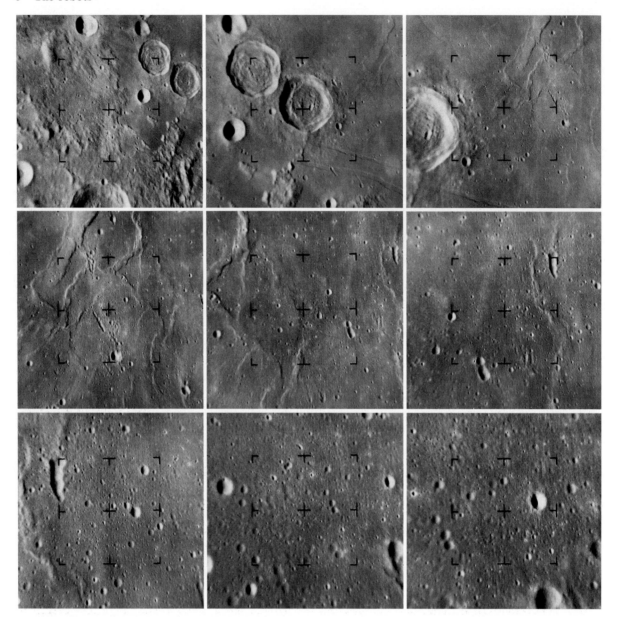

Ranger 8's slanting trajectory took it low over the 'raised floor' craters Ritter and Sabine *en route* to the southwestern section of Mare Tranquillitatis. This sequence starts at an altitude of 385 km.

everyone sees his own theories reflected in them''. The final frames showed objects on the surface which were just 0.3 metre in size. It was concluded that the floor of Alphonsus could be considered for one of the later Apollo missions.

Ranger had answered the fundamental question concerning the nature of the lunar surface: it looked as if it would support the weight of a spacecraft. Although the successful Rangers had returned excellent pictures of their targets, their viewpoint had rapidly shrunk to a scale which, while vital for assessing the condition of the surface, was too narrow to facilitate areal mapping.

LUNAR ORBITER

With Ranger, JPL had seized the initiative in the development of spacecraft for missions in deep space. In May 1960 it took on a far more adventurous project. It was to build two related spacecraft: one was to enter lunar orbit to map the surface, and the other was to land to determine the nature of the surface. A year later, when Kennedy initiated Apollo, the pace of these projects was accelerated. Unfortunately, not only did it soon become evident that Ranger's development would take longer than expected, it was also clear that JPL was overcommitted. Furthermore, the devel-

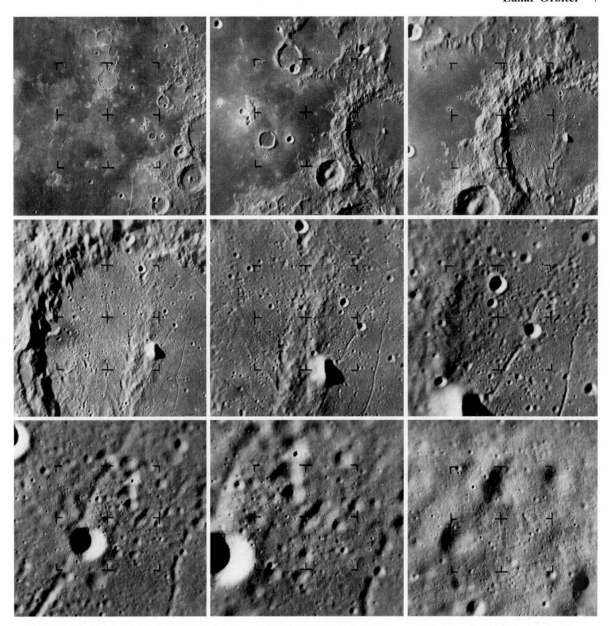

Ranger 9 dived into the large crater Alphonsus to investigate whether the central peak and various features on its floor were of volcanic origin. This sequence starts at an altitude of 1,600 km.

opment of the powerful Atlas-Centaur, which was to dispatch the new spacecraft to the Moon, was behind schedule. In coming to terms with Apollo's "before this decade is out" imperative, in 1963 NASA cancelled JPL's mapper and instead ordered the Langley Research Center in Hampton, Virginia, to develop a lightweight orbiter to ride the Atlas-Agena. This new spacecraft was to chart predetermined sites for their suitability as Apollo landing sites.

Ranger had yet to prove itself, but it was obvious that developing an orbiter would not just be a matter of fitting a motor to insert Ranger into lunar orbit. While JPL's TV camera was ideal for documenting a 20-minute plunging dive that would result in the destruction of the spacecraft, it was capable of providing the required high surface resolution only in its final few seconds, by which time its field of view was extremely constrained. To survey wide areas at similar resolution from an altitude of 50 km, Lunar Orbiter would expose film, and later develop this onboard and scan it for transmission. Because the orbiter had to be lightweight, the camera could not be shielded against cosmic radiation, so extremely 'slow' film (it was rated at 2 ASA) was needed, which in turn meant that the camera had to be able to compensate for the spacecraft's motion. A two-

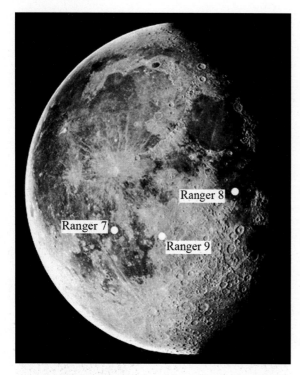

The locations of the targets of the three successful Ranger missions.

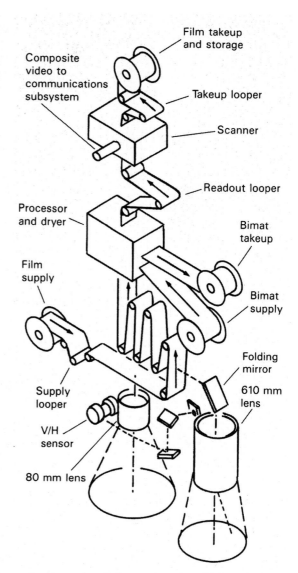

A schematic of the manner in which a Lunar Orbiter spacecraft processed its film.

lens system was used, with a wide lens providing context for a telephoto lens. As a result, Lunar Orbiter was a significant advance over Ranger. Luckily, NASA was able to adapt a camera built by Kodak for a reconnaissance satellite. In December 1963 NASA issued the Lunar Orbiter contract to Boeing. As with Ranger, Lunar Orbiter would employ 3-axis stabilisation, but a spacecraft's configuration is intimately related to its payload, and although it was possible to use many off-the-shelf systems, Lunar Orbiter turned out to be very different to Ranger. The budget allowed for five operational vehicles, plus a spare for engineering trials. It was expected that three successful flights would be sufficient to survey all the sites on the list for the first Apollo landing, which was as far into the future as NASA was looking at that time. To accomplish this, Langley devised three interleaved flight plans, which were designated 'A', 'B' and 'C'.

The spacecraft was to employ an elliptical orbit with a 1,500-km high point (apolune) above the Farside and a 50-km low point (perilune) on the Nearside. The orbital axis was selected to enable the spacecraft to take its pictures at a low Sun angle, so that the surface relief would be most visible. In fact, the perilune point would keep pace with the terminator, and take 10 days to move 2,500 km along the Apollo zone, documenting each site in ideal lighting. It was more complicated than that, of course. Although the candidate Apollo landing sites were in the equatorial zone, they were distributed across a strip 300 km wide. To reach sites near the edge of this 'box', it was necessary to tilt the spacecraft's trajectory with respect to the equator. The first spacecraft would fly with its perilune 11 degrees south of the equator; the second would have its perilune 11 degrees north of the equator; the third would be positioned as necessary to plug gaps, and to make follow-up investigations of the most interesting sites.

Lunar Orbiter 1 was launched on 10 August 1966 and achieved lunar orbit on 14 August. It sent its initial pictures, showing Mare Smythii on the eastern limb, on 18 August. In effect, this

A model of the Lunar Orbiter spacecraft.

sequence was just 'advancing' the film into the camera. In its 'mapping' mode, the camera took four telephoto shots for each wide-field shot, so that both provided contiguous coverage of the surface. Unfortunately, a malfunction in the motion compensator on the telephoto meant that its pictures were smeared. Although the flight controllers considered operating at a higher altitude in order to reduce the smearing, and so perform a modified mission which would map the *entire* Apollo zone with a resolution of 25 metres, it was decided instead to survey the planned sites using only the wide-angle lens.

On 23 August, in an impromptu experiment, the spacecraft turned just before it flew behind the limb and snapped the Earth. This was the first time that the Earth had been seen set against the lunar horizon, and the picture caused a sensation in the Press.

Although Ranger 7 had shown that the bright rays which radiate from fresh-looking craters are secondary impacts by ejecta from the primary, it was a matter of dispute why some large craters did not seem to possess rays. Shoemaker and Mike Carr had independently come to the

Once in orbit of the Moon, Lunar Orbiter 1 took this picture of the crescent Earth moments before the spacecraft passed behind the limb of the Moon on 23 August 1966. It was the first time that the Earth had been seen from so far away in space.

An oblique view of the crater Copernicus taken by Lunar Orbiter 2 on 23 November 1966. The 300-metre-high mounds in the foreground are the central peak complex. The terraces are the 100-km-diameter wall. The Carpathian Mountains form the horizon. To astronomers used to seeing craters from overhead, this 'pilot's eye' view caused a sensation, and the Press dubbed it the 'Picture of the Century'.

conclusion that the rays faded. Old craters had rays, but they had blended into their background terrain. As evidence, they charted the faint rays radiating from Eratosthenes, which stratigraphic analysis indicated to be older than Copernicus, whose rays were prominent. One of the Tranquillitatis sites which Lunar Orbiter 1 surveyed, labelled Site 'A3', was lightly rayed by Theophilus, off to the south. As Shoemaker had predicted, although the site was pocked, the profiles of these craters were eroded, and so they posed no problem to Apollo.

On 29 August, Lunar Orbiter 1 imaged the ninth potential Apollo landing site on its target list, and thereby completed its primary mission. In addition, it had taken a number of pictures while at apolune, to record wide areas of the Farside. Luna 3 and Zond 3 had imaged the Farside, but at low resolution, so these new pictures were a welcome bonus. The spacecraft transmitted telemetry for another two months in order that the degradation of its systems could be assessed, and then it was deorbited to clear the way for its successor, which started work on 18

November. In addition to documenting the remaining 11 candidate sites, Lunar Orbiter 2 was able to snap a number of sites of scientific interest that were of no immediate interest to the Apollo selectors. Although not under consideration for Apollo, area 2P-5 was important because it was Ranger 8's impact site, and it was hoped that the 'calibrated' crater would yield useful information about the mechanical properties of the lunar surface.

To follow up on the assessment of the extent to which rays mitigated against a landing site, Lunar Orbiter 2 documented a ray close to Copernicus, and found it to be too rough for Apollo. On 23 November, to prepare the camera for its next mapping sequence, the spacecraft turned to snap an oblique shot of the 100-km-wide crater itself, some 240 km north of its track. The result was not just a boon for the study of the crater whose ejecta blanket had served as the basis for Shoemaker's pioneering study in 1960 which demonstrated that stratigraphic analysis could be used to chart regional relationships, but the orbiting astronaut's perspective of its 300-metre-high multiple central peak and terraced walls prompted the Press to dub it the 'Picture of the Century'.

In addition to its own northern targets, Lunar Orbiter 2 provided high-resolution images of its predecessor's best sites, and area 'A3' (now called 2P-6) was given the green light. Of the new sites, 2P-2 seemed suitable, but 2P-4 was rejected for being too rough.

Lunar Orbiter 2 completed its photography on 26 November, having functioned perfectly. With all of the candidate sites imaged, the US Geological Survey (USGS) made terrain maps of the potential landing sites for the Apollo

planners and geological maps for their own use. In addition to photographing the most favourable sites from various angles in order to facilitate stereoscopic mapping for detailed analysis of the topography, Lunar Orbiter 3 was to chart landing approach routes. When the camera developed a fault, it was decided to replay the backlog of exposed frames, but 72 frames remained to be processed when the film transport motor failed. Despite this problem, the project had clearly achieved its objective.

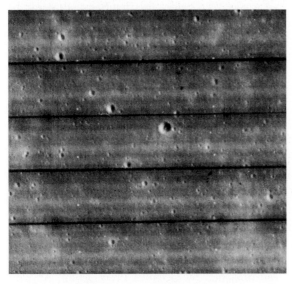

This view of area 2P-2 by Lunar Orbiter 2 showed it to be worthy of consideration as a potential Apollo landing site.

An oblique view of the Marius Hills area by Lunar Orbiter 2.

After being inspected by Lunar Orbiter 2, area 2P-4 was deemed to be too rough to be a potential Apollo landing site.

An oblique view of the crater Kepler by Lunar Orbiter 3.

Rather than terminate the series, NASA released the remaining two spacecraft to the scientists, who chose to fly them in near-polar orbits. The oblique shots of mare flow fronts, crater interiors and ejecta blankets had been informative, but the geologists yearned for a sense of *perspective*, so they sacrificed detail for areal coverage. On 11 May 1967, Lunar Orbiter 4 assumed an orbit with a 6,000-km apolune designed to give hemispherical coverage of the Farside, and a 2,500-km perilune (*50 times* higher than used to take 'close look' pictures of likely landing sites) for mapping of the Nearside. The 100-metre resolution, which was five times better than that of the best Earth telescope, would yield insight into regional morphology. Particularly welcome was the coverage of the limb, where the terrestrial view is foreshortened. By continuing around the limb, objects previously glimpsed only in libration were seen in context. The complexity of the multiple-ringed Orientale Basin, which had been hinted at by Zond 3, was dramatically revealed. Unfortunately, the camera suffered a spate of problems and only 133 frames were retrieved. This mapping was continued by Lunar Orbiter 5 in August, but in this case the perilune was lowered to 100 km in order to provide 2-metre-resolution stereo photography of sites of special scientific interest that might serve as 'geological feature' targets for advanced Apollo missions.

In the period of only a year, the Lunar Orbiters not only satisfied the objective of surveying likely sites for the first Apollo landing, they had also returned the first clear views of the Farside, tremendously advanced the state of Nearside regional geology, and identified more feature sites than were ever likely to be visited. Even after they had finished imaging, these spacecraft provided insight into the lunar *interior*. Although the first spacecraft was deorbited before the arrival of its successor, it was noted that the spacecraft's orbit was being perturbed, which meant that the gravitational field was uneven. To follow up, the other spacecraft were not deorbited until their propellant was almost exhausted, and with spacecraft in equatorial and polar orbits it was possible to 'map' the field in sufficient detail to show that the mare-flooded basins were the source of the strongest gravity. In view of the excess *mass* that was evidently *con*centrated in such basins, they were named 'mascons'. They would significantly complicate Apollo planning.

SURVEYOR

In May 1960 NASA assigned the Surveyor project to JPL, which was already developing the Ranger spacecraft. It was an ambitious plan to build a spacecraft to make a 'soft landing' on the Moon. In October 1962, the Apollo planners announced that because the mechanical properties of the lunar surface would influence the design of the Apollo spacecraft, the development of the lander ought to be given a higher priority than orbital photography, whose results would have only operational value. However, the Centaur rocket stage was late, and it was 1966 before the first Surveyor could be launched.

The Surveyor mission planners faced the same problem as their Apollo counterparts: where should they send their first spacecraft? Safety considerations effectively obliged them to select a mare site. In fact, not only was this consistent with the objective of characterising the surface in the Apollo zone, but the data would be scientifically significant, because the maria account for fully 30 per cent of the Nearside.

When Surveyor 1 was finally launched on 30 May, the 'old hands' at JPL might well have wondered whether they were in for a rerun of the teething troubles they had faced with Ranger, but the spacecraft touched down safely in the Flamsteed Ring, an old crater whose 112-km rim has been breached by Oceanus Procellarum. As with Ranger, the solitary instrument was a television camera. The first picture showed the vehicle's footpad resting on the surface, which was barely indented; so much for Gold's insistence that the spacecraft would sink into dust, and disappear. The camera then took a multitude of telephoto frames that were mosaicked to form panoramic views. There were small craters and rocks, but the surface was generally flat, and the horizon was essentially featureless. The overall impression was of a lava flow. Surveyor 1 continued to transmit panoramas in order to document how the surface looked under different lighting, and when the Sun set it went into hibernation for the fortnight-long lunar 'night'. Not only did it awaken with the return of the Sun, it did so each 'morning' for the rest of the year. Because Surveyor's landing accuracy was expected to be no better than 30 km, one of Lunar Orbiter 3's jobs was to find out precisely where the lander was located. It returned 64 frames which, once mosaicked, covered the entire area, and the landing site was eventually identified by virtue of the fact that the lander cast a distinctive shadow.

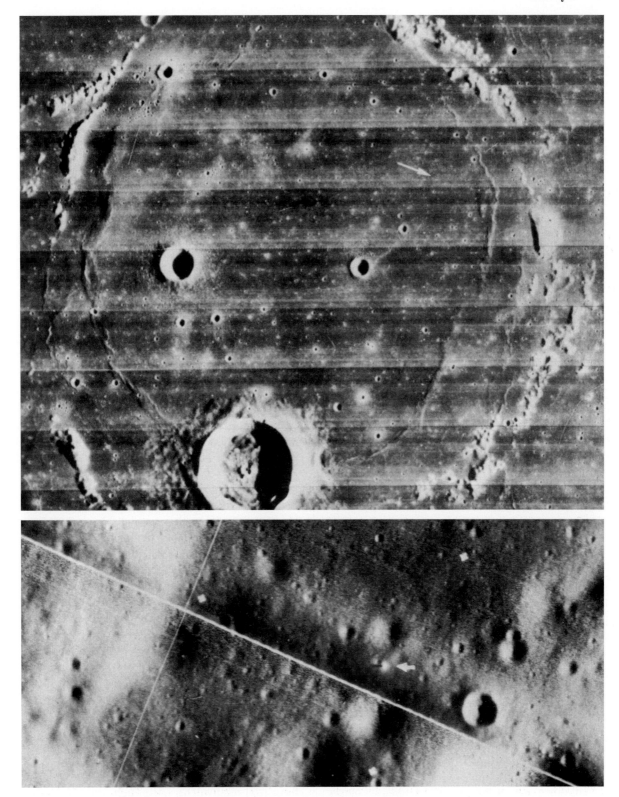

The landing site of Surveyor 1 is arrowed in the Flamsteed Ring, an old crater that has been flooded by Oceanus Procellarum (top). After an exhaustive search, Lunar Orbiter 3 found the lander on the surface (arrowed, bottom), casting a 3-metre-long shadow in the opposite sense to the craters.

Surveyor 1 photographed its own shadow, revealing there to be an intense 'zero-phase' glare of backscattered sunlight.

A telescopic photograph of the Lansberg area of Oceanus Procellarum, with the Surveyor 3 landing site indicated by a small circle.

Having achieved success at the first attempt, it came as a shock when Surveyor 2 started to tumble during a trajectory correction *en route* to the Moon and was lost. It had been intended to inspect a potential Apollo landing site in Sinus Medii.

On 20 April 1967 Surveyor 3 landed in Procellarum near the crater Lansberg, but it came down inside a 200-metre crater and bounced several times. In contrast to its predecessor, several hundred kilometres to the west, its view was severely limited. On the other hand, it conducted a thorough study of a sizeable mare crater, which would not have been possible if it had landed on the open plain. There was no sign of outcropping rock in the crater wall to indicate the thickness of the regolith, but the wall was pocked by smaller craters, one of which had excavated large rocks, so the regolith was clearly not hundreds of metres thick.

In addition to the camera, Surveyor 3 had a remotely controlled arm fitted with a scoop so that it could determine the mechanical properties of the surficial regolith, dig trenches to inspect the subsurface, lift rocks to weigh them, and roll rocks over to assess the extent to which their state of erosion was selective. It scraped out four trenches and found that although the surficial material was soft, it consolidated with depth, and 15 cm down was as hard as rock. Unlike its hardy predecessor, Surveyor 3 survived only one lunar night. Lunar Orbiter 3 also located this landing site.

Surveyor 4 was sent to Sinus Medii, but contact was lost while firing its solid-fuel retrorocket several minutes before it was scheduled to land.

Having sampled two western mare sites, and failed twice trying to reach the meridian, NASA dispatched Surveyor 5 to Tranquillitatis to sample an eastern mare. It landed 60 km from where Ranger 8 had hit, and 25 km from the 'A3' site which had passed its Lunar Orbiter inspection and graduated to the shortlist for the initial Apollo landing. However, it landed on a 20-degree slope in a 12-metre crater, and could barely see over the rim!

Instead of a scoop arm, Surveyor 5 had an instrument to study the chemical composition of the regolith. Once this boxy package had been lowered to the surface, it irradiated the soil with alpha particles from a radioactive source and then measured the extent to which these were backscattered. The energy of the reflected particles was proportional to the mass of the atomic nuclei off which they bounced. When data was integrated over many hours, the energy spectrum identified the elemental abundances. This 'beam' could not penetrate deeper than the surficial material, but the gardening of the regolith by micrometeoroid bombardment suggested that the chemical composition of the surficial layer would be typical. The results indicated calcium, silicon, oxygen, aluminium and magnesium. This suggested that the bedrock was basalt, but the high ratios of iron and titanium hinted that it was subtly different to its terrestrial equivalent. Although this supported Kuiper's concept of the mare as being volcanic extrusions, Urey pointed out that it did not rule out his hypothesis that the maria were formed simultaneously, and were a veneer of impact melt splashed out by the formation of the Imbrium Basin. A little data went a long way though, and

Lunar Orbiter 3 established that Surveyor 3 landed just inside the rim (arrowed) of a 200-metre-diameter crater.

John Gilvarry said that it supported his contention the maria were dried out marine deposits.

The Apollo planners were determined to sample Sinus Medii, so Surveyor 6 was sent to fill in for its forerunners, and landed without incident on 10 November 1967. It came down near the end of a 40-km-long ridge. Surveyor 2 had carried only a camera, and Surveyor 4 had a

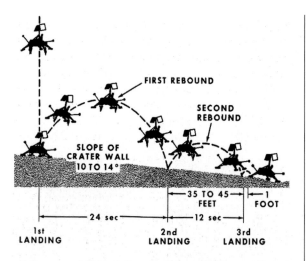

Surveyor 3 bounced several times before coming to rest.

Surveyor 3 used a 'pantograph' remotely controlled arm to excavate several trenches in the loosely consolidated lunar surface material.

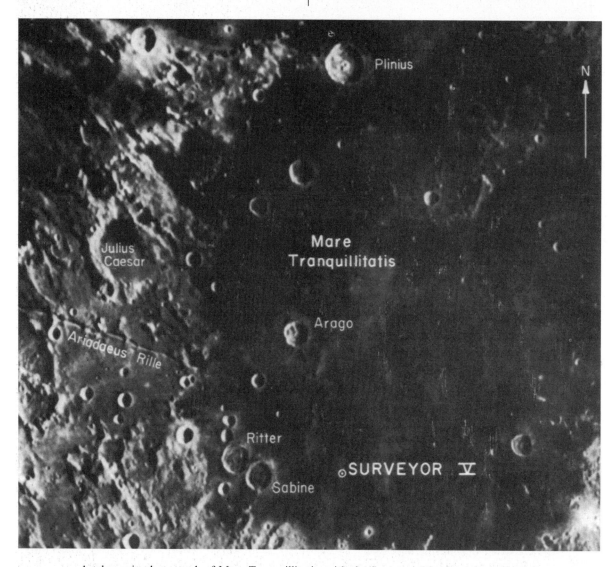

A telescopic photograph of Mare Tranquillitatis, with the Surveyor 5 landing site indicated.

scoop, but Surveyor 6 had the chemical analyser, and the results proved to be similar to Tranquillitatis: an iron-rich basalt.

Because the maria in the Apollo landing areas had turned out to be remarkably similar, NASA released the final Surveyor to the scientists, who promptly found themselves with the dilemma of deciding where to send it. The final Lunar Orbiter had scouted 'feature' sites for Apollo. Some argued that the final Surveyor should visit one of these sites, but others argued that it should be sent to somewhere that Apollo was unlikely to visit, not for being uninteresting but for being inaccessible. The consensus soon settled on Tycho, the bright ray crater in the Southern Highlands. It had been imaged by Lunar Orbiter 5, and its flank out to about 12 km beyond the 85-km rim was composed of irregular hills and valleys; the

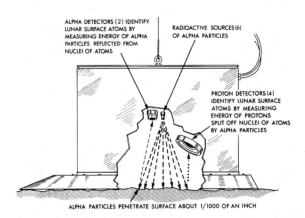

A schematic of the instrument designed to determine the elemental composition of the lunar surface material.

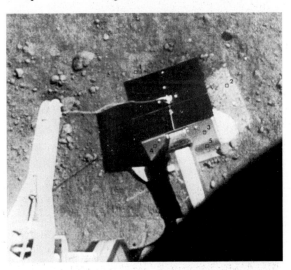

Surveyor 5's chemical analysis instrument after being lowered to the lunar surface.

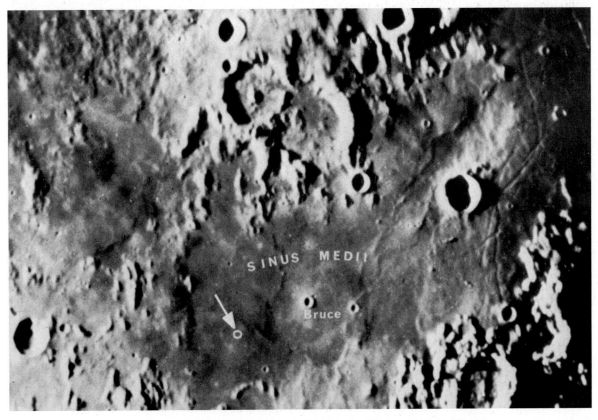

A telescopic photograph of Sinus Medii, with the Surveyor 6 site indicated.

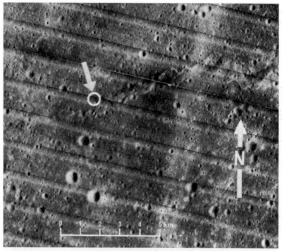

Surveyor 6 landed near the western end of a ridge.

Surveyor 6's view of the nearby ridge.

generally undulatory terrain from there out to about 40 km was etched with a network of radial ridges, each typically a couple of kilometres long and several hundred metres wide. Surveyor 7 was sent to the northern peripheral area, and on 10 January 1968 it not only landed safely but was within 2 km of the designated target, 30 km from the rim.

By cutting margins, JPL was able to install both the robotic arm and the chemical analyser, which proved fortunate because the boxy instrument became hung up, and if it were not for the arm nudging it free the scientific investigation would have been undermined. In addition, the arm was used to place the instrument onto a patch of excavated soil to check that this was the same as the surface, and also to stand it on a rock. The arm also scraped seven trenches for soil-mechanics tests. These exposed what seemed to be solid rock at a depth of only 10 cm, which suggested that the regolith was extremely shallow. No one knew how old Tycho actually was, but its prominent system of rays indicated that it was 'recent'. Upon seeing the elemental abundance data, some reasoned that the lunar highlands were an 'alumina-rich basalt', but Shoemaker concluded that the dominant rock in the Tycho ejecta, which had been excavated from deep within the crust, was anorthositic gabbro. This

feldspathic rock was clear proof that the Moon was thermally differentiated. It was evidently not Urey's 'cold' primordial body. John Wood later picked up on this, and argued that the Moon had been so hot in the final stage of its accretion that it had been molten to a considerable depth. The heavier elements in this 'magma ocean' sank to create a magnesium and iron ('mafic') silicate mantle. The lighter elements floated to the surface and cooled to create the first crust. Plagioclase feldspar is rich in aluminium. The rock that Tycho had dug up from a depth of many kilometres was evidently very rich in plagioclase.

MISSIONS ACCOMPLISHED

Work on Ranger and Surveyor began soon after NASA's creation, but before Kennedy decided to undertake Apollo. The "before this decade is out" imperative meant that both Ranger and Surveyor were redirected to support Apollo. Lunar Orbiter, which was initiated afterwards as an alternative to the orbital element of Surveyor, was also designed specifically to pave the way for Apollo.

By providing the first close-up views of the lunar surface, Ranger established:

- that the maria were probably lava flows;
- that they almost certainly had sufficient bearing strength to support a spacecraft;
- that although the maria were cratered and strewn with rock, they were nevertheless fairly flat and there were clear areas on which Apollo would be able to set down; and, significantly,
- that all maria were probably the same;

and, in doing so, it completed its redefined objectives.

Lunar Orbiter not only provided high-resolution imagery of all the candidate Apollo landing sites, but went on to perform extensive regional mapping and revealed that the lunar gravitational field is systematically irregular, and this gave insight into the relationship between the 'circular maria' and the basins they fill. Surveyor had been charged with determining the character of the mare regolith, both mechanically and chemically, and it not only did this but went on to provide a significant insight into the nature of the highlands. As a result, as Apollo prepared to venture out to the Moon, the planners were well-advanced in selecting a landing site.

The target selected for Surveyor 7 was a blanket of ejecta north of the rim of the crater Tycho (arrowed).

A mosaic of pictures by Surveyor 7 showing the litter of rocks and the ridges on the northeastern horizon.

SELECTING A LANDING SITE*

The Lunar Orbiters had photographed specific sites that had looked suitable on the basis of telescopic studies. Once the high-resolution images had been examined, the shortlist of candidate sites was filtered by stringent *operational* criteria.

Firstly, the Apollo flight dynamics team insisted that the prime site be east of the lunar meridian in order to allow room further west for one or two back-up sites that would be suitably lit if the launch were to be postponed by a few days. Launch opportunities ('windows') for any given site occurred only once per month, and it was thought better to go for a secondary site a few days late than wait a month for the prime site to present itself again. An easterly target was required because the Apollo spacecraft would fly in from the east. It had to be just after local sunrise at the site because the Sun had to be low on the horizon to cast sufficient shadow to reveal the topography. The Moon's axial rotation is tied to its orbit of the Earth, so it spins once a month. As a result, the Sun moves across the lunar sky at a rate of only 12 degrees in 24 hours. This in turn meant that the back-up sites had to be spaced about 24 degrees apart in lunar longitude so that the illumination would be right for each 2-day's delay in launch. On the other hand, the prime site could not be too far east or there would not be

---

*   The actual landing sites of all lunar missions can be found on page 365.

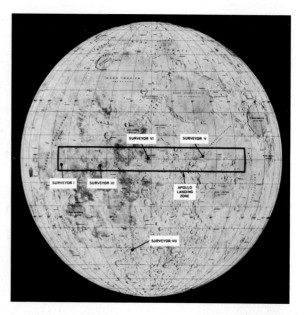

With the exception of Surveyor 7, which landed at a site of 'scientific interest' in the Southern Highlands, the Surveyors investigated the maria in the zone favoured by the Apollo planners.

sufficient time after coming around the limb to perform the navigational checks prior to initiating the powered descent.

Secondly, all the landing sites had to be in a narrow strip within 5 degrees of latitude of the lunar equator. A higher latitude site would involve a less propellant-efficient trajectory, and, for this first landing, economy was a priority. Furthermore, not only did all the sites have to be flat so as to minimise the need to manoeuvre in the final phase of the descent to avoid obstacles, the terrain during the run in had to be flat in order not to complicate the task of the landing radar.

Altogether, these safety constraints restricted the prime site for the first landing to one of the eastern maria on the equator, the back-up to Sinus Medii, and the reserves to the western hemisphere. Mare Foecunditatis, however, was too far east to provide a comfortable margin for the final navigational check. This left Mare Tranquillitatis, which contained two Apollo Landing Sites (ALS).

The most easterly site was ALS-1, east of Maskelyne Crater; it had been called 2P-2 by the Lunar Orbiter selectors. Apollo 8's trajectory in December 1968 was timed to view it under ideal illumination. Peering down from an altitude of 100 km, Jim Lovell described it as a smooth mare plain marred only by small craters. Bill Quaide and Verne Oberbeck studied the craters and inferred that the regolith was relatively deep. As only high-

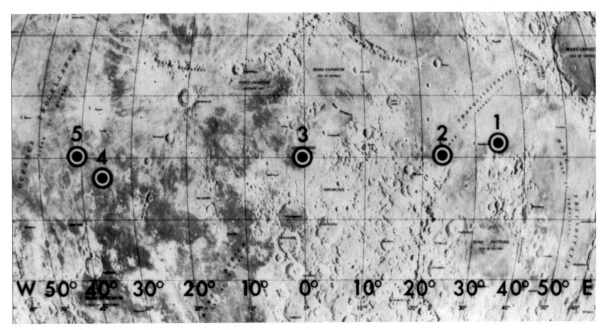

After being charted from lunar orbit, five sites were selected as candidates for the first Apollo landing. Operational constraints required them to be on the maria and near the equator.

energy impacts could penetrate the regolith to excavate bedrock, and the site selection process ensured that there would be no large craters in the vicinity, they argued that there would be few large rocks to threaten a landing. However, the fact that the mare was a little lighter than usual led Don Wilhelms to propose that this site might sit on a substrate of highland plains material. Although ALS-1 was a safe landing site, the fact that it was atypical of the maria made it less attractive for the first mission.

ALS-2 was in the southwestern part of Tranquillitatis. It had initially been called 'A3', then 2P-6. It would be easier to interpret its geological context because it was typical of the maria, but it was 25 km southeast of Surveyor 5's landing site,

and so had been subjected to alpha particle backscattering analysis and hence was regarded as "partially known". Nevertheless, it was one of the maria whose solar reflection was stronger toward the blue end of the spectrum than the red, and the cause of this spectral variation was unknown. In the end, therefore, the shortlist for the prime site for the first landing had only one entry. When Apollo 10 made a low pass over this site, Tom Stafford reported it to be generally suitable, but noted that the far end of the 'landing ellipse' was fairly rough. He warned that if the commander of a mission attempting to land at this site discovered that he was coming in 'long' and was running low on propellant, then his best option might be to abort.

2

Magnificent desolation

EAGLE'S DESCENT

The perturbation of the orbits of the Lunar Orbiters had shown that the distribution of mass within the Moon was surprisingly lumpy. Although the flight dynamics team believed they had understood this 'mascon' effect, Apollo 8 had been perturbed, and if a landing had been attempted then Frank Borman and Bill Anders would have found themselves descending on unfamiliar terrain. The Apollo 10 LM separated, inserted itself into the descent orbit, and followed through to PDI, at which point it shed the descent stage and rehearsed an abort. At this lower altitude, the perturbation was greater, and if Tom Stafford and Gene Cernan had tried to land they too would have come down off target.

By the time of Apollo 11, the issue was under control. Nevertheless, as Neil Armstrong made the final navigational check prior to PDI he realised that they were heading for the rough end of the target ellipse. Indeed, when Eagle pitched upright for the final phase of the descent, and Buzz Aldrin began to recite the angles for the Landing Point Designator that indicated to him where the computer intended to land, Armstrong saw that it was heading for a large crater surrounded by a blanket of debris. He didn't really know where he was, but it didn't matter because it was not important that he set down at a *specific* point. The imperative was to find a *safe* spot, so he took a degree of manual control and cut his rate of descent in order to prompt the computer to overfly the unfortunately located crater at an altitude of barely 100 metres. This 180-metre-wide pit had been called West Crater because it was west of the planned landing site. Once he was beyond the crater, Armstrong took full control and started to ease Eagle down. As he sought a flat spot, he was no doubt haunted by Stafford's stark warning about straying downrange: "If you don't have the hover time, you're going to have to shove off." And

Armstrong had little choice but to continue to fly downrange, because there was a 25-metre crater (later dubbed 'Little West') immediately beyond the fringe of West Crater's ejecta blanket. With literally seconds to spare, Armstrong set Eagle down on clear ground some 45 metres beyond the smaller crater, then casually reported his success: "Houston, Tranquility Base here. The Eagle has landed."

The Apollo 11 mission patch.

As he flew overhead in Columbia, Mike Collins used a telescopic sextant to look for Eagle. He hoped to recognise it because it would cast a shadow in the opposite sense to the craters, but was unable to locate it because he did not really know where to look. In fact, the actual landing point would be identified only via the film from the 16-mm movie camera that had been mounted in Aldrin's window. The computer had aimed for a point almost 8 km beyond the designated site, and Armstrong had landed 400 metres further downrange.

Neil Armstrong, Apollo 11 commander.

Michael Collins, Apollo 11 command module pilot.

Buzz Aldrin, Apollo 11 lunar module pilot.

As they worked through the post-landing checks, Aldrin remarked on the strange illumination. It was impossible to see surface detail when looking straight ahead, which was due-west, because there were no shadows when viewing down-Sun in the 'zero phase'. Looking cross-Sun, however, the definition in the airless environment was truly amazing. In his first out-of-the-window geological report, Aldrin judged there to be "every variety of rock". It looked as if they would have their work cut out collecting a suite of representative samples. As he bent forward to peer out, he realised that the colour of the rocks was dependent on the viewing angle with respect to the Sun, which would complicate formulating descriptions.

MOONWALK!

"Okay, Houston," Armstrong called several hours later, "I'm on the porch."

"Roger, Neil," replied EVA CapCom Bruce McCandless. Armstrong pulled a lanyard to release the Modular Equipment Stowage Assembly (MESA) on the descent stage under Aldrin's window, which aimed a camera towards the ladder as it hinged down horizontal. "And we're getting a

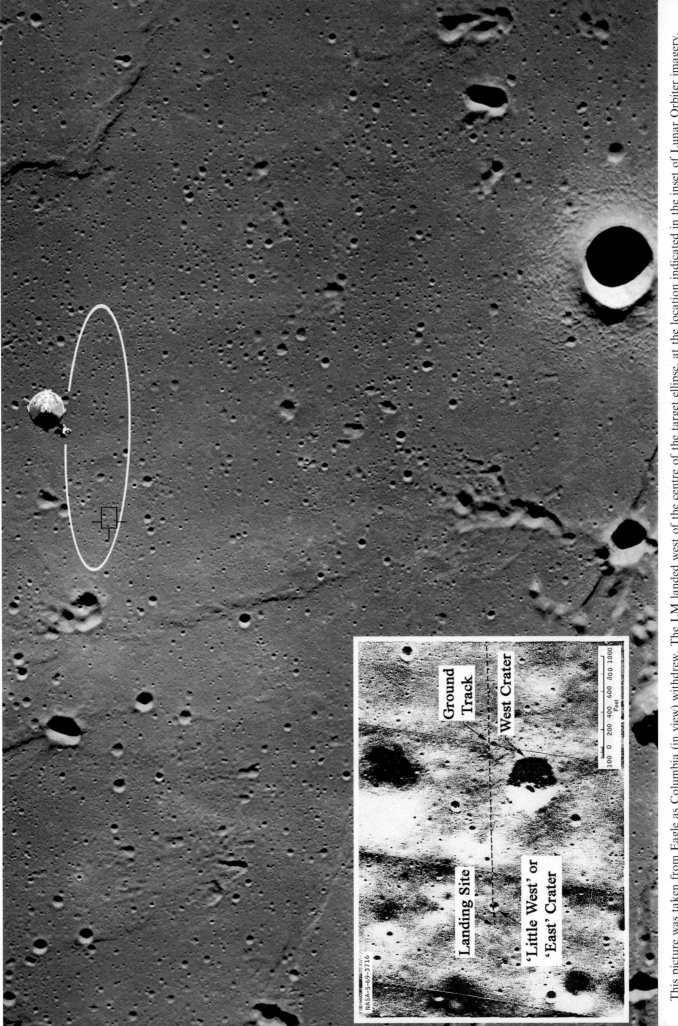

This picture was taken from Eagle as Columbia (in view) withdrew. The LM landed west of the centre of the target ellipse, at the location indicated in the inset of Lunar Orbiter imagery.

Ground
Track

West Crater

Landing Site

'Little West' or
'East' Crater

100 0 200 400 600 800 1000
Feet

NASA-S-69-3716

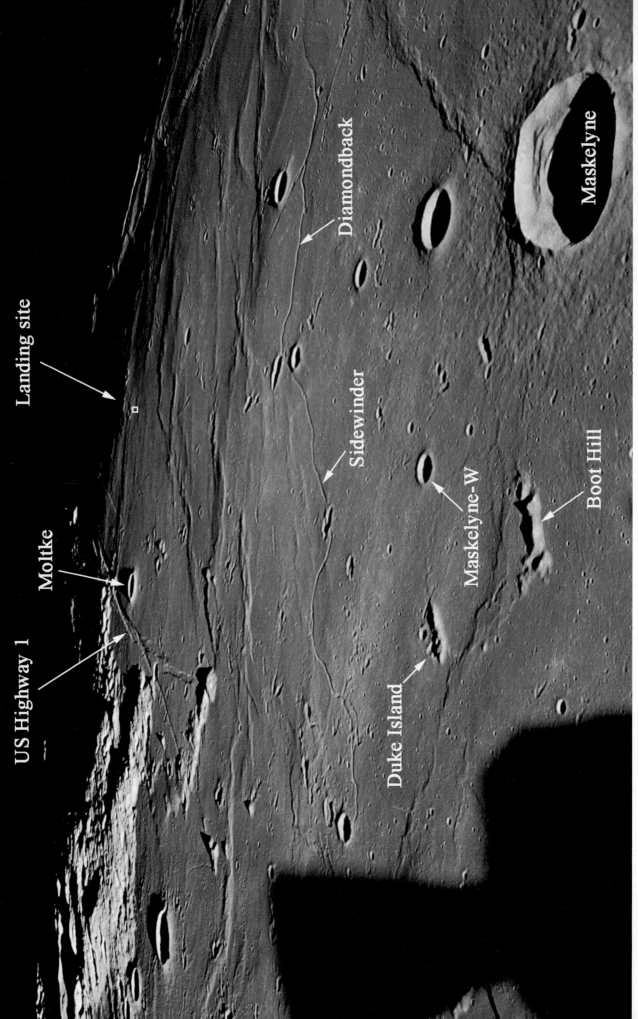

A westward-looking view of the Apollo 11 landing site taken from orbit just after local sunrise. The profile on the left of the frame is the silhouette of one of Eagle's attitude control thrusters.

A TV camera in the Modular Equipment Stowage Assembly gave 'live' coverage as Neil Armstrong descended Eagle's ladder.

The version of the Hasselblad camera developed for use on the lunar surface.

picture on the TV!" said McCandless. The picture was pretty contrasty, and it was initially upside down, but it was relayed by networks around the world with the tag 'Live From The Moon'.

"I'm at the foot of the ladder," Armstrong reported. Standing on the 1-metre-wide footpad, he looked at the soil; it was "almost like a powder". As a pilot, the highlight of the mission had been the landing. But for the public, Man would not have reached the Moon until Armstrong's foot was on the surface. He had given a lot of thought to what he might say to mark the historic event – the first time that a human being had *stood* upon the surface of another planet. "I'm going to step off the LM now." Keeping a firm grip of the ladder with one hand, he turned and set his boot on the ground. "That's one small step for a man," he observed, and then after a moment's reflection, "one giant leap for Mankind." He raised his foot, and noted that the soil held a sharp imprint of his boot. He scraped the loose soil with his toe, and saw that it coated his boot "like powdered charcoal". Remaining in the LM's shadow, he stood back a few metres to assess the landing site. "We're essentially on a very level

place." He was referring to the site on which he had set Eagle down, but this applied to the plain itself – apart from the small craters and rocks in the near-distance, the horizon was essentially featureless.

Once Aldrin had lowered a Hasselblad camera by lanyard, Armstrong mounted it on a bracket on his chest. This done, he stepped about 5 metres out to take the 'contingency sample'. He had a folded-handle scoop in his pocket. Although the soil was easily scuffed by his boots, it became very compact a few centimetres down, so he scraped the bag across the surface to fill it up with soil and a few small rocks, and then detached the bag and put it in his pocket. Now, even if he was obliged to curtail the EVA, he would not leave without a piece of the Moon. As an experiment, he poked the end of the handle into the ground. It penetrated only about 15 cm before coming to a halt. Either he had struck a rock, or there was a layer of consolidated material just below the surface. This impromptu soil-mechanics experiment over, he threw the handle away. "You can really throw things a long way up here!"

As Aldrin made his way out, Armstrong photographed him. The low rung of the ladder had been placed high up the strut just in case a 'hard' landing compressed the strut, but the LM had barely indented the surface. On turning around, he was awed. "Beautiful view."

"Magnificent," Armstrong agreed.

"Magnificent desolation," Aldrin announced, having finally found an adequate description.

Aldrin's first task was to assess his stability. With the Portable Life Support System (PLSS) on his back, he had to lean far forward in order to balance his centre of mass. This done, he joined

A depiction of the commemorative plaque on the main strut of the Eagle's landing gear.

This picture of Buzz Aldrin's boot print in the regolith shows that although it was almost a powder the material was sufficiently cohesive to retain the imprint of the boot's tread.

Armstrong at the ladder to unveil the plaque on the strut.

"Here," Armstrong read, "men from the planet Earth first set foot upon the Moon." There was a depiction of the Earth on the plaque, together with the signatures of the participants. "We came in peace for all Mankind."

Moving to the MESA, Armstrong retrieved the TV camera so that it could be mounted on a tripod 20 metres northwest, where it would be able to watch them at work. The whole area was peppered with small craters, and as he walked he noticed that in the middle of one fresh-looking pit there was what seemed to be black glass.

Meanwhile, Aldrin unpacked the Solar Wind Collector (SWC). This sheet of aluminium foil was to be 'hung' facing the Sun. On being returned to Earth, it would be 'cleaned' in order to isolate the solar wind ions which had penetrated the metal. He discovered what Armstrong had already noticed: there was a solid substrate just beneath the surface, but the lightweight staff remained standing.

With the TV deployed, Armstrong unpacked Old Glory. This national flag did not constitute a territorial claim, of course; this was prohibited by treaty. But the flag underlined the Cold War motivation that had prompted the decision to land on the Moon. As a flag could not 'fly' on the airless surface, the cloth was supported by a telescoping horizontal strut.

Armstrong went to unpack the Sample Return Containers (SRC) at the MESA. Aldrin undertook a systematic mobility evaluation, and was soon running back and forth between the LM and the camera. The only complication was the need to position himself to accommodate his elevated centre of mass as he changed direction. As Aldrin finally drew to a halt, McCandless called to announce that President Richard Nixon wanted to speak to them, so the astronauts took a few minutes off to reflect upon their achievement, and then it was back to work. While Armstrong ran back and forth with a scoop collecting rocks for the 'bulk sample', Aldrin borrowed the Hasselblad and took a series of pictures which included the imprint of his boot in the soil, the state of the LM's landing gear and several panoramic views of the landing site.

By this point, they were running about 30 minutes behind schedule, but there had not been a plaque or the flag in the 'integrated' simulation that had been used to refine the timeline, and there had certainly not been a Presidential call. However, their PLSSs held oxygen and coolant water

for at least 4 hours, so this slight delay was not a matter for concern. In fact, for this first mission, only 2 hours of surface activity had been intended. It had once been planned to set up a full 'scientific station' of instruments that would monitor the lunar environment once they had departed, but this had been reduced to two instruments. The Early Apollo Scientific Equipment Package (EASEP) comprised the Passive Seismic Experiment (PSE) to report on seismic activity, and the Laser Ranging Retro-Reflector (LRRR) mirror for geophysical research. These instruments were stowed in the Scientific Equipment (SEQ) Bay, on the left-rear side of the LM, and it was Aldrin's job to unload them. While this was being done, Armstrong took a series of extremely close pictures of the surface using a stereoscopic camera. Aldrin lifted the packages, one in each hand in order to maintain his balance, and carried them to the deployment site, 20 metres southwest of the LM. Armstrong emplaced the LRRR, then resumed taking pictures. The LRRR was totally inert; it had only to be oriented facing the Earth. The PSE, however, had to be set up with solar panels oriented to catch the Sun. It also had to be levelled, which proved a little tricky. Once activated, it began to transmit data to Earth, and was revealed to be sufficiently sensitive to detect the vibrations caused by the two men walking around.

The original plan had called for 30 minutes of 'documented sampling', including driving in a core tube and photographing rocks before they were retrieved, so that their context would be known, but McCandless warned that there was only 10 minutes available for this work. When he had taken the bulk sample, Armstrong had taken care to collect a suite that included both typical and exotic rocks, so he knew it would not be a disaster if they left without a documented rock. When the sampling plan had been drawn up, it had been generally assumed that they would be far from a large crater, but the pit that he had been obliged to fly over to reach a clear site was only 45 metres away, so, without saying anything, he ran to this crater's southern rim and took a panorama across it, including the LM for context. Gene Shoemaker had predicted that the regolith would be 3 to 6 metres deep, and this impact had only just managed to reach the basement. As Armstrong would later put it, the crater had some "pretty good-sized" rocks on its floor. He did not venture in to collect one, though.

Meanwhile, Aldrin went to get the core sample alongside the SWC. However, no matter how

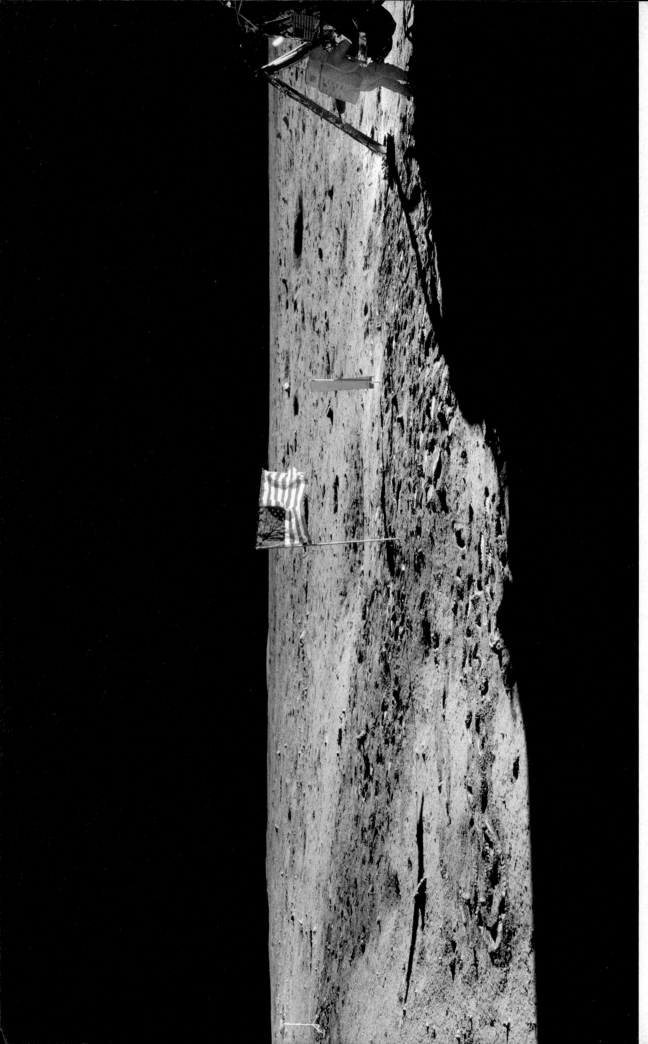

A northward-looking view showing the television camera, the US flag, the Solar Wind Collector and Neil Armstrong working at the Modular Equipment Stowage Assembly.

Buzz Aldrin works to unload the two instruments of the Early Apollo Surface Experiment Package from the Scientific Equipment Bay.

Buzz Aldrin stands alongside the deployed Passive Seismic Experiment, with the Laser Ranging Retro-Reflector in position just beyond.

Monitoring the signal from the Passive Seismic Experiment.

MOONROCKS

Ever since Kennedy committed NASA to "landing a man on the Moon", the agency had been planning to study the samples that the astronauts would bring back, and it built the Lunar Receiving Laboratory (LRL) alongside its Manned Space-craft Center. The handling procedures would not only protect the lunar material from the Earth's atmosphere, but would also protect the Earth against any 'bugs' sufficiently hardy to survive on the radiation-soaked lunar surface. The risk of back-contamination was exceedingly low, but no chances were being taken for this 'first contact' situation.

On 25 July, the SRCs arrived. There was a tremendous sense of expectation as the Preliminary Examination Team assembled. There was almost 22 kg of rocks and soil. Once it was all catalogued, samples were sent to 150 Principal Investigators at institutions all around the world for analysis. With samples to examine, it was time for the geochemists and petrologists to play their part. Most astronomers ignored the Moon, just as they did the rest of the Solar System, preferring to study the Universe at large. The authority of the International Astronomical Union (IAU) in controlling the naming of lunar surface features had been undermined by astronauts naming the objects that they would see during their descent and on the surface. But as the Moon was transformed from being a light in the sky into a place for humans to explore in person, this was inevitable.

The 'cold' Moon theory that Harold Urey had advocated was badly undermined by the first large rock to be examined in the LRL. It was basaltic, and it implied that the mare plain was a lava flow. To Urey, the maria were the result of impact-melting on an enormous scale. While the astronauts had been out on the surface, he had been encouraged that they had *not* reported finding the "frothy vacuum lava" proposed by Gerard Kuiper. Having dismissed Armstrong's report of vesicles, Urey was sure that the evidence would confirm the Moon to be undifferentiated and, apart from impact damage, essentially a pristine object of ultrabasic material. Isotopic dating indicated that the basalt had crystallised 3.84 to 3.57 billion years ago. Compared to terrestrial basalt, the lunar form was rich in titanium, which was the reason for it being so dark. Given the Moon's low bulk density, it was clear that the body could not be made entirely of such rock. It argued for an endogenic rather than an exogenic origin. This proved that the Moon had thermally

much he hammered it, the tube would not penetrate the substrate. He tried again 3 metres away, with the same result, so he settled for what he had in the tube and then stowed the tube on the MESA, together with the rolled up SWC, ready for Armstrong to pack away. In the expectation that the lunar soil would be loosely compacted, the tube's penetrometer had been fitted with an inner flange to prevent material from tricking out as the tube was withdrawn, but this feature had made the tube impossible to drive in. After several prompts from McCandless, Aldrin climbed the ladder back into the LM.

Armstrong hastily used a pair of tongs to lift rocks, which he popped into a large collection bag, describing one as "really vesicular". He topped up the box with 6 kg of regolith to ensure that the rocks were packed tight. Once the two boxes were sealed, he used a lanyard on a pulley to hoist them one at a time to Aldrin, and then left the surface without comment.

Having taken manual control in the final phase of Eagle's descent, Neil Armstrong set down 45 metres beyond a 25-metre-diameter crater, later dubbed 'Little West'. Towards the end of the moonwalk he sprinted back to take a panoramic sequence of the crater's interior.

A view out of Buzz Aldrin's window after the moonwalk, showing the flag, the TV camera and lots of boot prints.

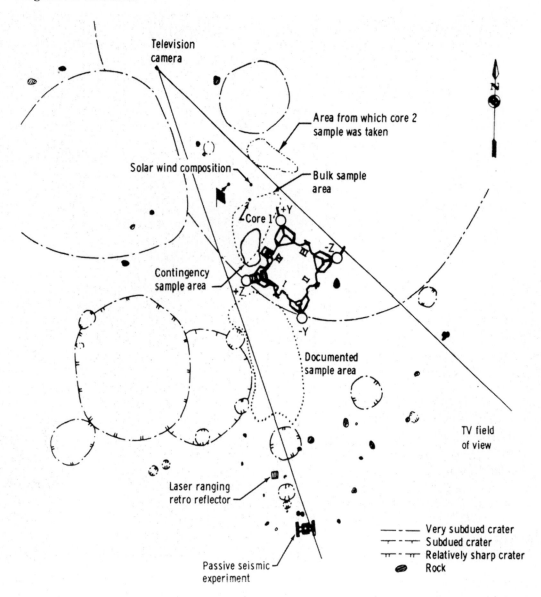

Television camera

Area from which core 2 sample was taken

Solar wind composition

Bulk sample area

+Y

Core 1

-Z

Contingency sample area

+Z

I

-Y

Documented sample area

TV field of view

Laser ranging retro reflector

Passive seismic experiment

——— Very subdued crater
—·—┬ Subdued crater
—┬┬— Relatively sharp crater
⬤ Rock

A post-flight analysis of the area surrounding Eagle in which the moonwalk was conducted.

differentiated, with the lightweight aluminous minerals migrating to the surface and then being overlain by an extrusion of lava from the interior. This effectively ruled out Urey's hypothesis. The titanium-bearing mineral, which was new to science, was dubbed 'armalcolite' by combining the first letters of the three astronauts' surnames. The most striking discovery was the total absence of hydrous minerals. The lunar basalt was also deficient in volatile metals such as sodium. The low-alkali (that is, sodium-depleted) lava would have had an extremely low viscosity, which explained why it had flowed so far and left so few 'positive-relief' features. The Moon accreted from the solar nebula 4.6 billion years ago, and the basins were excavated during the first billion years.

In the case of Tranquillitatis, the dark plain that we see today was built up by episodic volcanism over *several hundred million years*. The fact that *two* forms of basalt were identified implied either that there had been several sources or that one source had undergone chemical evolution over an extended period.

Armstrong's 'packing' soil produced a wealth of data. The majority of the lithic fragments were pulverised basalt. There was little material of meteoritic origin. This was consistent with the gardening process in which a large impact dug up bedrock which was progressively worn down by smaller impacts to yield the seriate regolith. Many of the discrete samples turned out to be consolidated soil, made by the shock of an impact

With no time left for 'documented sampling', at the end of the moonwalk Neil Armstrong collected some 20 'grab samples' massing a total of 5.5 kg. Here they are viewed in the Sample Return Container in the Lunar Receiving Laboratory.

compressing the regolith. The 'glassy material' that Armstrong observed in the middle of what was evidently a fresh crater had come as a surprise, but the geologists soon had several candidate explanations. It was argued by Jack Green that the crater marked the spot where a semi-molten 'volcanic bomb' fell after being ejected from an explosive vent, but this supposed that the Moon had recently been active and that there was a local vent, neither of which suppositions had been established. Tommy Gold proposed a typically creative model of which the least said the better. In fact, the glassy material was just regolith fused by the shock of a high-speed impact.

A small residue in the regolith was startlingly different in character. On the basis of his analysis of data from Surveyor 7, which had landed on the flank of Tycho in the Southern Highlands, Shoemaker had posited that 4 per cent of the Tranquillitatis regolith would be minuscule fragments of light-coloured rock, and this was the case. This rock was plagioclase feldspar. Although this is one of the commonest minerals in the Earth's crust, terrestrial plagioclase is rich in sodium; the lunar variety was depleted in sodium but enriched with calcium, making it calcic-plagioclase. A few of the fragments turned out to be sufficiently pure to be anorthosite (this being the name for a rock that is at least 90 per cent plagioclase), but most were diluted by a variety of mafic minerals and so were more properly called anorthositic gabbro, as at the Surveyor 7 site.

Some of the lunar material returned by Apollo 11. Top left: sample 10049 (0.193 kg); lower left: 10018 (0.213 kg); top right: a piece of highly vesicular basalt 10072 (0.447 kg); and lower right: a 'glass bead' of melt 0.25 mm in diameter that was in the regolith and was deformed by several small cavities, one of which is surrounded by a fragmented area that was produced by the shockwaves of the small impact.

Shoemaker's rationale for the presence of highland material in the mare regolith was based on the way in which the most recently formed highland craters sprayed out rays of material. As regards the highlands, it was now clear that the Moon's primitive crust was composed of anorthositic rock. Nevertheless, the geologists were eager to find evidence of the 'upland fill' that they believed made the 'light plains' in the highlands, but there was no sign of this in the Tranquillitatis regolith. Further study of the anorthositic fragments yielded an insight into the absence of europium (one of the rare earth elements) in the mare basalt. It turned out that the plagioclase had an affinity for europium, and was enriched in it. The original anorthositic crust had soaked it up so efficiently that the lava that extruded from fractures in the floors of the basins was from a reservoir that was deficient in this element.

When the Apollo 11 results were formally presented at the First Lunar Science Conference in Houston in January 1970, they confirmed the idea that the lunar crust formed as 'scum' on a magma ocean.

A SENSE OF PERSPECTIVE

By 1969, studies of meteoroids had shown that the Solar System formed 4.6 billion years ago. However, the oldest known terrestrial rocks were 3.5 billion years old. The first billion years of our planet's history was a *total* mystery, and *very little* was known of the ensuing 3 billion years.

As it happened, geology itself was undergoing rapid development in the late 1960s. The study of magnetic anomalies on the sea floor had demonstrated that the Earth's crust is composed of a set of 'plates'; mid-ocean ridges are constructive 'margins' where plates are forced apart; and ocean trenches are where they are subducted into the mantle. The theory of 'plate tectonics' explained why the quest for the ancient crust had been so fruitless; most of it had been 'recycled'. That which

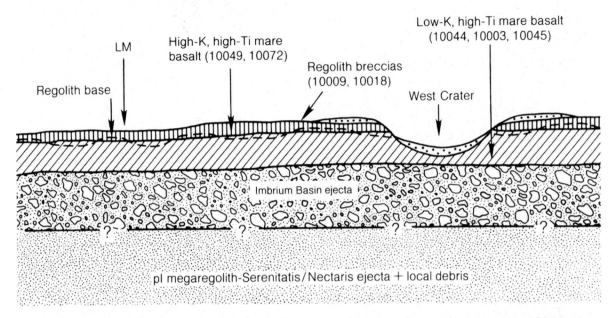

A schematic east–west geological cross-section through the Apollo 11 landing site showing at least two basalt lava flow units overlying older ejecta from major basins (modified after Beaty and Albee, 1980). The presence of the Imbrium ejecta is inferred. Numbers refer to specific collected samples representative of the various units inferred to be present. The base of the regolith (dashed line) locally penetrates into the Low-K mare basalts. The abbreviation 'pI' stands for pre-Imbrian in the megaregolith. Courtesy the Lunar and Planetary Institute and Cambridge University Press.

survived had been intensely metamorphosed and eroded. One reason that geologists were eager to study the Moon therefore, was to gain insight into the early history of the Earth. One inescapable conclusion was that even though the evidence had been destroyed long ago, the bombardment that excavated the vast lunar basins must also have pummelled the Earth.

3

'Pin-point' landing

RELAXING THE PACE

After so many years in development, the Apollo program* had sprung forth from the starting block with a flight in cislunar space at the end of 1968 and a schedule for 1969 which called for missions at roughly two-monthly intervals. The driving force behind this hectic pace had been the desire to satisfy John F. Kennedy's challenge of a landing on the Moon before the decade was out. Apollo 10's rehearsal of lunar orbit activities to the point of PDI cleared the way for Apollo 11 to try for a landing. If this was achieved, there would be enough Saturn V rockets for a total of ten expeditions to the surface. Even as Eagle landed, and Armstrong took his "small step", it was expected that the flight rate would be sustained, and that, consequently, the expeditionary phase of lunar exploration would be concluded in 1971 to clear the way for a new program of space station

The Apollo 12 mission patch had a nautical theme reflecting the all-Navy crew.

development. Upon Apollo 11's triumphant return, however, NASA announced that Apollo 12 would be postponed from September to November, and that the subsequent pace would be reduced to a 3-monthly basis. This slow-down did not overly concern the scientists, because, by enabling the results from one mission to feed through to the next, it would enhance the overall coordination of the program.

In terms of the alphabetic labels used to describe the build-up of the program's capability, the initial landing ('G') was to be followed by four 'H'-class missions. Unlike the first landing, in which all the surface activity was undertaken in a single day, these new missions were to involve an overnight stay so that there could be two 4-hour EVAs, with a full scientific station being deployed on the first outing and site exploration on the second. In contrast to Apollo 11, on which, apart from Neil Armstrong's impromptu inspection of 'Little West' Crater, he and Buzz Aldrin had remained close to Eagle, on the 'H'-class missions it was envisaged that the crews would eventually be allowed to venture out a kilometre or so while undertaking a meticulously planned traverse to study the geological 'feature' that had attracted the site selectors.

THE NEED FOR PRECISION

Apollo 11 had proved the LM's ability to land on the Moon, but the fact that it had strayed so far off target was embarrassing, and the flight dynamics team were eager to demonstrate that they could deliver a 'pin-point' site. The ability to land within a kilometre of a given location was a prerequisite to being able to follow a specific

---

* Details of the Apollo missions are given on page 366.

geological traverse. The flight dynamics team had devised a simple method for the LM to correct for trajectory errors. The Primary Guidance and Navigation System (PGNS) would be told to head for a 'different' landing point, although in reality this redirection would take it to the intended target. They were so sure that this would work that they reduced the size of the target ellipse. Also, given the relaxation of the pace, it was decided to reduce the requirement for back-up sites from two to one.

There had been five sites on the shortlist for the historic first landing. The easterly ALS-1 and 2 sites in Tranquillitatis had been backed up by 3 in Sinus Medii, with 4 and 5 in Oceanus Procellarum in reserve in case of a prolonged launch delay. After Apollo 11 went to ALS-2, it would have been natural to send Apollo 12 to one of the other sites, but the geologists were eager to sample something more interesting, and, inevitably, this meant landing as *close* as possible to a sizeable crater. In fact, even before Apollo 11 flew, the site selectors had drawn up a list of interesting craters for this eventuality, starting by re-examining the sites that had been rejected for the first landing owing to the *inconvenient* proximity of a crater.

Although a number of pin-point sites were nominated in the vicinity of ALS-5, they were pretty bland. A site in Hipparchus Crater was scientifically interesting, but was deemed to be insufficiently documented. Hal Masursky's suggestion of the hummocky terrain to the north of Fra Mauro Crater was rejected as being too demanding. Relaxing the constraints enabled ALS-6 to be reinstated. This was in the Flamsteed Ring, which was almost totally swamped by Procellarum. The regional stratigraphy made it Eratosthenian in age. Its geological context was both well understood and significant, so samples would yield absolute dates to refine the timescale of a wide area. However, the fact that ALS-6 was not far from where Surveyor 1 had landed prompted the flight dynamics team to propose making *this* the target. Of course, it would be humiliating if Apollo 12 missed, but to set down within walking distance of another spacecraft would be a powerful demonstration of precision. Unfortunately, this site was so far west that it left no room for a back-up. But Surveyor 3 was located a little further east and could accommodate a back-up. The Lunar Orbiter team had designated this site 3P-9 and, using their con-

Apollo 12 commander Pete Conrad and lunar module pilot Al Bean in a LM simulator.

servative criteria, had ruled it too rough for the first Apollo landing. Once it was reinstated, the geologists eagerly endorsed it because it was crossed by a ray from Copernicus, just 370 km to the north. The prospect of dating Copernicus swung the selectors behind this site, and it was designated ALS-7.

In addition to scouting the blandest sites for the first Apollo landing, the Lunar Orbiters had photographed many 'feature' sites. Nevertheless, this program had been so dedicated to locating wide open sites for the first landing that the coverage of more demanding sites was too cursory to certify them. It was evident, therefore, that if Apollo was to exploit the relaxation of the flight dynamics constraints, the early missions would have to inspect 'out of area' sites for their successors. The selection of a site for Apollo 12 would therefore significantly influence the remainder of the program. ALS-7 offered the benefit of surveying both the important Fra Mauro area and Davy Rille, a chain of craters believed, by some, to be volcanic vents on the line of a fault. Once pinpoint accuracy had been demonstrated, it would be feasible to assign 'feature' rather than 'area' targets.

Dick Gordon, Apollo 12 command module pilot.

On 10 April 1969, when Pete Conrad was given command of Apollo 12 and a launch slot in September, there was a distinct possibility that he would become the first man to walk on the Moon. Apollo 10 had yet to execute its rehearsal in low orbit to clear the way for Apollo 11 to attempt a landing. If either mission failed, he and Al Bean would inherit the task, and the ALS-2 landing site, so they began training with this in mind. When Apollo 11 returned in triumph, and Apollo 12 was slipped from September to November, they switched to a more demanding 'H'-class mission at a more adventurous site. Conrad had flown with Dick Gordon, the CMP, on Gemini 11, and they had served on the Apollo 9 back-up crew, along with Bean. As they were an all-Navy crew, they named their vehicles Yankee Clipper and Intrepid. A commander sets the mood for his crew. Armstrong, burdened with the pressure of being 'first', had been low-key and all business, but Conrad intended to have fun.

LANDING BY THE SNOWMAN

"Hey! There it is!!" Conrad called excitedly when Intrepid pitched and he gained his first look at where the PGNS was heading. The 200-metre crater that Surveyor 3 had set down in had been dubbed Surveyor Crater. From Conrad's point of view, it and a smaller crater beyond resembled the body and head of a snowman respectively. To his delight, the line of the LPD indicator ran through the target. "Son-of-a-gun!" he chuckled. "Right down the middle of the road."

"42 degrees," Bean called, a few seconds later.

Conrad had once jokingly suggested that the computer be programmed to aim for Surveyor Crater. In fact, it had been told to aim for a point to the north of the crater. But the 42-degree mark on the LPD was precisely on Surveyor Crater. "It's heading straight for the centre of the crater!" he exclaimed. "Look out there." The LMP's job was to 'stay inside' and read the instruments, but this was so amazing that Conrad wanted his friend to see it too. "I can't believe it."

Bean stole a glance for himself. "Fantastic!" he agreed.

"Go for landing," confirmed CapCom Gerry Carr.

Once it became clear that the PGNS was indeed heading for Surveyor Crater, the interior of which was in deep shadow, Conrad intervened and nudged it a little north, in order to be ideally placed for the preferred traverse. However, there seemed to be rocks everywhere. At an altitude of 100 metres, he reduced the rate of descent and made a hovering tour of Surveyor Crater's northern rim in search of a clear spot.

"You're really manoeuvring," Bean observed upon seeing the wild gyrations of the inertial reference '8-ball' as Conrad tilted the LM this way and that to impart and then to cancel horizontal motions.

"Yeah," Conrad confirmed. He was too busy to talk.

"Come on down, Pete," Bean prompted.

Conrad continued to hover, slowly manoeuvring around Surveyor Crater, until he saw a clear spot near the northwestern rim. Once he was over the desired position, he cancelled all of his horizontal components and made a vertical descent. As he passed through 30 metres, the DPS plume stirred up so much dust that he could not see the ground. Denied visual cues, he was obliged to 'come back in'. He knew that he had to land before he started to drift, because Surveyor Crater was somewhere behind him.

"Slow the descent rate," Bean urged upon seeing the sink rate.

"30 seconds," Carr warned.

"Contact light!" called Bean.

Conrad hit the Engine Stop button, and the LM settled.

"Good landing, Pete," Bean congratulated.

"We're in good shape," Conrad announced. "We're in a place that's a lot dustier than Neil's. That was an IFR landing."

"It's beautiful out there," Bean enthused, finally able to gawp out of the window.

"Congratulations from Yankee Clipper," Gordon called just before he went below the horizon. "Have a ball."

With Intrepid confirmed to be healthy, Conrad and Bean began to power down its systems so that they could start the first EVA. "Man, I can't wait to get outside," Conrad chuckled.

SCIENTIFIC STATION

"Whoopie!!" Conrad yelled as he jumped backwards off the ladder's lowest rung. "Man, that might have been a small one for Neil, but it was a long one for me." He stepped off to one side in order to sneak a peek into the crater behind Intrepid, and saw Surveyor 3's mast-mounted solar panel poking out of the shadowed far wall. He was delighted. Not only had he had landed on target, but the scientists has specified the correct crater.

Even in a community of mission-oriented test pilots, Conrad was an achiever. The carefully worked out timeline was designed to facilitate 'ticking all the boxes' on the checklist. His motto was: 'Get ahead, then stay ahead.' His curiosity satisfied, he 'ticked' the first box by scooping up 2 kg of regolith for the contingency sample while Bean made his way out. While Conrad erected a dish antenna, Bean transferred the TV camera from the MESA to a tripod. Unfortunately, as Bean carried the camera away from Intrepid, he inadvertently aimed it directly at the Sun and burned out the vidicon detector. This was all the more frustrating because the low scan-rate monochrome camera used previously had been superseded by a colour camera.

The objective of the first day's excursion was the deployment of the Apollo Lunar Surface Experiment Package (ALSEP). This 'scientific station' incorporated a more sophisticated seismometer than that left by Apollo 11 (which had been damaged by the intense cold of the lunar night). The package also had a magnetometer, a solar wind spectrometer and instruments to study the tenuous 'atmosphere' of exotic ions. The instruments had to be arrayed in a circle around a Central Station that provided communications and fed power from a plutonium Radioisotopic Thermal Generator (RTG). This power supply would keep the electronics warm during the lunar night. The instruments were on a pair of pallets stored in the Scientific Equipment (SEQ) Bay on the left-rear quadrant of the descent stage. Bean employed pulleys to lower the pallets to the surface. Everything went well until Bean tried to 'load' the power unit. As a precaution, the fuel element for the RTG was carried in a separate flask mounted on the rear of the LM. The flask had to be hinged down, the domical end cap unscrewed, and the cylindrical element eased out and transferred to the power unit. To Bean's dismay, the element stuck fast half way out of the flask. Determined not to slip behind on the timeline, Conrad got the geological hammer and delivered a sharp blow to the flask to shake the element free. "Never come to the Moon without a hammer," he chuckled. Ed Gibson, serving as the EVA CapCom, dutifully noted down this empirical gem for future crews. Bean fastened one pallet to each end of a crossbar, and then carried the 'dumbbell' a hundred metres west to a flat spot that Conrad had selected. Removing the instruments from the pallets was no trivial task, because they were bolted on. These were 'fast release' bolts, but required a long tool to be used, and it was

Pete Conrad just after exiting Intrepid's hatch.

Al Bean descends Intrepid's ladder.

Pete Conrad set Intrepid down just outside the 200-metre-diameter crater in which Surveyor 3 had landed (that spacecraft is out of frame). Conrad is working at the Modular Equipment Stowage Assembly. A large dish antenna was erected to broadcast the TV back to Earth.

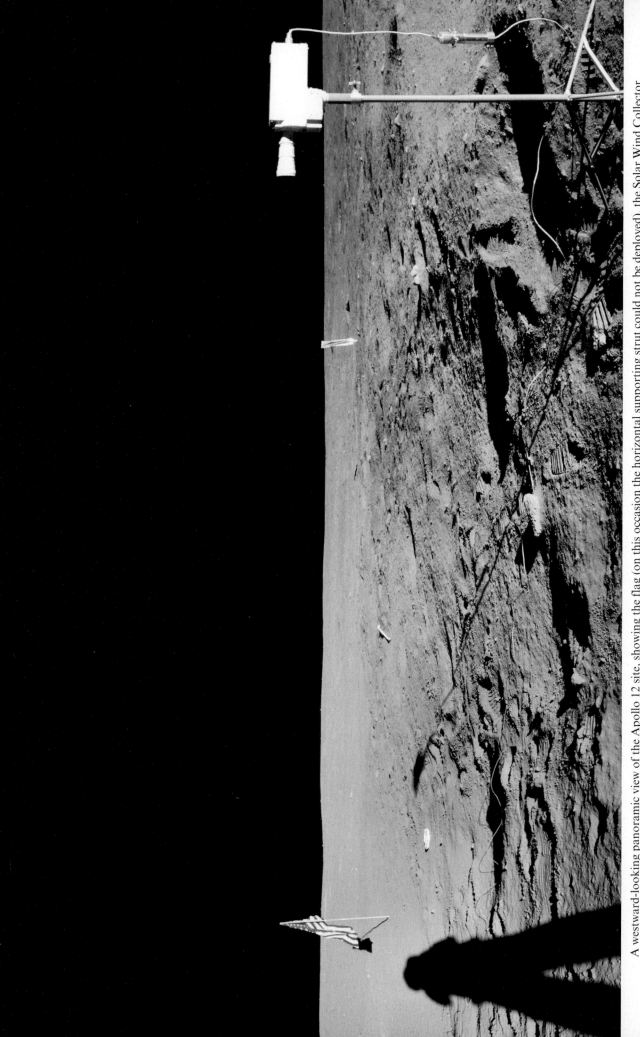

A westward-looking panoramic view of the Apollo 12 site, showing the flag (on this occasion the horizontal supporting strut could not be deployed), the Solar Wind Collector and the TV camera – which was unfortunately 'burned out' by the Sun entering its field of view while it was being carried by Al Bean to the deployment position.

impossible to see the bolts in the absolute darkness of the shadows. With so many pieces of apparatus to set up, Conrad and Bean had small spring-bound books strapped to their cuffs. The pages were laminated and the binder followed a curve, so the pages could readily be flicked over, yet remain open at the selected page. This cuff-checklist was such a success that all future crews employed it. Armstrong and Aldrin had had a single page sewn onto their gloves.

With half-an-hour to spare after completing the deployment of the instruments, Conrad and Bean set about 'selected sampling'. They had spotted a pair of conical "mounds" about 1 metre across the base, one just outside Head's northern rim and the other 50 metres further out. The geologists' ears perked up when the astronauts described the mounds as being "like volcanoes", but they proved to be nothing more exotic than piles of loosely consolidated ejecta. Nevertheless, they took a sample.[1]*

As they were already half-way there, Gibson suggested that they go out another 75 metres in order to take a look at Middle Crescent. On their way, they stopped to take a few samples[2] from a small fresh-looking crater. Standing on the eastern rim of the 350-metre-wide Middle Crescent Crater, which Bean said looked "rather old", their shadows extended far across the floor, and the opposite wall was washed out with backscattered zero-phase glare. When Conrad interpreted the boulders on the floor of the crater to mean that it had exposed "bedrock", Bean felt obliged to point out, "We don't see any outcropping." When Gibson hinted that they ought to leave, they lifted a rock[3] that was clearly a fine-grained basalt, and then literally ran home. Although breathless by the time they got back to Intrepid, they were delighted that their first day had gone so well. With a few minutes to spare, they sampled several more rocks[4] from beside the spacecraft.

Armstrong and Aldrin's 22 hours on the lunar surface had included a sleep assignment, but sleep had proved elusive. Conrad and Bean, however, *had* to rest, because they were to spend two days on the Moon. Nevertheless, even after a full day involving a 4-hour excursion, it was no simple matter to 'wind down'. Although they were allowed to remove their helmets and gloves, it had been decided that they should remain suited.

This was not so much out of concern for Intrepid's continuing integrity, as for the zippers and seals of the suits, which might well deteriorate if they became clogged with lunar dust and leak during the second excursion.

GEOLOGICAL TRAVERSE

The run out to Middle Crescent at the end of Apollo 12's first moonwalk was trivial in comparison to the traverse planned for the second excursion. Although this was to take the form of a loop around the nearby craters, involving a total trek of 1.5 km, it was important that Conrad and Bean never stray more than about half a kilometre from home, because if they were obliged to rely on their emergency Oxygen Purge System (OPS) this was the greatest distance they could be expected to run and still have time to scramble back into the LM. The duration of the moonwalk was set by the rate at which the PLSS consumed oxygen and coolant water. The rates at which Armstrong and Aldrin had consumed their supplies had provided a calibration which allowed the rules to be eased sufficiently to extend the EVA duration to 4 hours. Nevertheless, from time to time, Houston sought an "EMU check", whereupon the astronauts reported the readings on their gauges. Although the consumption during the ALSEP deployment had enabled the first EVA to be extended by half an hour, the rules for the geological traverse were conservative since it was better to be safe than sorry.

A number of traverse routes had been charted, each to suit a different landing point within the general vicinity of Surveyor Crater. As it happened, Conrad had set down right on one of these routes. Middle Crescent was actually on the route that they would have taken if they had landed downrange and north of target. Being on target, however, their route would be the preferred one: around Head's northern rim, south to the near rim of Bench, out west to Sharp, back east past Bench's southern rim to Halo, then northeast to Surveyor, where they would visit that spacecraft and a crater called Block, after which they would return home. Apart from Surveyor and Head, the craters had been named for their distinctive morphological characteristics: Sharp had a sharp-looking rim; Bench appeared to have an inflection in its wall (a feature that the geologists referred to as a bench); Halo seemed to be surrounded by lighter material; Block, which was on the northeastern rim of Surveyor Crater and

---

* Information pertaining to each sample is given on page 369.

Al Bean retrieves the plutonium core of the Radioisotope Thermal Generator with which to power the instruments of the Apollo Lunar Surface Experiment Package.

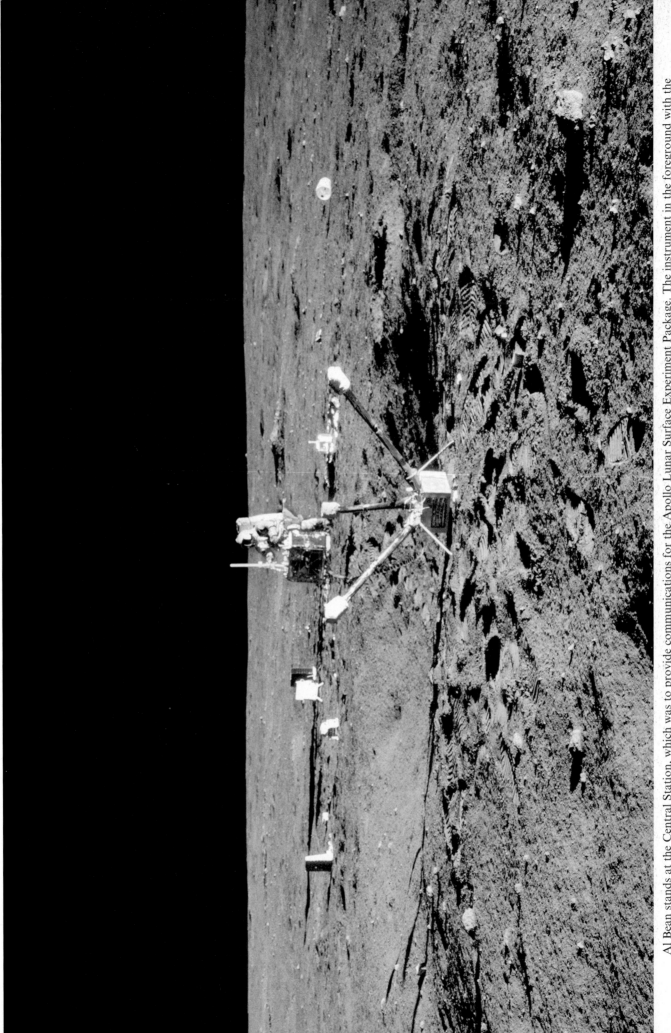

Al Bean stands at the Central Station, which was to provide communications for the Apollo Lunar Surface Experiment Package. The instrument in the foreground with the three 'arms' is the Lunar Surface Magnetometer.

With a few minutes to spare at the end of the first day's activities, Pete Conrad and Al Bean went to inspect the crater named Middle Crescent. In addition to the long shadows resulting from the low Sun angle, this view shows the disadvantage of taking pictures from directly up-Sun owing to the brilliant glare of backscattered sunlight.

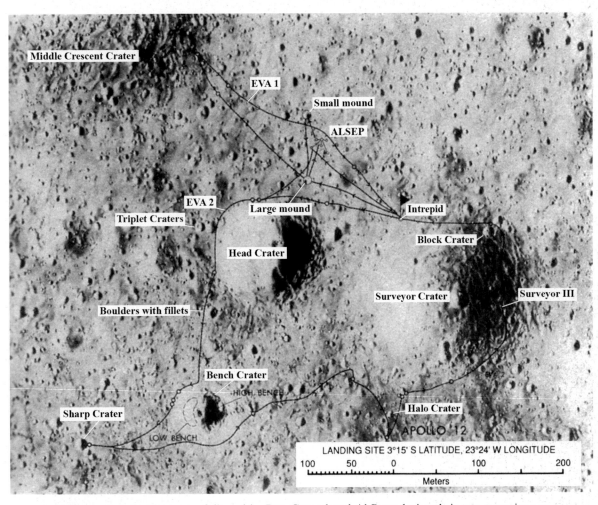

Middle Crescent Crater

EVA 1

Small mound

ALSEP

EVA 2

Triplet Craters

Large mound

Intrepid

Block Crater

Head Crater

Surveyor Crater

Surveyor III

Boulders with fillets

Bench Crater

HIGH BENCH

Halo Crater

Sharp Crater

LOW BENCH

APOLLO 12

LANDING SITE 3°15' S LATITUDE, 23°24' W LONGITUDE

100   50   0   100   200

Meters

A map showing the routes followed by Pete Conrad and Al Bean during their two excursions.

had been photographed by that spacecraft, was littered with blocks of rock. These names had the merit of being easy to remember. None of them had been known to astronomers prior to the space age, and if Surveyor 3 had not landed nearby, they wouldn't have attracted the attention of the site selectors. For the first landing, the selectors had avoided craters, but the craters were the real attraction for this follow-up mission for the reason that they had excavated the mare plain. To sample the rocks on a crater's rim was equivalent to drilling a hole. As events transpired, Armstrong had set down within walking distance of a sizeable crater, but he had not really had time to exploit that fact. Conrad and Bean were to tour their craters, comparing their characteristics, and taking pictures and samples. It was a lucky coincidence that there was such a variety of crater morphologies within the half-kilometre radius imposed by the 'walkback' constraint. On the way, they were to dig a trench, drive in a core tube and crack rocks with their hammer. Their

portable Hand Tool Carrier (HTC) held tools in a peripheral rack and rocks in a bin in the middle. They had trained to take 'before and after' pictures during their sampling, with a three-legged gnomon providing scale, local vertical and an illumination chart, and to store the samples in individually numbered bags rather than tossing them all into the same box.

On the first day, with its eastern wall in shadow, Surveyor Crater had appeared to be a real pit, and Conrad had doubted their ability to venture in, but the Sun had risen and now that its shallow floor was visible the crater looked much less imposing. Conrad had no doubt that they would be able to reach the spacecraft half-way down its wall. He was also keenly aware that because the spacecraft was towards the end of their traverse, if they fell behind they might have to miss it out, which would mean that, from the public's point of view, his mission would be seen to have failed; so he was determined to stick to the timeline.

"It's a heck of a lot deeper than it looked,"

Conrad said as they moved around the northern rim of 120-metre-wide Head. He threw a rock in, to attempt to stimulate the seismometer a hundred metres away, but Gibson relayed negative results. Conrad was struck by the fact that the rock did not roll, it just slithered to a halt.

They paused at a 1-metre crater that was littered with glass-coated fragments, but when Conrad grasped one with his tongs the 'rock' disintegrated. He bagged[5] as much of it as he could. Appearances could be deceptive on the lunar surface: what had appeared to be a solid rock had turned out to be a clod of soil that had been compacted by the shock of the impact that made the crater. In a sense, of course, it was rock, and it was subsequently characterised as 'regolith breccia'. The glass was soil that had been so severely shocked that it had melted. This had then splashed out and coated the clods. Being friable, such rock is soon fragmented and gardened into the regolith by the rain of micrometeoroids. Its presence in a crater is an indication that the crater is young.

As Bean followed Conrad around Head's rim, he noticed that his commander's boots were breaking through the surficial regolith to expose a lighter material. This provoked howls of delight from the scientists in Mission Control's 'Backroom'. One of the reasons for selecting Surveyor 3 as the target was that one of the bright rays radiating out from Copernicus crossed the site. Could this subsurface be the ray? Conrad was surprised, because they had not exposed such material in deploying the ALSEP nearby. "Let's trench this!" he decided. After taking a surface sample,[6] they scraped out a 15-cm-deep hole and sampled[7] the floor. Reluctant to use formal geological terms, Conrad referred to everything as "stuff". For good measure, he added to the bag a small fragment of rock that he dug out of the trench. Subsequent analysis of the isotopic ratios in the material showed it to be 810 million years old, and although the link is tenuous, this is generally accepted as dating Copernicus. In providing an absolute date for this stratigraphic feature, Apollo 12 achieved one of its objectives.

Having sampled the light material, Conrad was keen to recover time, so he set off apace towards Bench. However, on the way they paused first to photograph one of a trio of small craters – together dubbed Triplet – and again when Bean picked up a rock[8] that glinted in the Sun. Although they collected several rocks[9] on the northern rim of Bench, they often neglected to take 'after' pictures and instead of placing the rocks into individual bags they stuffed them into the 'saddle bag' that Bean had strapped to his thigh. This so complicated the task of the 'receivers' in the LRL that it was difficult to identify every rock's context. Indeed, the largest rock[10] of the mission, a 2.4-kg monster, was not documented at all. Although most of the rocks were basalt, the lax sampling procedures made it difficult to relate the various different types to any bedding. Bench was of such interest to the geologists because the inflection in its wall hinted that it had not only penetrated the regolith to excavate bedrock, but also that this stratigraphy was in outcrop. Conrad and Bean confirmed that there was a distinct bench, and reported a small mound in the centre of the crater, but when Gibson suggested that they venture in to try to chip samples from *in situ* rock, Conrad refused. In fact, the wall looked much steeper than it really was; it was only 15 degrees.

Conrad set off again (running between sites allowed more time to be devoted to a site) in search of Sharp. Although this 12-metre crater displayed a prominent rim in overhead imagery, it was 120 metres west of Bench and proved elusive in the zero-phase glare. "I'm not sure this is it," he admitted as a suitably-sized pit opened up before them.

"Let's use it anyway," Bean suggested. The crater was interesting because its half-metre-high raised rim was covered with the lighter material.

Conrad turned to check his map, taking bearings on the Sun and Intrepid, some 350 metres away. "It's got to be it," he decided. "It's awful soft," he observed as he scrambled up onto the rim and his boots dug deep into loosely compacted material. "Look at the bottom of that!" The wall was scoured with a radial pattern that looked just "like blast effect". Despite being 3 metres deep, the crater had not excavated rock.

"This soil must be of a different make-up," Bean mused as Conrad scraped another trench. Not only was the lighter material exposed on the surface, the fact that the trench did not pass through it indicated that it was a substantial deposit. This time they put the soil into a Sealed Environment Sample Container (SESC) in order to preserve it for analysis. Bean drove a short core tube through the floor of the trench to sample any layering below. After Aldrin's difficulty taking a core, the tip of the 46-cm-long tube had been redesigned. It penetrated much more readily this time, but that was in part due to the softness of the soil.

Gibson now turned them back east: to make up

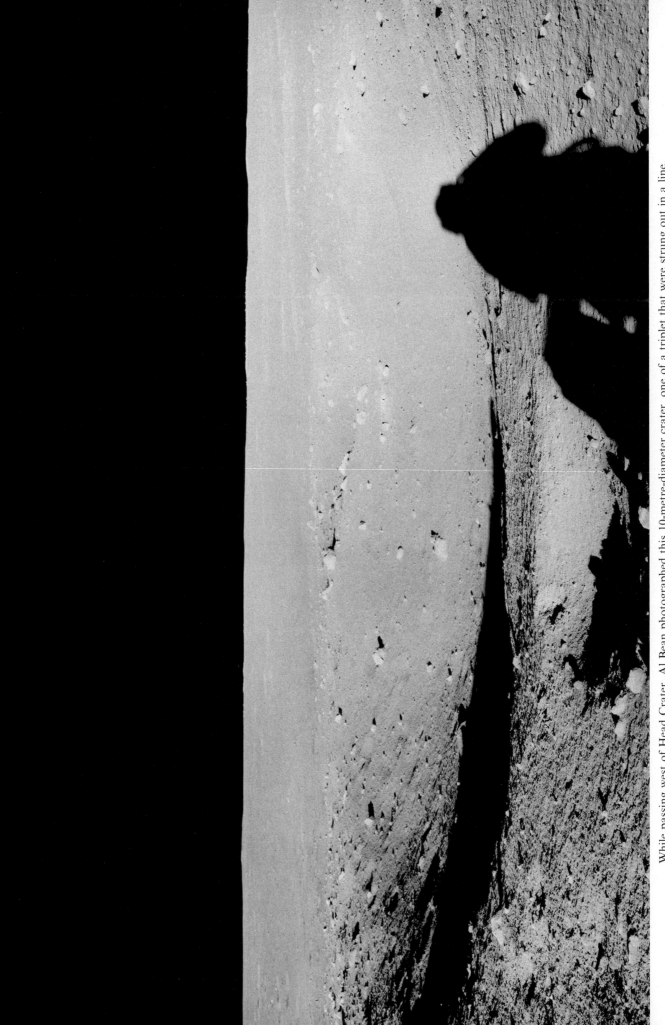

While passing west of Head Crater, Al Bean photographed this 10-metre-diameter crater, one of a triplet that were strung out in a line.

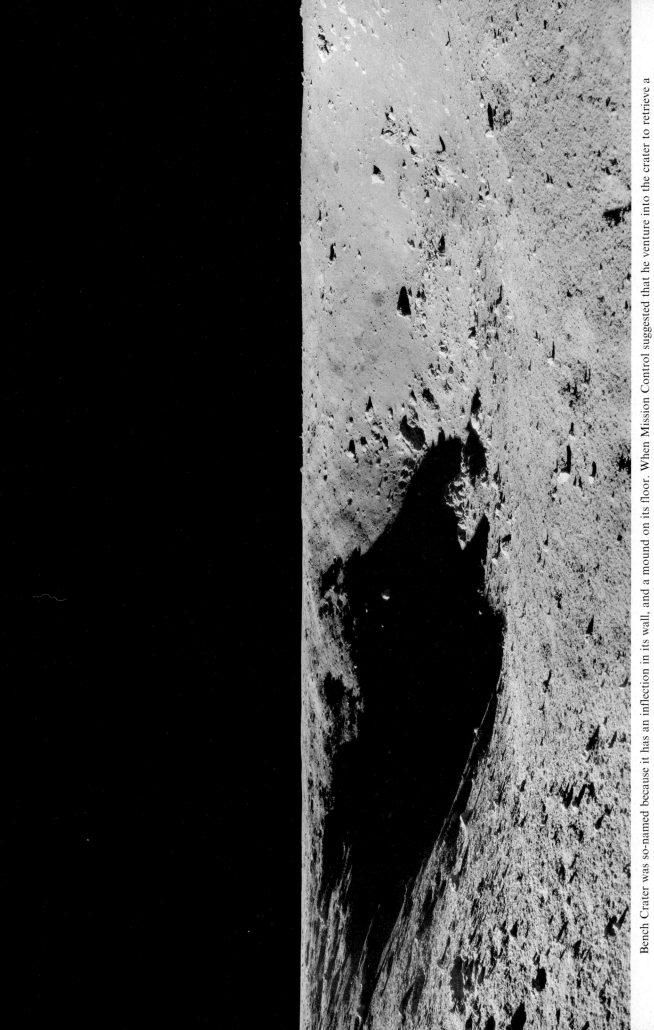

Bench Crater was so-named because it has an inflection in its wall, and a mound on its floor. When Mission Control suggested that he venture into the crater to retrieve a rock, Pete Conrad refused.

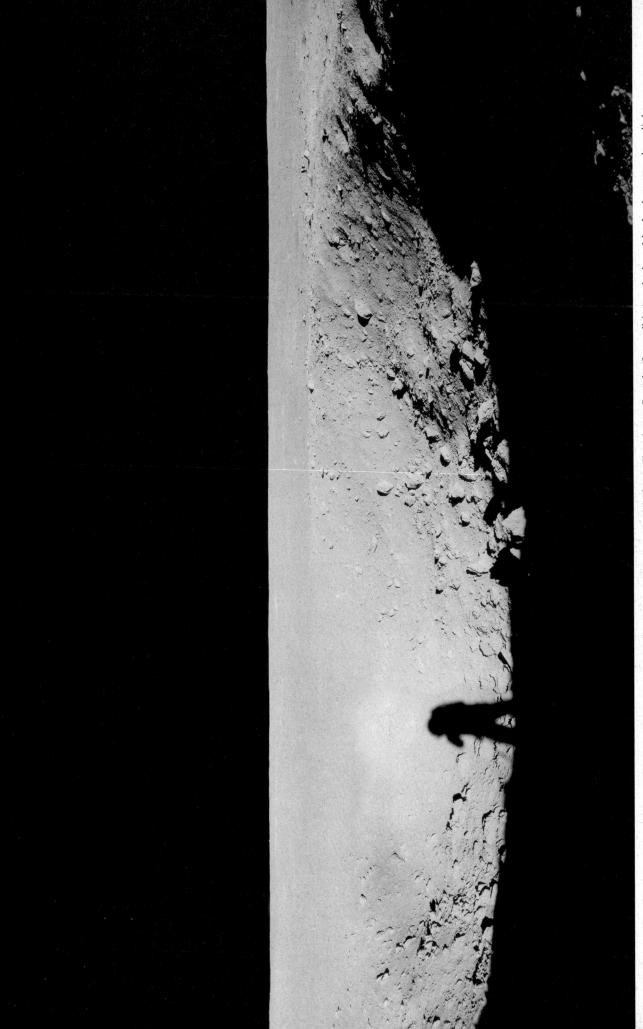

The interior of Sharp Crater. Unfortunately, because Pete Conrad photographed it from directly up-Sun the far wall is 'washed out' by backscattered sunlight.

Pete Conrad sampling in the vicinity of Halo Crater. The three-legged gnomon was positioned alongside the sample. Also in view is the framework Hand Tool Carrier.

time they were to bypass the rocks on Bench's southern rim and head straight for Halo. After criss-crossing the area where this 6-metre pit was supposed to be, Conrad decided to push on to Surveyor. However, a few seconds later, he found a crater that looked as if it might be their objective.

"I'm not sure we're in the right place," Bean said.

"It's your call," Gibson emphasised.

In fact, as Conrad struck off 40 metres south in order to sample among a cluster of small craters, he passed close to Halo without recognising it. They were to take a double-length core. This was the first time that this had been attempted, and there was no guarantee that a long tube (assembled by screwing together two short tubes) would go in, but there was so much soft soil that Bean felt lucky. He pushed the first 15 cm in by hand, and then hammered the rest of it in without difficulty. After pulling it out, he was astonished to see that the soil at the bottom of the tube was the same as on the surface.

With yet another box 'ticked', Conrad and Bean moved northeast to sample their 'dessert', in the form of the Surveyor spacecraft. They walked 100 metres along the crater's southern rim until they were due south of their target, then they crossed the subdued rim and made their way down to its level, approaching it cross-slope. It had been believed that the crater's wall would be thick with dust, and that there might be a danger of the spacecraft sliding, so they had been told not to stand down-slope of it. In the event, the surface inside the crater was as firm as it had been on the plain, away from the fresh craters, so they had no difficulty on the slope. As Conrad had observed earlier, now that the crater was illuminated it looked quite tame. To their amazement, the spacecraft was no longer white, it was a dull tan. They speculated that the paint might have been discoloured by exposure to the harsh environment, but it turned out to be a thin coating of dust that had been stirred up by their landing.

Surveyor 3 had bounced three times, progressively moving down-slope before it came to rest, so Bean photographed the marks that it had left, as well as the soil that was piled up against the footpads. After the structure had been documented, Conrad used bolt-cutters to sever the struts of the scoop that had excavated the first lunar trench. However, the real prize was the TV camera, because it contained electronics, optics and complex mechanical systems that the engineers were keen to study for indications of deterioration.

Before leaving the mutilated Surveyor, they retrieved a "brick-like" rock[11] that the spacecraft's camera had studied in minute detail, so that the scientists could assess their analysis.

The final objective of this pioneering traverse was to be the blocky crater set just inside the northeastern rim of Surveyor Crater. As Conrad and Bean started across to it, Gibson warned that they would have time only for a single sample.

"It's a pretty fantastically interesting crater," Conrad countered, "with big chunky rocks."

"Very angular," Bean added. In fact, they were the largest and most angular rocks that they had seen. Because it was within the larger crater, which had conveniently cleared away much of the regolith, this smaller impact had been able to excavate a lot of rock, most of which had remained in the immediate vicinity, which made it an excellent sampling site because the rock was 'in context'.

Although Bean suggested cracking open a few blocks to see what was beneath the cover of dust, Conrad refused; they were running late. After retrieving several small fragments,[12] Bean offered to get some of the lighter subsurface material, but Conrad refused again. "I'll just pick up this big rock," Bean persisted. He lifted a rock, and then chased after his commander. "I bet you everything is basalt," Bean observed presciently.

HURRY UP, AND WAIT

After racing back to the LM to secure their pre-launch margin, Conrad and Bean promptly found themselves idle for 2 hours. This was frustrating, because their suits could readily have supported at least another hour of activity in the immediate vicinity of the LM, time that could have been spent collecting rocks, but the flight surgeon welcomed this slack because it enabled the astronauts to relax. At nearly 3 hours, the geological traverse had taken a little longer than planned, but the 'boxes' had been 'ticked' and Mission Control was satisfied with the operational success. The fact that the men had taken each other's picture standing by the Surveyor would make their traverse a public success too because, for the first time, there had been a tangible objective.

Both the flight dynamics team and the Public Affairs Office were delighted with the pin-point landing. The geophysicists were pleased to have the ALSEP operating. Only the geologists were disappointed – firstly by the fact that the astro-

Al Bean in the vicinity of Halo Crater. He had a 'saddle bag' on his left thigh for samples.

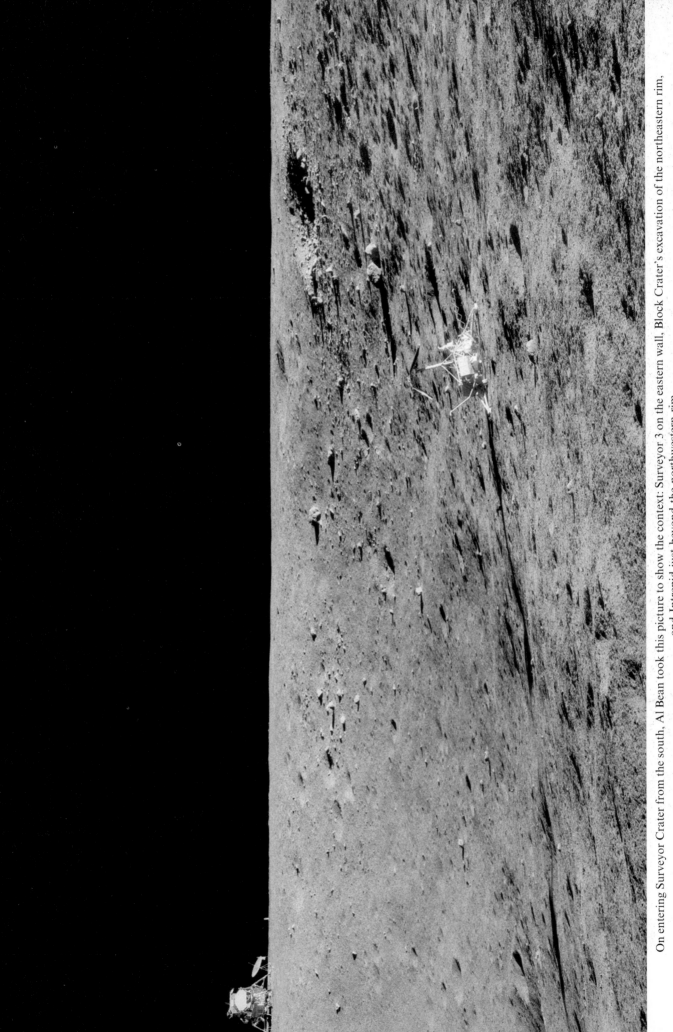

On entering Surveyor Crater from the south, Al Bean took this picture to show the context: Surveyor 3 on the eastern wall, Block Crater's excavation of the northeastern rim, and Intrepid just beyond the northwestern rim.

Pete Conrad stands beside Surveyor 3, one hand on the TV camera and the other on the 'pantograph' remotely controlled sampling arm.

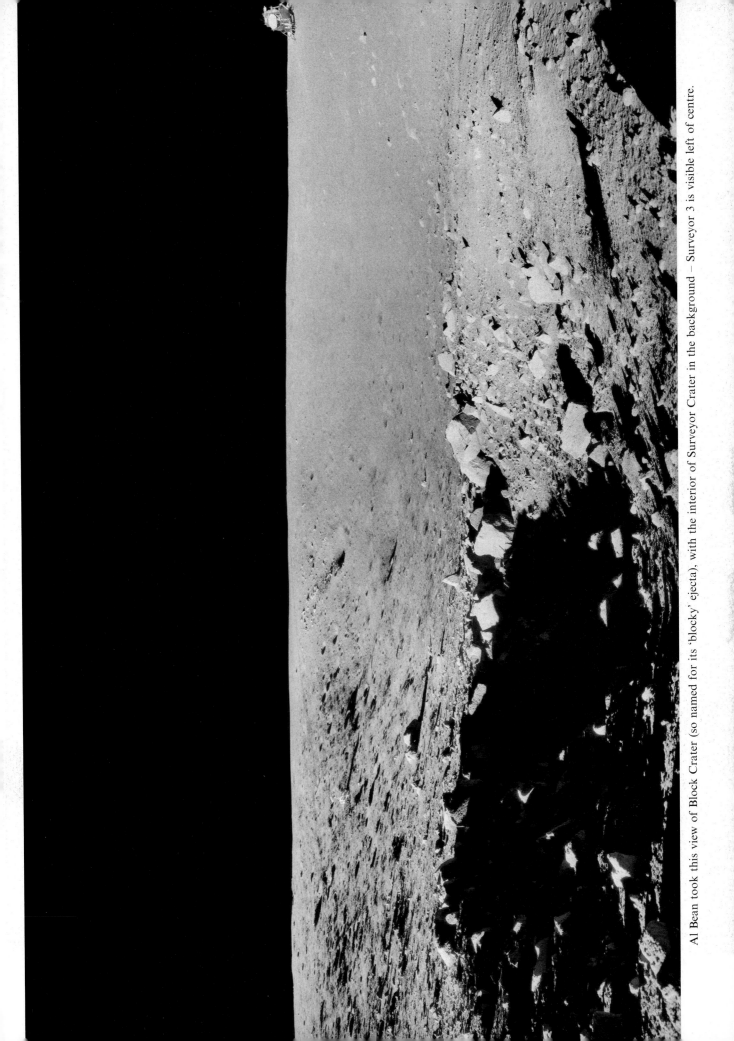

Al Bean took this view of Block Crater (so named for its 'blocky' ejecta), with the interior of Surveyor Crater in the background – Surveyor 3 is visible left of centre.

nauts had been able to sample so little of what was there, and secondly that the sampling had been undertaken so casually that it would prove difficult, and sometimes impossible, to reconstruct a rock's context.

Conrad and Bean had been outside four times as long as their predecessors; at 35 kg, had taken 60 per cent more lunar material; and at 400 metres, had ventured much further from the LM than had Armstrong on his dash to 'Little West' Crater. In fact, because they had been obliged to carry their tools and samples, Conrad and Bean had been operating close to the limit of what *could* be achieved during a simple walking traverse. If they had planned to stay out longer, and had carried portable instruments, their route would still have been restricted by the walkback constraint. Although the precision of Intrepid's landing made practicable the more demanding 'feature' sites that the scientists were keen to visit, future crews would require some form of wheeled transport to carry their tools and samples.

Yet, Conrad and Bean's exploits also dramatically showed how conservative Apollo 11 had been. Of course, the rationale for selecting a bland target was fully justified, but imagine if Armstrong had set Eagle down in the belly of the Snowman and he and Aldrin had snipped off Surveyor's camera.

SECOND 'ROCK FEST'

When the LRL 'receivers' unpacked Apollo 12's SRCs, they found 34 individual rocks. In contrast to the Apollo 11 samples, in which breccias and basalts had been represented equally by number, only two of the new rocks were breccias; the rest were crystalline.

An analysis of crater morphology had prompted Gene Shoemaker to predict a relatively thin 2-metre regolith, so, as expected, most of the craters Conrad and Bean sampled had excavated bedrock. The shallowness of the regolith, the low abundance of solar wind particles in the soil, and the lack of pits in the rocks from micrometeoroids (these would later be dubbed 'zap-pits') all indicated that this was a relatively young site.

The crystalline rock was coarser and more texturally diverse than that from Tranquillitatis. As before, it was low in volatiles, but it was also low in titanium. On reflection, therefore, it seemed that the Tranquillitatis basalt might be *enriched* in this element. Significantly, four kinds of basalt were identified in the Procellarum samples: olivine

basalt, pyroxene basalt, ilmenite basalt and feldspathic basalt. This indicated that there had been several distinct flows in this local area. However, the crystallisation dates clustered within a fairly narrow window. The combination of the visually distinct patchwork of flows in this region, and the chemical variation in the basalts from different flows within a short interval, suggested that the extrusions resulted from partial melting of pockets of rock by heat from the interior rather than a succession of flows from a single evolving reservoir situated deep in the mantle. Although the geologists had been keen to visit Surveyor 3 to date Copernicus via its ray, they had known that the context of this site would complicate the exploitation of the basalt ages. Nevertheless, the dates from Procellarum had profound implications for lunar history. The first isotopic ratio measurement yielded an age of 2.7 ($\pm$0.2) billion years. This meant that fully *a billion years* had elapsed between the Tranquillitatis and Procellarum extrusions. Even the strongest critics of the proposal that the maria had formed simultaneously were taken aback by such an extended period of volcanic activity. The next result pushed the age up to 3.4 billion years, but as the data came in it converged on 3.2 billion years. The Apollo 12 basalt proved to be the youngest returned from any site sampled. Coming so soon in the program, this 500 million year span in the ages of an eastern and a western mare did not simply confirm that the maria had not formed simultaneously, it also indicated that the driving process had been both endogenic and well established. The Moon was definitely not a 'cold' unevolved body that had been splashed by melted rock from a massive impact soon after it accreted. It was a differentiated planetary body. In fact, the degree of chemical differentiation that it had undergone was remarkable.

Geochemist Paul Gast made a surprising discovery in the Apollo 12 basalts: an abundance of potassium, phosphorus and a variety of rare earth elements. Linking their chemical symbols, he coined the label 'KREEP'. After attempting to isolate this material, he realised that it was not present as a distinct mineral. Consequently, the term did not indicate a new type of rock; it was an *adjective*, and it is more correct to say that the Apollo 12 basalts are KREEPy. By way of an 'instant science' explanation, he suggested that the additive might have been picked up from the material underlying all the basin ejecta, through which the basalt would have passed on its way to the surface, and so might even represent the 'non-

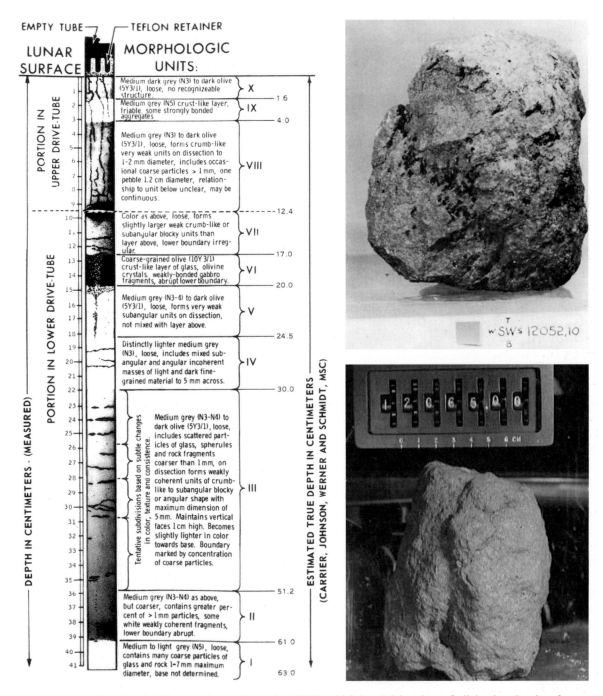

An analysis of an Apollo 12 core tube; and samples 12052, which is a 1.9-kg piece of olivine basalt taken from between the craters Head and Bench; and 12065, a 2.1-kg pigeonite porphyry, consisting of plagioclase, pigeonite and ilmenite.

mare' volcanism that some people thought was common in the highlands. However, when the KREEPy basalts proved to be rich in radioactive elements, in particular thorium and uranium, it was realised that this material could not be common in the crust, as the associated radiogenic heating would have prevented the crust from cooling sufficiently to halt volcanism. The mys-

tery of the KREEPy additive was not resolved until more such rocks were returned by later missions.

After being discarded, Intrepid's ascent stage was sent crashing onto the Moon; it hit at a point 75 km east-southeast of the ALSEP's seismometer, which recorded the lunar crust "ringing" for almost an hour with a signature unlike a terrestrial

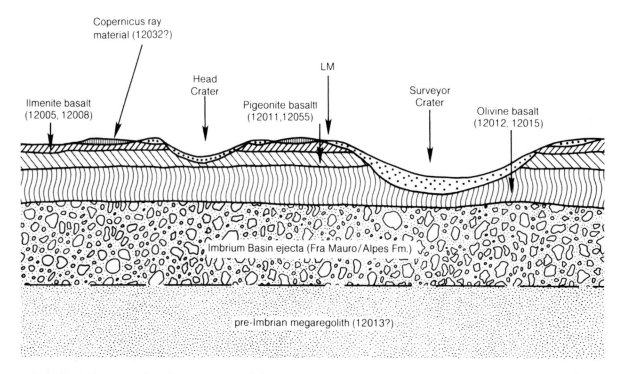

A schematic east–west geological cross-section through the Apollo 12 landing site showing several basalt lava units overlying older ejecta from major impact basins (modified after Rhodes *et al.*, 1977; Wilhelms, 1984). The presence of the Imbrium ejecta layer is inferred. Numbers refer to specific collected samples representative of the various units inferred to be present. Courtesy the Lunar and Planetary Institute and Cambridge University Press.

signal. At the First Lunar Science Conference, in January 1970, Gary Latham, the Passive Seismic Experiment's Principal Investigator, reported the Apollo 11 results and the early data from Apollo 12. It had initially been difficult to distinguish between a moonquake and an impact, but Intrepid's impact had served to calibrate the system, and it turned out that surprisingly few of the 150 events on record were true quakes. It seemed that the lunar crust was brecciated to a depth of about 30 km. In effect, it was a thick 'megaregolith'.

As the seismic data accumulated, patterns were revealed. Although the *average* rate of events was about one per day, there was a spate when the Moon was at perigee. This indicated that these particular events were triggered by the tidal forces of the Earth's gravity distorting the Moon. Furthermore, these events seemed to come from a fairly localised area south of the detector. Once other such instruments were deployed, it would be possible to triangulate the epicentres.

Another conclusion was that there were fewer impacts than expected. The seismometer was sufficiently sensitive to detect a grapefruit-sized meteoroid striking within a radius of 350 km. It detected only one such impact per month, on average. These could also be expected to arrive in batches. The Earth/Moon system orbits the Sun, and in doing so it periodically passes through swarms of meteoroids left behind by expired comets. It was evident, however, that there had not been an impact anywhere on the Moon comparable to the crash of Apollo 12's ascent stage. To probe the deep structure of the crust, it was decided that on future missions the spent S-IVB stage of the Saturn V launch vehicle should be sent crashing down onto the Moon.

Amazingly, when the TV camera from Surveyor 3 was taken apart, streptococcus bacteria were found. If not post-flight contamination, it was ironic that although the Apollo crews were put into biological isolation on their return in order to safeguard the Earth from infection by any lunar 'bugs', the only microorganism returned was of terrestrial origin.

4

Knowledge from the Moon

SATISFYING THE GEOLOGISTS

Having made a precision landing with Apollo 12, the flight dynamics team felt sufficiently confident to further reduce the target ellipse to enable the next mission to aim for a more confined site in rougher terrain. To escape the confinement of the equatorial zone, it was necessary to relax the propellant margins. The requirement for a back-up site was also deleted, and a launch delay of up to three days would be accommodated by making the landing at the higher Sun angle. This relaxation of the constraints did not 'open up' the Moon, because high-latitude sites were still out bounds, but it was a welcome degree of flexibility. It was the contraction of the ellipse, and the elimination of the rule that the general area be free of terrain relief, which allowed more interesting sites to be considered. In anticipation of a successful landing by Apollo 11, the site selectors had met on 10 June 1969 to plan Apollo 13. Some geologists were eager to visit a large crater. Hipparchus and Censorinus would have met the projected operational constraints, but the consensus was in favour of the terrain to the north of the crater Fra Mauro.

The Fra Mauro Formation is a hummocky morphological unit that dots much of the periphery of the Imbrium Basin. In fact, it is the most expansive stratigraphic unit on the Nearside. Dating it would 'lock in' many other structures. Contemporary understanding of the early history of the Moon was derived from the manner in which the Imbrium ejecta had splattered across thousands of kilometres, sculpting ruts and groves. As far as the geologists were concerned, dating the Imbrium impact was the single most important item on the agenda.

The trick was to find a crater in the hummocky terrain that had a clear line of approach from the east, had a landing site within a kilometre or so, and had a *very* blocky rim. A 370-metre pit on the crest of a north-south trending ridge some 40 km

north of Fra Mauro was chosen. Because it was precisely the kind of terrain that had been avoided while seeking 'safe' sites, the selectors found themselves faced with the task of certifying a site using just *four* high-resolution frames that had been taken by Lunar Orbiter 3 for 'scientific' interest. However, because it was on the eastern shore of Procellarum, the site was well illuminated when Apollo 12 was in orbit, and Dick Gordon was able to take excellent pictures. As a result, on 10 December 1969 Fra Mauro was confirmed as the target for Apollo 13. In view of its shape, the 'drill hole' crater that Jim Lovell and Fred Haise were to visit was named Cone.

The best terrain for a landing was the relatively flat plain about 1 km west of the ridge, but this was judged to be too close to the fringe of the crater's debris field for comfort, so a spot another kilometre downrange was chosen as the computer's aim point. Lovell, however, had the option of taking over if he thought he could land short. By this point in the program, landing accuracy was being assumed. So vital was Cone to the geological objective that there would be little merit to an excursion in the event of the LM straying beyond walking distance of the crater.

A major refinement to the descent profile was to use the SPS engine of the main spacecraft to place the LM into the descent orbit, so that the LM would have more propellant for the powered descent. As missions headed into ever rougher terrain, it was important to increase the time available to hover to select a safe spot on which to touch down. The flight controllers were still haunted by Charlie Duke's call to Neil Armstrong that he had less than 30 seconds of propellant left as he hovered in the 'dead zone', below the altitude at which it was safe to shed the descent stage to perform an abort, yet high enough to wreck the vehicle if its engine shut down and it fell to the surface. This margin would hopefully guarantee Lovell sufficient time to hover, if the plain proved

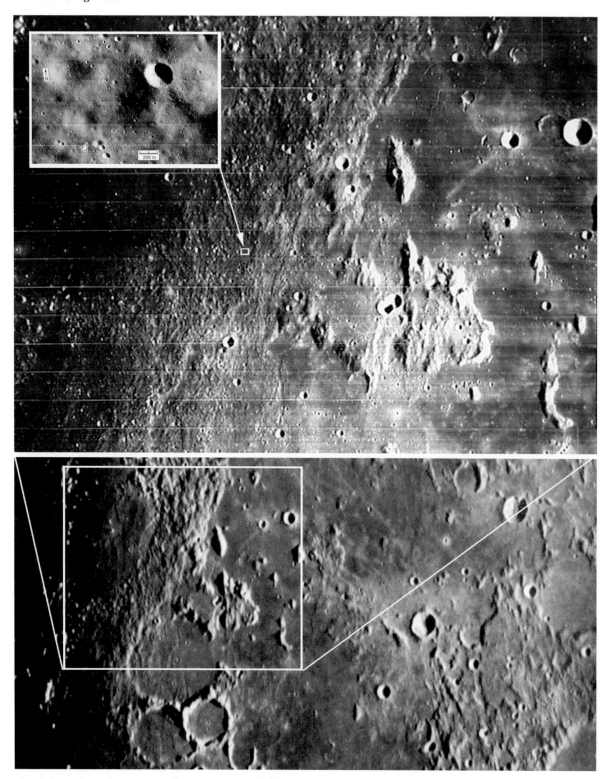

After two landings on mare plains, NASA decided to investigate the hummocky formation to the north of Fra Mauro, the large eroded crater near the terminator in the telescopic photograph (bottom). The area had been imaged by Lunar Orbiter, with the location of Cone Crater indicated in the inset.

to be rougher than expected. A month before they were scheduled to do so for real, Lovell and Haise undertook a high-fidelity rehearsal by trekking over a mock-up lunar surface made by their geologist tutors in the Verde Valley in Arizona.

Fra Mauro matched the single-feature 'H'-mission criteria perfectly. While it was a key site because it could be related to the larger-scale structure of the Nearside, all that would be needed to sample it would be a 'radial' through Cone's ejecta blanket to chart the vertical structure of the Fra Mauro Formation in which the crater was embedded. If the crater had punched all the way through the Imbrium ejecta, there would be material from the local basement sitting on its rim. This was not a key objective, but it would be a very welcome bonus. Following the 'H'-mission format, the ALSEP was to be set up on the first 4-hour EVA, and the second day would be devoted to the geological traverse. Although Cone was not as large as Meteor Crater in Arizona, Lovell and Haise expected to be rewarded with an awesome sight once they completed the ascent of the ridge.

CUTS

The first Lunar Science Conference opened on 3 January 1970 to present the preliminary results of the first two lunar landings to a spellbound audience. Ironically, while the conference was in session, NASA announced that one of the eight Saturn V rockets had been requisitioned to launch the Skylab orbital workshop; an Apollo mission had been cancelled. Furthermore, because the space station was not scheduled to fly until 1972, it had been decided to further relax the pace of the lunar missions to one flight every 6 months. This put Apollos 13 and 14 in 1970, Apollos 15 and 16 in 1971, and Apollo 17 in early 1972. Skylab would host several crews over an 18-month period, and be followed in 1974 by Apollos 18 and 19. By placing the final two lunar missions out 'on a limb', this plan virtually invited the budgetary axe.

APOLLO 13'S ORDEAL

A week before Apollo 13 was due to launch, Charlie Duke, the back-up Lunar Module Pilot, contracted the measles. Medical tests found that Lovell and Haise were both immune, but Ken Mattingly was not. The incubation time was such that it was not feasible to determine before the launch window opened whether Mattingly had the

disease. If he was permitted to fly and the launch went ahead on schedule, he could well develop the symptoms while in lunar orbit, which was not desirable. It was decided that Jack Swigert should move up from the back-up crew and fly in Mattingly's place. If Lovell or Haise had been at risk, Apollo 13 would have needed to be post-poned for a lunation. Swigert was fully capable of looking after the CSM while the LM was on the surface, and he was amply familiar with the observational work which Mattingly had planned. He doubtless thought that '13' had become his lucky number.

Unfortunately, while Apollo 13 was in transit to the Moon an oxygen tank in the service module exploded, crippling the CSM. The LM became the 'lifeboat'. After an ordeal involving looping around the Moon, Lovell, Haise and Swigert splashed down safely within sight of the recovery forces.

DEEP SOUNDING

Acknowledging that supreme piloting was still essential, but was no longer the point of the exercise, Lovell had chosen the motto 'Knowledge from the Moon' for his mission, which was not only to have sampled the geological unit most crucial to understanding the formation of the Imbrium Basin, but was also to have attempted to assess the state of the lunar interior. Even though the astronauts did not reach the lunar surface, their spent S-IVB stage was sent crashing down.

At a depth of 60 km the Earth is 1,000°C. This is sufficient to melt rock, and is why the Earth is so volcanically active. The transition to a fluid state is detectable by seismic soundings, and is called the 'low velocity' zone. Geologists had interpreted a variety of lunar surface features as being volcanic in form, and hence that there had to be (or, more properly, there must once have been) a fluid mantle. It was hoped that the shockwave from the impact of the 15-tonne vehicle would reflect off a discontinuity, and thereby prove that there was still a fluid mantle. The S-IVB hit 44 km from Apollo 12's seismometer. The signal persisted for almost 4 hours. This not only confirmed the announcement at the First Lunar Science Con-ference that the outer crust was intensely brec-ciated, but also showed that this material extended to a depth in excess of 100 km. Such a thick crust meant that any surviving fluid was unlikely to be able to feed magma to the surface. This implied that the Moon is now effectively inert. The Moon

The Apollo 13 patch (top left) includes the motto: 'Knowledge from the Moon'. Ken Mattingly, left behind owing to a medical issue, looks on in Mission Control (centre left). "There's a whole side of that spacecraft missing," exclaimed Jim Lovell as the Service Module was jettisoned (top centre). After a suspenseful radio blackout, a TV camera with the recovery forces showed the welcome sight of the Command Module Odyssey descending by parachute (top right). The crew of Lovell, Fred Haise and Jack Swigert waved as they exited the recovery helicopter (centre right). Only once the astronauts were safely on the aircraft carrier did flight director Gene Kranz light his cigar (bottom left). On his way home, Lovell took a little time out to catch up on recent news (bottom right).

is evidently structured differently to the Earth. But how different? If Lovell and Haise had been able to emplace the thermal probes of their ALSEP's Heat-Flow Experiment they might have been able to find out.

APOLLO GROUNDED

The near loss of a mission had been turned into a magnificent success for impromptu flight operations, but the program was stalled until the cause of the oxygen tank's explosion was tracked down and rectified. It was during this period of reflection that the budgetary axe fell: funding for the final two Apollo missions was denied. However, the announcement, on 2 September 1970, contained a revision of the existing mission sequence. The 'H'-class single-feature mission that had been planned for Apollo 15 would be deleted, and this flight would be upgraded to exploit the new spacecraft and surface transport that were under development for the 'J' missions to recover some of the lost scientific yield of the program. The Saturn V rockets had been bought and paid for, so cancelling their use saved only the operating cost, which was of the order of $20 million per flight. A fortnight after losing this funding battle with Congress, Tom Paine resigned as NASA's administrator.

In 1967, when visiting the Apollo 7 crew in training to make the first test of the CSM, Lyndon Johnson was heard to say of the effort being devoted to Apollo, "It's too bad, but the way American people are, now that they have all this capability, instead of taking advantage of it, they'll probably just piss it all away." This prophesy was now coming true.

REVISED PLANS

When the crews for Apollos 13 and 14 had been selected in August 1969, the plan had been for Apollo 13 to land at Fra Mauro, which was known to be Imbrium ejecta, and for Apollo 14 to visit a volcanic site. The 'dark mantle' near the eastern shore of Mare Serenitatis in the vicinity of Littrow Crater, for example, appeared to be a layer of pyroclastic ash. Was this evidence of 'recent' volcanism? There were rilles in the same area. Had these spewed forth the pyroclastic? This area was to have been photographed by Apollo 13. Without this documentation, Littrow could not be certified as a landing site. But this was not a matter

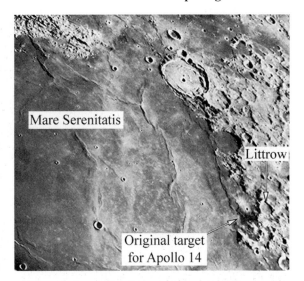

The 'dark mantle' site on the eastern shore of Mare Serenitatis that was originally assigned as Apollo 14's target.

The Apollo 14 mission patch.

of immediate concern, as Fra Mauro was more important, and on 7 May 1970 this was formally assigned to Apollo 14.

Within weeks, the schedule was revised. The remedy for the failed oxygen tank was to revise the electrical system and to install a reserve tank in a different part of the service module. Because modifying the already-built vehicles would take time, Apollo 14 was slipped to early 1971.

By this point, the program was tightly coupled. As more difficult sites were considered, the more lamentable was the focused Lunar Orbiter coverage. In retrospect, it may have been better to have built more of these marvellous spacecraft and interleaved them with Apollo surface explorations.

Al Shepard, Apollo 14 commander.

Stu Roosa, Apollo 14 command module pilot.

Ed Mitchell, Apollo 14 lunar module pilot.

ONCE MORE MOON-BOUND

The launch of Apollo 14 on 31 January 1971 was 40 minutes late due to excessive cloud, but the translunar coast trajectory was altered to recoup the delay in order that the operations in lunar orbit would occur 'on time'. The transposition and docking manoeuvre went perfectly until Stu Roosa eased Kitty Hawk, the CSM, back in to extract Antares, the LM, from the expired S-IVB. In the spirit of the 'right stuff' school of astronautics, he completed the transposition by using the bare minimum of propellant. However, the latches on the tip of the docking probe failed to engage when it slipped into the drogue. After several unsuccessful attempts, he withdrew and flew in formation while Houston assessed the situation. All that manoeuvring had made a mockery of his propellant economy. It was decided that the latches on the tip of the probe must have frozen. The recommendation was to insert the probe into the drogue as during a 'soft' docking, allow time for the two vehicles to align, and then command the probe to retract while Kitty Hawk fired its thrusters to force the main docking collars together in the hope that the primary latches would engage. It worked. This was fortunate, because another abort in transit to the Moon may well have prompted the cancellation of the rest of the Apollo program on the basis that it had already achieved its *original* objective.

Kitty Hawk made first the lunar orbit insertion burn and then the descent orbit insertion burn, to release Antares. Immediately following separation, Shepard and Mitchell undertook a rehearsal of the procedure for PDI, and were shocked to find that the Abort Guidance System (AGS) was triggered; if they had been attempting to land, this spurious signal would have automatically jettisoned the descent stage and initiated a rendezvous with Kitty Hawk. After studying the telemetry, the engineers in Houston decided that the abort switch must have a short circuit. While the LM was behind the Moon, a software 'patch' was written to allow the system to ignore the manual abort switch. The update was radioed up as soon as the spacecraft came around the eastern limb for its nominal landing pass. Mitchell worked against the clock to key in the modified program, which made the AGS passive; in the event of an abort he would have to key a command sequence into the computer. Meanwhile, tracking verified that the trajectory was correct, so, with just a few minutes to go, the flight director authorised PDI. However, the landing radar failed to supply altitude to the PGNS. Without altitude and rate of descent, the mission rules mandated an abort. Fortunately, there was time to troubleshoot the problem. In fact, the radar fault was a side-effect of installing the abort switch patch *after* the radar had been powered on. The remedy was to cycle the radar's circuit breaker to re-initialise the system, so as to re-establish the link with the PGNS, but it took several minutes to analyse the problem and figure out the solution, and during this time Shepard and Mitchell let Antares continue to fly a nominal descent profile. What they would have done if the radar had not come on-line is a matter for speculation.

Formalising Lovell's intention to land as 'short' as possible in order to reduce the length of the trek to Cone Crater, the aiming point had been moved 500 metres east. As he descended, Shepard observed that the plain beyond Cone Ridge was "a little rougher" than he expected. He was not concerned, however; he had plenty of hover time, so he reduced his rate of descent and flew over to the flattest site he could find. "We made a good landing," he announced casually as he set Antares down a mere *50 metres* from the designated target.

CHORES

Immediately after stepping onto the surface,

Shepard stepped to his left in order to peer around Antares at Cone Ridge, and reported it to be "a very impressive sight".

The first EVA was devoted to deploying the ALSEP. To minimise the effect of the dust that would be stirred up when Antares lifted off, it had been decided to set up the package at least 100 metres west of the LM. Not only did the rough terrain make it difficult to find a level site, but transporting the pallets gave a preview of the difficulty that the astronauts would face during their traverse up Cone Ridge. The surface *appeared* flat, but it was undulatory with a 2-metre vertical amplitude, and even although their spacecraft was nearby the astronauts sometimes lost sight of it.

The ALSEP included the laser reflector and the passive seismometer in order to expand these networks and several new instruments had been added, but Apollo 13's Heat-Flow Experiment was not carried forward. One of the new activities was the Active Seismic Experiment (ASE). For this, Mitchell emplaced geophones and then walked along the line with a 'thumper' which he placed on the ground at 3-metre intervals to fire an explosive charge, with the seismic data from the geophones being transmitted to Earth. In effect, this reran the fieldwork conducted in Antarctica for the International Geophysical Year in the late 1950s, when a similar device revealed that the southern ice cap was considerably thicker than had been thought. However, when Mitchell hit the trigger, the first charge misfired and he had to switch to the next charge in the sequence. Although 8 of the 21 charges failed, the experiment was successful. If the Fra Mauro Formation had indeed been laid down by the impact that excavated the Imbrium Basin, then it was an old structure and could be expected to have a mature regolith. The ASE data revealed the regolith to be 8.5 metres deep, and the fact that the surface undulated across a 2-metre range made this an average. This was in agreement with the pre-flight estimate of 6–12 metres. The experiment indicated that the Fra Mauro Formation was only 75 metres thick, but with Cone on the crest of a 100-metre-tall ridge, the crater had almost certainly not excavated the underlying material.

CONE RIDGE

To enable Shepard and Mitchell to return from Cone with a heavy load of rocks, they were provided with a small trolley that could be pulled

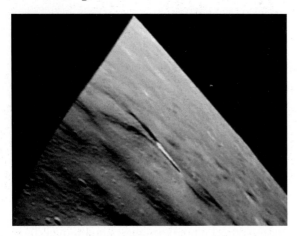

A frame from the 16-mm movie camera in the right-hand window of Antares as it passed low over the ridge on which Cone Crater is situated.

along like a golf cart; in typical NASA jargon this was called the Modular Equipment Transporter (MET). The first task on the second day's outing was to configure the MET. If they had been walking unaided, they would have taken the minimum of tools, but the availability of the cart had led to their being assigned a substantial load. In addition to the basic sampling kit of a hammer, a pair of tongs, a scoop, a gnomon and a dispenser for storage bags, they had tubes for taking core samples and two SESC cans, the 16-mm sequence camera, the close-up camera, a magnetometer and a 30-metre-long tether. After half-an-hour, a few minutes late, they took the MET in tow and set off east. Once they walked out of frame, having left the TV camera staring up-Sun, the few networks that were showing the moonwalk promptly switched off.

The trek up Cone Ridge had not been expected to pose any significant problem, particularly in the low lunar gravity, but the MET bounced around rather more than had been expected, and dragging it required the astronaut to adopt a slow and rather terrestrial gait that was inconvenient in an environment where a hop-skip was more efficient.

To document the transition from the plain to the flank of the ridge, samples were to be taken at two sites, designated 'A' and 'B'. The geologists had selected specific points and Mitchell, whose task it was to navigate, tried to check the terrain features against his photographic map. Site 'A' was to be 350 metres east of the landing site, about one-third of the way to Cone Crater. However, within 5 minutes he discovered that the way the hummocky terrain masked the features on his map made it difficult to navigate. "I don't know exactly where we are," he warned. Of course, he was not lost, because their destination was visible when-

ever they crested a local rise, but he needed cross-bearings to gauge their rate of progress. His task was not helped by the fact that most of the landmarks on the map were *depressions*. Upon cresting the next rise, they saw a 100-metre-wide, 5-metre-deep valley cutting across their path. This turned out to be the first of a series of troughs that etched the plain in parallel with Cone Ridge.

While Shepard pushed on, Mitchell paused to study his map. He finally spotted an eroded crater with a trio of small pits within it. This distinctive morphology meant that it was the crater they had called Weird. "If we head north of that," he noted, "we're in business." It was evident that finding the specified sampling points would be both difficult and time consuming.

"There seems to be quite a few rocks up to 2 or 3 feet in size," Shepard noted as he dragged the MET across the trough. They were not taking pictures as they went, so their commentary would be the only record of the geological setting.

Fred Haise, having been denied the opportunity to make this traverse for himself, was now serving as CapCom. He asked whether they saw any change in the regolith as they entered the outer fringe of Cone Crater's ejecta blanket.

"It all looks the same, Fredo," Mitchell replied.

"That's what I was afraid of," Haise observed dryly. He asked whether the rocks that they had noted were in the form of a ray radial to Cone.

"They're fairly generally scattered," Shepard replied.

As they crested the next rise, Shepard suggested that they declare it to be Site 'A'.

"Well, it's pretty close," Mitchell agreed, "but I don't think it's exactly right." He paused to check out the landmarks as Shepard descended into the next valley. "That large crater to your right, Al – just beyond that." There were three 20-metre craters, with the ones to the north and south fresh-looking and the one to the west subdued; it looked about right. Despite this apparent match, it would later be calculated that they were 150 metres west and 100 metres south of where Mitchell estimate them to be. Shepard stopped and reported that although the regolith was fine-grained dust, as at the landing site, there were more pebbles and the larger rocks were more plentiful.

In retrospect, it might have been more productive for the geologists to have told the astronauts what kind of terrain to look out for, then let them forge head directly for Cone, halting only when they saw something relevant; even if they didn't know where they were, a photographic panorama would identify the site. This was the first substantial lunar

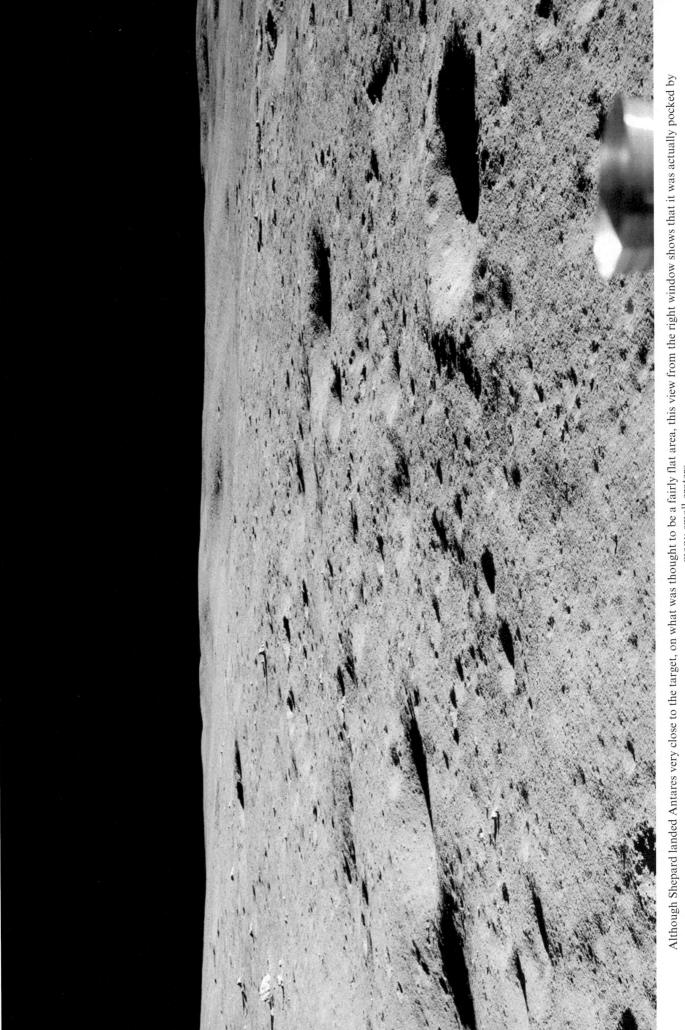

Although Shepard landed Antares very close to the target, on what was thought to be a fairly flat area, this view from the right window shows that it was actually pocked by many small craters.

As soon as he was on the lunar surface, Al Shepard stepped around the north side of Antares and raised a hand to block the Sun as he inspected the ridge on the crest of which Cone Crater was situated.

As this northward-looking view of Antares shows, to a man standing on the surface the site looked deceptively level.

For the Active Seismic Experiment, Ed Mitchell laid out a line of geophones and then walked along the line with a 'thumper', firing explosive charges in order to probe the structure of the Fra Mauro Formation.

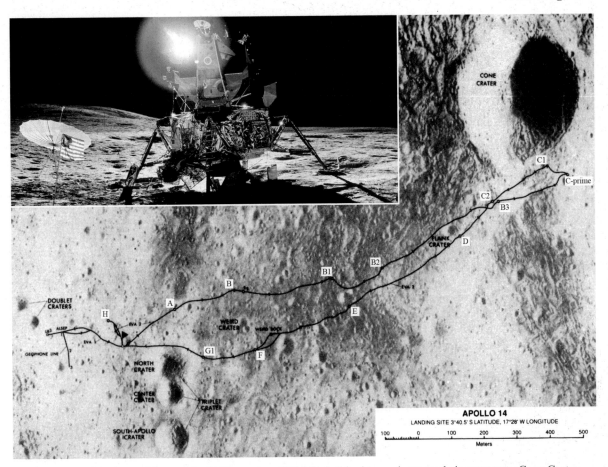

A photographic chart of the terrain Al Shepard and Ed Mitchell had to navigate on their traverse to Cone Crater, with the actual sampling sites marked. The inset is an eastward-looking view of the ridge from the landing site.

geological traverse, however, and the planners were on a learning curve. They had been too specific in designating sampling sites. The astronauts, being pilots, considered navigation to be part of their trade. As a result, performing these secondary tasks was jeopardising the primary objective.

At Site 'A', Mitchell was to measure the magnetic field. The Lunar Portable Magnetometer (LPM) had been introduced to follow up on the Apollo 12 report of a surprisingly intense magnetic field. The field was thought to be remanent magnetism in the rocks, rather than a dipole in the Moon's core. In order to establish the variability of this magnetic field, it had been decided that they should take the LPM with them to Cone, where they would jettison it to accommodate rock samples. It was a tricky piece of apparatus to operate. It had to be set on a tripod well clear of the MET and properly orientated and levelled. The control panel was on the MET, and connected to the instrument by a cable. There were three sensors, mounted one in each of the cartesian axes, and to cancel any sensor bias Mitchell had to take *three* sets of data, reorientating the sensors each time. Furthermore,

to ensure consistency against fluctuations, he had to sample each X:Y:Z dataset three times. It took him 4 minutes to set up the instrument, 10 to take the measurement, and 5 to stow it – this latter task being hindered by the fact that as soon as he let go of the winding crank, which did not have a lock, the ribbon cable promptly unwound and draped itself in loops down the rear of the MET, but since it did not reach the ground he left it in this state. The field was *fossil* magnetism locked into the rocks, so if the rocks were jumbled their fields cancelled out, but if they were mutually aligned they reinforced. The field at Site 'A' proved to be fully three times that measured by the Apollo 12 magnetometer. Upon finishing with the LPM, Mitchell took the pan, and then chose a rock from the rim of one of the craters as a documented sample. As he lifted it with his tongs it broke apart, so he popped a few of the pieces[1]* into the bag.

---

* Information pertaining to each sample is given on page 369.

Shepard's task was to take a core sample. The tube went in easily until it was on its second section, and then a few hefty taps with the hammer drove it almost all the way in, and he secured a 1-metre sample, which was the deepest vertical section yet taken of the lunar regolith.

By the time they were ready to leave, this site had consumed fully 35 minutes. The loading of the MET alongside Antares had taken almost half-an-hour and the first leg of the route was 8 minutes, so they were well into their allotted time. They set off believing that they were one-third of the way towards their objective, but they were actually only 175 metres from Antares, and Cone was still a kilometre away! The next task was to find Site 'B', which the geologists had set on the southern rim of a 20-metre crater, 230 metres beyond the intended first sampling point. Finding a specific crater that *small* in such terrain was going to be difficult. As they crested the next rise, Mitchell informed Haise that there was evidence of a ray of ejecta in the shape of a string of boulders off to the south. The two men then stood with the map between them, trying to relate the terrain to the map. There *was* a 20-metre crater in the valley ahead, but was it their target? "This is the one that's half-way between 'A' and 'B' isn't it?"

"Yeah, I think so," Shepard agreed.

They needed a cross-reference. "We should be able to spot a little chain of craters, just to the south." They turned to look.

"That little chain?"

"Kind of small," Mitchell mused, sceptical of the identification.

The Site 'B' crater was supposed to be in a field of boulders, northeast of Weird. There were boulders off to the south, but where was Weird? There was a ridge due south, and Shepard suggested that the crater was behind it. This would place them due north of Weird, in which case the crater ahead was the one whose rim they were to sample. But Mitchell thought Weird was the large crater visible to the southwest, which put them beyond Shepard's estimate.

"Al and Ed," Haise began tentatively, "I don't think you have to worry too much about the exact position. If you're close, that should be good enough."

The moonwalkers continued to debate for another minute or so, and then decided to sample their current position as a substitute for the assigned site. They were about 200 metres from where they made their first stop, and the presence of boulders met the geological requirement to sample the fringe of Cone's ejecta blanket. Site 'B'

called for a 'grab' sample, which involved simply lifting a rock without bothering to document its context. They chose their rock,[2] and took a pan. The effort devoted to looking for (and failing to find) a specific location was hardly justifiable simply for a grab sample. They had spent almost 4 minutes standing still debating their position before deciding that they should sample where they were. It would have been much better for Mitchell to pause to grab a rock whenever he saw something geologically relevant, while Shepard kept going with the MET. A pan needed only a minute to take. Even if they both stopped, one could be taking the pan while the other sampled. In addition, it would have been advantageous if Mitchell had paused from time to time to take a pan simply to record the lay of the land, so that the variation in the rock coverage as they approached Cone could be analysed *post-flight*. But lunar field geology was in its infancy, and the lessons had to be learned the hard way. It would later be determined that they made their second stop close to where they should have stopped for Site 'A'.

"We're almost 15 minutes behind in the time-line," Haise warned, once they were ready to depart.

"We'll pick it up later," Mitchell promised.

"We'll see about that," Shepard noted more sceptically.

"The next stop is the top!" Mitchell insisted.

They were to pass just north of Flank Crater, which was near the summit of the ridge, with Cone just over the crest. The nature of the terrain changed dramatically once they crossed the next rise. In addition to an increase in the coverage of rocks, the regolith was much coarser, and there were so many extremely eroded craters that the surface, until now undulatory but smooth, became much rougher. "There's not 10 yards between craters," Mitchell reported. "There's hardly a level spot anywhere." They had penetrated into Cone's ejecta. This was the transition which they thought they had reported on (negatively) earlier. Obliged to follow a winding path, they had to check their bearings every time they climbed out of a hollow because, with the horizon being so close, it was all too easy to emerge pointing in the wrong direction.

Upon encountering a crater that was littered with debris, Shepard decided to grab a sample. Opting not to bother retrieving his tongs, he dipped onto one knee to grasp a rock[3] by hand. In fact, this was a difficult manoeuvre, because the waist of the suit was rigid (a flexible suit was under

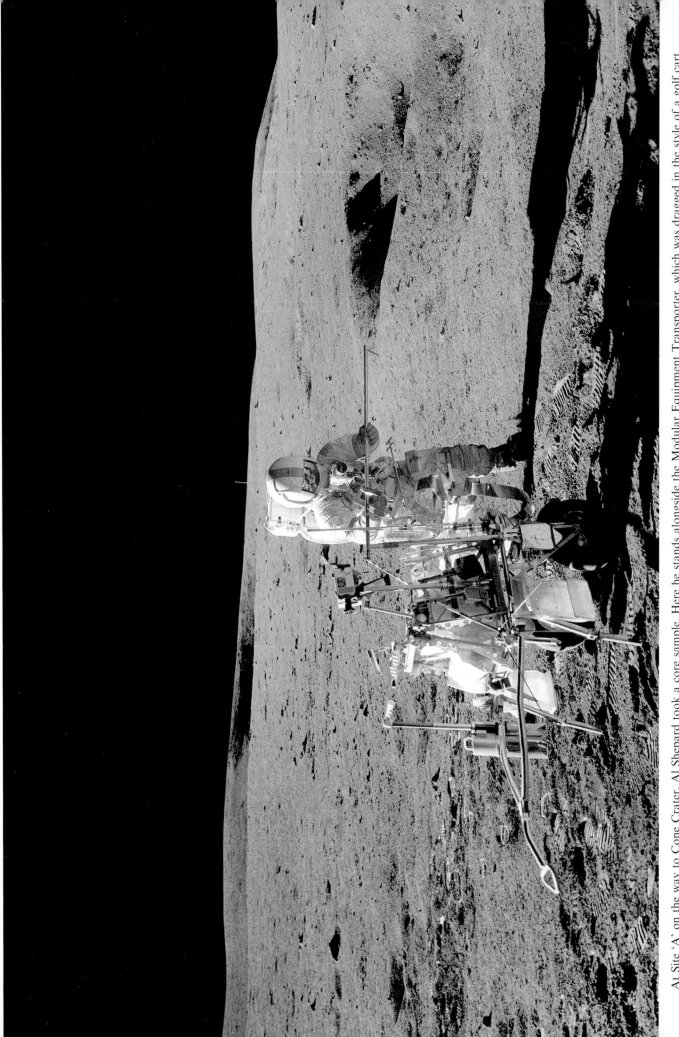

At Site 'A' on the way to Cone Crater, Al Shepard took a core sample. Here he stands alongside the Modular Equipment Transporter, which was dragged in the style of a golf cart.

development) and his knee hit the ground with such force that it dug "3 or 4 inches" into the soft soil on the crater's rim. "Do you know where we are, Fredo?" he asked. Haise assured them that he had a fair idea, so there would be no need to take a pan to document the site.

"We're starting uphill now," Mitchell informed Haise a few minutes later.

"Flank should be right over there," Shepard opined.

"Just out of sight, you mean?" Mitchell retorted cynically. "Let's go over and see."

As they pushed on up the hill, the slope increased to 10 degrees and they began to huff and puff. "Why don't we pull up beside this big crater?" Mitchell suggested, to give their suits time to cool off; Shepard agreed. Haise proposed that they take a pan to document their ascent of the ridge. They tugged the MET a little further up-slope so that they could set it in a crater, where it would be safe. "I can't really spot this crater on the map," Mitchell admitted, "but I think I know where we are." There was a large crater off to the right.

"Okay, Houston," Shepard called to Haise. "We're passing to the north of Flank."

In fact, their rest stop was 700 metres east of the landing site and, ironically, they were not far from where they should have stopped for Site 'B'. Because they took a pan from this spot, it was later designated Site 'B1'.

As the coverage of small rocks increased, the MET became rather animated. To cross one particularly rough spot they hoisted the cart between them and carried it. The back-up crew had placed a bet that the MET would not make it to Cone's rim, so this news prompted wry smiles in Houston. But the point of the MET was to enable them to return with a full load of rocks, and to leave it behind to save time would be counter-productive.

About 150 metres beyond the rest point, they paused at a partially buried rock. It was the biggest rock that they had seen thus far. "I think it's worth taking a picture," Shepard said. "It's very old, very weathered. There's some crystal shining through the fractures." It had a substantial up-slope fillet, but the upper surface seemed to be fairly clear of dust. The fact that dust was piled up around the side, but not on top was of interest to the geologists because it provided information on micrometeoroid rates. A fillet implied that a substantial amount of dust had been thrown around by nearby impacts. This dust would not only have piled up around the rock, it would also

have settled on top of it. The bare upper surface implied that micrometeoroid impacts had swept the rock clean. An additional factor was the extent to which the fillet was composed of material shed by the rock, and the asymmetry enabled this to be assessed. Mitchell took a pan in order to enable the boulder's position to be located; it was later designated Site 'B2'.

"Just over this ridge," Shepard prophesised once they resumed their climb.

"We got fooled on that one!" chuckled Mitchell upon cresting the rise. There was another ridgeline ahead. Surely this must be the raised rim of Cone crater? On the other hand, it was strangely asymmetric. "Wait a minute!" called Mitchell on realising that he was looking at Cone Ridge's distinctive southern shoulder, where it fell steeply. If that was so, then they had strayed south of their route. He said they should swing northeast.

"That's at least 30 minutes, up there," Shepard pointed out. They were already 2 hours into the EVA. There was a field of blocks nearby. "We'd probably do better to sample those boulders."

Mitchell was frustrated. "Let me look at that map."

"What I'm proposing," Shepard argued, "is that we use that field as the turning-point. It seems to me that if we continue, we'll spend a lot more time in-traverse and we won't get many samples." It was a sure bet that such large rocks were from Cone and sampling them would be almost as good as sampling the rim itself.

Mitchell could sense the decision being made, and he didn't like it. "Let's keep going." He held the map so that Shepard could see it. "Let's head for *that* boulder field." He indicated a map location, apparently just over the next rise, where debris was marked as littering the crater's rim.

"I don't think we have *time* to go up there," Shepard retorted.

"Oh, let's give it a whirl!" Mitchell pleaded. "Gee whiz. We can't go back without looking into Cone Crater!" Outcrops in the wall of the crater might expose the vertical structure of the Fra Mauro Formation, and possibly even the under-lying material. "Fredo, how far behind are we?"

"About 25 minutes down," Haise replied immediately.

"We'll be an hour down, by the time we get to the top," Shepard warned.

"Well I think we're going to find what we're looking for up there." The rocks from the rim would be from the deepest part of the excavation, so the rim was not to be abandoned lightly.

"Ed and Al," Haise interjected. "The word

In taking a panorama at the 'rest stop', Site 'B1', Al Shepard caught Ed Mitchell inspecting the map.

At Site 'B2', Al Shepard (shown here) and Ed Mitchell encountered a large rock with a well developed fillet of eroded material that implied the rock had been in place for a considerable time.

from the Backroom is to consider your current location the edge of Cone Crater."

"I think you're finks!" Mitchell countered. The issue was *time*. He saw one last option. "Why don't we lose our bet, Al, and leave the MET."

Shepard ignored Houston's recommendation. "We'll press on a little farther."

"We have a 30-minute extension," Haise called. Their PLSSs were not consuming coolant water as rapidly as feared, so there was a little flexibility in the margin.

"Okay!" acknowledged Shepard.

They halted to allow Mitchell to take a pan at what would later be designated Site 'B3'.

Haise relayed a prompt from Deke Slayton that *he* would cover the bet if they reckoned that the MET was jeopardising their chances of reaching the rim.

"The MET's not slowing us down," Shepard replied. "It's just a question of time."

To Mitchell's amazement, when his commander set off he resumed the easterly trek. This was because Shepard had not been convinced by Mitchell's argument that Cone was to the left.

"We're in the middle of the boulder field," Shepard told Haise as they crested the rise. "But we haven't reached the rim yet." He was working on the expectation that the crater would have a raised rim. They pushed on until they reached the summit of the ridge. Although this presented a clear field of view, there was no sign of their objective! Unfortunately, since the 370-metre-wide pit was simply embedded in the ridge, it did not possess a raised rim. As Mitchell had surmised, the crater was off to the north. In fact, it was now *below* them, but they could not see it because the ridge sloped away to the north and the crater's far rim was lower than its near rim, placing it below their local horizon. However, the crater's proximity was evident from the density of the field of boulders, which covered "as much as a square mile". Shepard called another halt, and asked

Mitchell to see if he could identify the debris on the map.

"We're on the southern rim," Mitchell said confidently. The map showed several distinctive boulders. He looked around but, without a clear sense of scale, he did not realise that one particular white rock corresponded to the bright blob on his picture. If he had recognised it, he would have realised that that they were now southeast of the crater.

"Ed and Al," Haise broke in. "We've already eaten the 30-minute extension. I think we'd better proceed with the sampling." They were clearly near their objective and the sampling site could be worked out later from a pan.

"Okay, Fredo," Mitchell acknowledged. The ascent from Site 'B' had taken 48 minutes; far longer than planned. But he was satisfied that they had given Cone Crater their best shot. After lifting a rock[4] from alongside the MET, he set up the magnetometer.

Shepard, meanwhile, gave a geological commentary. "Most of these boulders are the same brownish grey that we've found, but we see one that's definitely almost white." This was the rock that was prominently displayed on the map. Because it turned out to have a distinctive shape, it became known as 'Saddle Rock'. "Beneath the dark brown surficial regolith there is a very light-brown layer." Since arriving on the summit, their boots had been scraping this up and, looking back, they could trace their route by the albedo change. As soon as he had taken a pan, he drove a single-length core into the soft soil to sample the layering. However, when he withdrew the tube, the coarse white material from the deepest part of the sample trickled out; he scooped it into a bag.[5] To complement the soil sample, he popped a few pebbles into another bag.[6]

With a pan, a rock and a core sample achieved at what would later be designated Site 'C-prime', it was time to move on and sample another part of the boulder field. "Let's go over there," Shepard prompted, indicating the prominent white rock.

Having taken the magnetic field reading, Mitchell had abandoned the LPM *in situ*, because it was now surplus to requirements, so he took the hammer and set off, leaving Shepard to drag the MET through the litter of rocks. On the way, he saw a rounded boulder that had split open.

"It's further than it looked," Mitchell noted on reaching Saddle Rock, which was designated Site 'C1'.

"That's the order of the day," Shepard observed dryly.

"Okay, Fredo," Mitchell reported. "I'm right in the midst of a pile of very large boulders." All the rocks were breccias with large clasts in a coarse matrix. This was hardly surprising, as the Fra Mauro Formation was a blanket of ejecta. Furthermore, they were distinctly two-toned "white and brown", with most being more-brown, a few being more-white, and several displaying banding at the contacts between the light and the dark parts. Mitchell took some pictures, then looked around for a piece of the lighter material but, frustratingly, all the fragments were too large to fit into a bag. "They're all so darn big!" he complained. He hit the brown part of one of the boulders with the hammer, and was rewarded with a reasonable chip.[7] The scar was much lighter than the weathered rock. In true field geologist's fashion, he rested the hammer on the rock to provide a scale, and photographed his work. After inspecting the boulders for himself, Shepard picked up a football-sized rock[8] that had the virtue of seeming to be typical of the site. In fact, this was not only the largest rock of the entire mission, it was also much larger than anything returned by any earlier crew.

"We have about one more minute," Haise announced as they manhandled the rock into the MET's storage bin. To allow Shepard and Mitchell a few minutes more in the boulder field, the Backroom had traded away two sampling sites assigned to the descent of the ridge.

As Shepard had feared, in continuing to push for the rim they had consumed their extension, which left them with very little time for sampling when they were obliged to call a halt. However, the situation was not as bleak as it appeared. They had been within 100 metres of the crater's rim ever since Haise had granted the extension, so they had been on-site for almost 30 minutes, although they had spent most of this on the move – they spent 6 minutes at the LPM site and 5 minutes at Saddle Rock. A study revealed that the field of two-tone rocks extended to the very lip of the crater, with Saddle Rock being just 15 metres from it. If they had had another 10 minutes, one or other of the men would almost certainly have strayed sufficiently far north to see the far wall of the crater poke over the near horizon. They had, in essence, achieved all of their objectives except photographing the interior of Cone. However, their load of samples (a core, a rock, a few pebbles, a chip, and a football-sized rock) fell far short of expectation. In fact, Shepard's rock accounted for fully 90 per cent of their 10-kg total, and if he hadn't been seduced by it they would have left the ridge with a paltry

Upon cresting the ridge, Al Shepard and Ed Mitchell were surprised to find no sight of Cone Crater, but the litter of rocks indicated that it was nearby. This position was designated Site 'C-prime'. In fact, the near-horizon in this northward-looking view was the rim of the crater.

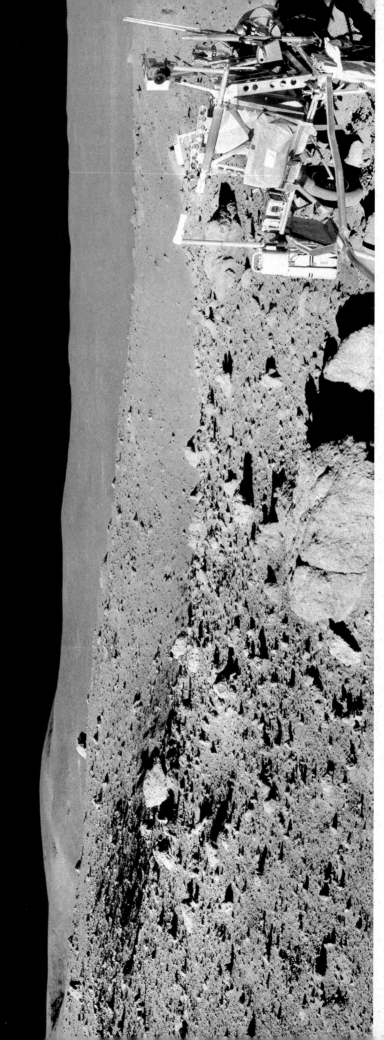

The Modular Equipment Transporter at Site 'C-prime' on the crest of Cone Ridge.

'Contact rock'   'Saddle rock'

In making their way through the rocks on the summit of Cone Ridge, Al Shepard and Ed Mitchell observed a split boulder. Two rocks that they would later inspect are arrowed.

0.96 kg of samples. Nevertheless, by virtue of its location it was material from deep within the Fra Mauro Formation, and was precisely what they had travelled so far to retrieve.

Within minutes of starting down off the ridge, they found an extremely eroded boulder "with tremendous fillets", so they decided to pause to take a sample. "Oh, man, that's hard!" Mitchell reported after delivering a firm blow with the hammer and achieving nothing. He examined the rock a little more closely. "Look at the *melt* in it." He persisted, and managed to break off several fragments.[9] Later designated Site 'C2', this was an excellent example of grab sampling.

"It looks like you're making better time going down, than up," Haise pointed out when told that they thought they were "about 10 minutes" from Weird Crater.

"The slope's a different way," Mitchell replied dryly. Moving downhill was easier going, but the MET was far more animated. They took turns, with one dragging the cart and the other trailing behind to retrieve things that bounced off. While moving west offered the advantage of being able to see the illuminated faces of the rocks, and their texture could be reported more readily, going both down-slope and *down-Sun* needed care because craters were hidden in the zero-phase glare. An impressive crater opened up directly ahead. "This is a *big* crater," Mitchell told Haise, "40 to 50 metres across; it has a fairly sharp crater in its southern rim."

Haise had been tracking their descent on the map by estimating their speed, which was why he had made the remark about their rate of progress. "That might be the one at 'E'," he offered helpfully.

The saddle-shaped rock was designated Site 'C1'.

Antares

Also at Site 'C1' was an angular boulder with an albedo variation indicating that it was a breccia that contained the 'contact' between two rock types.

Although Site 'E' had been deleted to make up time, Shepard decided to exploit the fact that they had stumbled upon it. "Why don't we grab a couple?" Any sample was better than no sample. They bagged[10] a "very crumbly rock" and one which had "flashing crystals" right off the rim.

Once they were off the ridge, they used a large rock 100 metres short of Weird Crater as a navigational point. "It sure is a big old boulder," Shepard observed. Their track took them south of the rock, but he ran over to snap a picture of it.

Mitchell pushed straight on with the MET, in search of Weird crater itself. "This country is so rolling and undulating, Fred," he reflected. "You can be by a fairly good-sized crater, and not even recognise it." The full irony of this observation was not evident though, because they had no idea just how close they had really been to Cone's rim. Halting south of Weird, which was Site 'F', he

bagged several rocks[11] while Shepard took a pan. Although time was running out, they were now using it very efficiently. This was how they should have done the sampling on the outbound leg.

"What else, Houston?" Shepard prompted Haise.

"We'd like you to proceed to the North Triplet for 'G'. One crater's diameter short, take a core and dig the trench."

The Triplet was directly ahead, a few minutes away. The record double-length core taken at Site 'A' was to be followed-up by a triple-core at Site 'G', but when it was about a section's length into the soft regolith the tube stopped, and no matter how heavily Mitchell struck it with the hammer it would not budge. "Solid rock!"

"Pull it up," Haise urged. "Move over, and try again."

The second attempt penetrated no deeper than the first. "I could beat it for 10 minutes," Mitchell

On the way to Site 'F' heading home, Al Shepard made a diversion to photograph a large rock near Weird Crater.

grumbled, "and not get another inch." He settled for what he had.

Meanwhile, Shepard took the scoop and went to dig a trench on the rim of a 6-metre crater 15 metres further west. "I'm going through three layers," he reported. "The fine-grained surface is dark brown; then there is a layer of what appears to be quite a bit of glass; and then some light material." The top layer was 5 cm thick. The glassy material, which was composed of small pebbles, was concentrated in a 2-cm layer. "I should be able to sample all three." Like the core, the trench revealed the vertical structure, but also extended the view horizontally, and the properties of the wall yielded soil-mechanics information. Taking a rest-break, he surveyed his location in more detail. Spotting an "interesting rock" in the crater, he went to retrieve it as a grab sample.[12] "It's dark brown. The dark part is fractured, and the fracture face is light grey with very small crystals." It was far too large for a bag, so he dumped it straight into the MET's bin. Upon resuming the trench, the deeper he dug, the more the walls tended to slump, polluting the separation that he had hoped to record, so he called a halt at half-a-metre's depth. The checklist called for a

SESC sample off the floor. Although he suggested that with the slumping this would not be worth doing, Haise told him to go ahead anyway.

When Mitchell said that he was going to get a few more rocks, Haise warned that they had only 3 minutes left, and he relayed a request from the Backroom that they document a partially-buried rock. After taking one rock,[13] Mitchell went to a group of 25-cm rocks just outside the rim of North Triplet, took prior-to-sampling pictures, then discovered that his chosen rock was deeply embedded. "Goddamn, it's bigger'n we thought." While Mitchell dug out the rock,[14] Shepard took the locator shot with Antares, 100 metres away, in the background.

During the first day's excursion, Mitchell had reported a cluster of rocks about 150 metres beyond where they had positioned the TV camera. The Backroom had decided that he should take a sample. The largest, a dark breccia[15] with substantial light clasts in it, had a protrusion on top that bore a striking resemblance to a turtle.

Meanwhile, Shepard had his own agenda. As the 'Old Man' of the astronaut corps, he had yearned more than most to walk on the Moon. He had in mind a moment of fun to celebrate his

At Site 'H', near Antares, Ed Mitchell inspected a boulder that had a protrusion on top of it which resembled a turtle.

success. "Houston," he called, as he stood in front of the TV camera. "You might recognise what I have in my hand as the handle of the contingency sampler. It just so happens to have a 6-iron on the bottom of it. I also have a little white pellet familiar to millions of Americans." He let the golf ball fall to his feet. "Unfortunately, I can't do this with two hands, but I'm going to try a little sand-trap shot." He swung and missed, swung again and this time chipped the ball.

"That looked like a slice to me, Al," Haise observed.

Shepard swung again, and this time struck the ball cleanly. "Miles and miles," he lied. With that, he set about the minutiae of packing up.

"I'm starting up," Mitchell announced. He took a rock box in his left hand and, using only his right hand to steady himself, climbed the ladder.

"Okay, Houston," Shepard announced a few minutes later. "The crew of Antares is leaving Fra Mauro Base."

"Roger, Al," Haise acknowledged. "You and

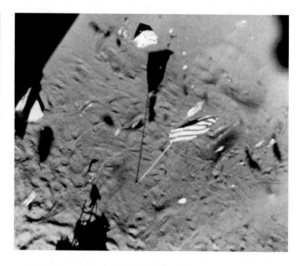

This frame from the 16-mm movie camera mounted in Ed Mitchell's window shows the blast from the ascent engine vigorously blowing the flag.

Ed did a great job." And then he added, with a rich sense of irony, "I don't think I could have done any better myself."

Apollo 14 samples 14047 from Site B (top); 14321, the 3.4-kg football-sized breccia from near the rim of Cone Crater; and Al Shepard and Ed Mitchell in the Lunar Receiving Laboratory inspecting their hoard.

THE FRA MAURO ROCKS

There were 33 individual rocks in the 43 kg of lunar material returned by Apollo 14. Their analysis turned out to be rather more complicated than had been the case for those from earlier missions, because the Fra Mauro rocks were consolidations of shattered precursors. Although one objective was to date when the fragments had been bound together in order to date the impact that applied the shock, this task was complicated by the fact that if a rock is melted, this 'resets' the isotopic 'clocks' used to yield the formation date. This was not an issue for the mare basalts, but the analysis of breccias involved dating the individual clasts. It was soon found that the samples tended to cluster into two age ranges, one spanning 3.96–3.87 billion years and the other 3.85–3.82 billion years. It was therefore inferred that the breccias were formed in the younger interval. The older dates applying to some of the clasts were the formation ages of the rocks that had been shattered by that impact. Although the small number of samples off Cone's rim meant that the error-bars were larger than hoped, it seemed clear that the Fra Mauro Formation had been laid down as a splash of ejecta 3.85–3.82 billion years ago. This was the 'ground truth' that Apollo 14 had been sent to determine. Although the primary objective had been achieved, a great many subsidiary issues remained unresolved.

It had been hoped that samples from Cone's rim would characterise the material underlying the Fra Mauro Formation. This putative highland basement was widely expected to be volcanic. At first, several intriguing samples did look as if they might represent ancient volcanism, but they turned out to be the first instances of another type of breccia. In fact, it was some time before it was realised that there were several kinds of breccia. That impact-formed breccias should prove to be poorly understood was a consequence of their being rare on Earth; they are found only in large craters, which are rare. Whereas clumps of consolidated regolith had been found on the mare plains, the Cone rocks were clasts of shattered rock bound in a matrix of pulverised rock. The term 'regolith breccia' was coined for the con-solidated-regolith type, and 'fragmental breccia' for the clasts-in-matrix type. Actually, as more examples were studied, it was found that some-times fragments of minerals were incorporated into a breccia, so not all clasts are lithic. Also, because breccias were caught up in impacts, it was possible to get 'breccias of breccias' in which the clasts in one breccia were bits of earlier breccias, and the terms 'one-rock' and 'two-rock' were introduced to accommodate this complex history. The 'Turtle Rock' was one of the first examples. A study by Ed Chao showed that it contained two types of clast: one dark and dense, and the other light coloured and friable. The dark fragments outnumbered the lighter ones in Mitchell's sample, but his photographs of the parent showed it to be more light than dark, indicating that it was a friable matrix hosting fragments of a much denser rock. The samples that were initially taken to be ancient basalt turned out to be yet another form of breccia. The "interesting" rock[16] that Shepard recovered near North Triplet had clasts containing a higher proportion of plagioclase feldspar than did the mare basalts. This was dubbed 'Fra Mauro Basalt' because it was thought to be igneous, but when small inclusions were found it was realised to be a breccia. In fact, it had been so heated as to be thoroughly melted. Impact-melt resembled basalt to the extent that it was a solidified melt, but a basalt is homogeneous. Despite the violence of the shock-melting, the resulting breccias were found to

contain some extremely delicate crystals which could only have built-up by diffusion processes as mineral-rich gas escaped from the ejecta. Although this crystallisation process was similar to sulphur encrustation at a terrestrial volcanic vent, the gas derived from the rubble itself, rather than the material on which this rested. The discovery of these crystals, some of which were metallic iron, indicated that the rubble had been hot when it was laid down, and had fused as it cooled. Not all clasts were metamorphosed. A few were igneous, and one,[17] at 3.96 billion years old, was not only the oldest sample yet found, it also showed there had been volcanism prior to the basin in-fill period and that this earlier lava was considerably more aluminous than the dark maria; this gave rise to the term 'non-mare' basalt.

One intriguing aspect of the Fra Mauro impact-melt breccias was that they were KREEPy, meaning that they were rich in both radioactive and rare earth elements. A study of the Fra Mauro samples revealed that the KREEPy material was originally a gabbro which solidified in the final phase of the fractionation of the magma ocean. During crystal-

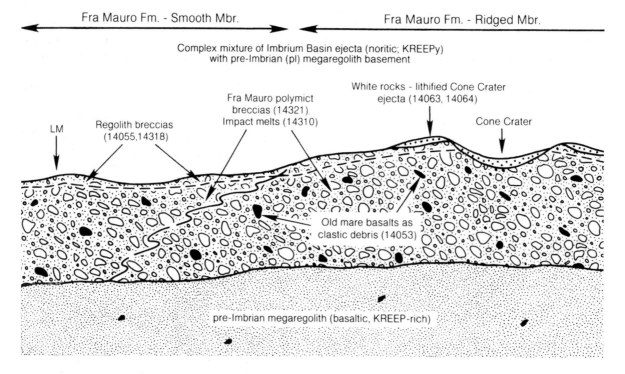

A schematic east–west geological cross-section through the Apollo 14 landing site showing the contact between the "smooth" and "ridged" members (Mbr) of the Fra Mauro Formation, which is an assemblage of complex impact-produced breccias (after Swann *et al.*, 1977). Numbers refer to specific collected samples that are representative of the units inferred to be present. 'LM' shows where Antares landed. The dashed line below the surface represents the depth of the regolith. Courtesy the Lunar and Planetary Institute and Cambridge University Press.

lisation, certain elements are either accepted or rejected according to whether they fit into the regular crystalline structure. The trace elements tend not to participate in mineralisation, so they would have remained in the melt as the various compatible elements were extracted, and so would have become progressively more concentrated. The radioactive elements at depth would have helped to maintain this concentrated reservoir molten, and then been locked in when it finally solidified. It was later excavated by the largest basin-forming impacts.

Analysis of the regolith's exposure to cosmic rays indicated that Cone Crater was formed a mere 25 million years ago. By striking the crest of the 100-metre ridge, it had no chance of reaching beneath the Fra Mauro Formation. If it had hit a few hundred metres to the west it may well have punched through the 75-metre blanket and, with easier access and more time for sampling its rim, what might Shepard and Mitchell have found?

THE BIG PICTURE

On 4 February, just after Apollo 14 settled into lunar orbit, its S-IVB hit the Moon some 175 km southwest of the Apollo 12 site. The data from this seismometer confirmed that the lunar crust is brecciated to a depth below that corresponding to the base of the terrestrial lithosphere. Although this was bad news for those who believed the Moon still to be volcanically active, they were encouraged by the Cold-Cathode Gauge Experiment's detection of a cloud of gas that was 100 times as dense as in cislunar space. On the other hand, this represented a better vacuum than that which was attainable in any laboratory on Earth. Unfortunately, the instrument did not measure the chemical composition of the gas. Although it was *possible* that this gas had belched from a volcanic vent, the thickness of the crust made this unlikely. The low escape velocity meant that only heavy gases such as argon and radon would linger near the surface. To follow up this positive detection, it was decided in future to install remote-sensing apparatus to enable a CSM to survey its ground track.

Although the Apollo 14 site was only 180 km from that of Apollo 12, the addition of a second seismometer meant that the moonquake epicentres could now be refined, and the data confirmed that the seismicity followed a highly regular pattern. "You can set your watch by it," joked Gary Latham. The repeatability suggested that the tremors were a fault system adjusting to tidal forces at perigee. It was also noted that there were distinct zones of activity, with 80 per cent of the events being produced in *one* zone, generally south of the Procellarum site and southwest of the Fra Mauro site. As the data accumulated, this zone was suspected to be related to an extensive network of rilles in the northern part of the Humorum Basin. The most surprising discovery was the *depth* at which the events occurred: 600 km. Although this imposed severe constraints on the possibility of there being a magma zone, the mechanism suggested for the quakes was magma periodically welling up through a series of deep faults. This conclusion was based on the similarity between the seismic signature of the deepest lunar quakes and those thought to be due to magma moving in a reservoir deep below the Hawaiian volcanoes. It would not require much lunar magma to flow to make events as weak as those observed. The fluid at that depth would have little chance of extruding onto the surface, however. It seemed that either the Moon had undergone dramatic cooling which had substantially thickened its crust, or the many features suspected of having been created by 'recent' volcanism were nothing of the sort.

Another issue that was causing consternation was the thickness of the maria. They account for 16 per cent of the lunar surface (an average of 30 per cent of the Nearside and 2 per cent of the Farside). If they were thick accumulations of lava then they represented a major proportion of the crust, but if they were no more than a thin veneer they were volumetrically insignificant. The favoured contemporary view was that they were 25 km thick. This implied a sustained process of extrusion. What was the timescale of the evolution of the Moon's 'heat engine'?

OPERATIONAL LESSONS

When Fra Mauro was assigned to Apollo 13, it had been believed that it represented an *ideal* target for an 'H'-class mission, but it had turned out to be just barely manageable, not for the difficulty of finding the planned landing site, which was not a problem, but for the undulatory character of the plain. The trek to Cone and back was a round trip of 3 km. If Lovell and Haise had landed at the more westerly site, it is likely that they would not have attained the crest of the ridge. Although Shepard and Mitchell did climb the ridge, the crater eluded them. The fundamental limitation was *time*. Another hour on Cone's rim would have fully justified the overhead of reaching it. In fact, the PLSS was rated for 7 hours, but this included the pre-egress time and a 2-hour post-ingress troubleshooting margin, and, as with Apollo 12, the rules obliged the crew to re-enter the LM rather than spend an hour or so sampling at the landing site. In effect, therefore, the time at Cone's rim was constrained by the need to preserve the *margin*. Although Apollo 14's visit to Fra Mauro had achieved its primary objective, it was evident that foot-traverses were no way to explore the Moon. The planners had seriously underestimated the difficulty involved in walking so far, and climbing so high, in such a brief time. As Mitchell put it while awaiting lift-off: "It was a darned hard climb to do rapidly." Certainly, the 'H'-class format would not do justice to the rich multiple-objective sites which were being proposed by the geologists for the remaining missions. It was impractical to devote so much of the limited time to travelling to and from the sample site; what was required was a *vehicle* that would not only minimise the transit time, and hence maximise sampling time, but would also enable more tools to be carried and more rocks to be collected, all of which would serve to increase the overall productivity of a traverse.

5

The wonder of the unknown at Hadley-Apennine

With two maria and the Fra Mauro Formation sampled, the geologists were eager to attempt a highland site, but none of the 'feature' sites had been surveyed sufficiently to certify them for a landing. However, there was no shortage of attainable alternatives. In addition to the possibly volcanic Marius Hills and Davy Rille, there was the 'dark mantling' on Mare Serenitatis that Apollo 14 had been set to sample before it was redirected to Fra Mauro. But the 2 September 1970 decision to recast Apollo 15 as the first advanced 'J'-mission made these seem rather tame.

The catena of Davy Rille, suspected by some of being volcanic vents along a deep fault, as photographed by Apollo 12.

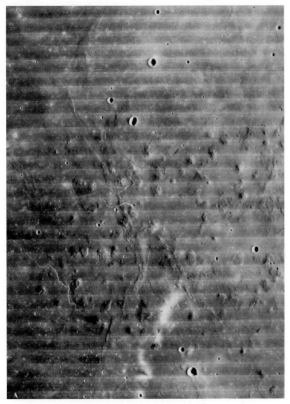

The seemingly volcanic Marius Hills, as photographed by Lunar Orbiter 4.

The enhanced capability offered by the Lunar Roving Vehicle (LRV) really required a 'multiple-feature' site. The favourite was a new contender: Hadley Rille, on the eastern rim of the Imbrium Basin. It would be demanding to reach, however, as it would require flying low over the Apennines, whose peaks are among the highest on the Nearside, and then descending steeply to a mare-flooded valley. In the selection meeting, the mission commander, Dave Scott, attracted by the site's majesty, insisted that a landing was feasible,

The target for Apollo 15 was Hadley Rille, a sinuous feature on the plain of Mare Imbrium, running close alongside the Apennine mountains. Note that the trend of the hummocky terrain is perpendicular to the arc of the mountain chain and hence radial to the Imbrium Basin. The largest crater in this view is Archimedes, and the lighter-toned area between it and the mountains is the Apennine Bench Formation.

and so it was decided to send Apollo 15 to explore Hadley-Apennine.

Of two types of lunar rille – linear and sinuous – Hadley Rille is one of the most impressive sinuous ones. It 'starts' in an arcuate cleft in the Imbrium-facing side of the Apennines and exploits a system of radial and peripheral clefts near the mare shore to follow the basin rim north for 100 km. Upon drawing parallel with Mount Hadley Delta, a secondary peak south of Mount Hadley, it cuts across the mouth of the embayed valley before resuming its sinuous path until finally petering out on the mare. The International Astronomical Union had long-ago given this patch of mare the unfortunate moniker Palus Putredinus (Marsh of Decay), but it promised to be a rich venue for Apollo's investigation of the basin which served as the basis for lunar stratigraphy.

The Apollo 15 mission patch.

In this westward-looking view from orbit Mount Hadley Delta is on the left, with the crater St George on its flank and the craters of the South Cluster at its base; Hadley Rille cuts the mare-embayed valley; Bennett Hill is top left; Hill 305 is top right; the hills and craters of the North Complex are right of centre; and the landing site is this side of the rille, between the North Complex and the South Cluster. Mount Hadley is out of frame to the right, and the shadow at the bottom was cast by one of the peaks of the Swann Range, which forms the backbone of this section of the Apennines.

SKIMMING THE MOUNTAINS

The landing site was beyond the backbone of the Apennine Range, so the LM could not employ the shallow 15-degree trajectory flown by earlier missions; to clear the mountains and land on the confined plain beyond, a 26-degree approach would be needed. To reach this site, it had been necessary to further relax the flight dynamics criteria. For a start, the rule that the track of the final approach must be covered by high-resolution Lunar Orbiter imagery had to be deleted because, although the area had been imaged, it was as a site of scientific interest. In fact, at best, the resolution at the landing site itself was 20 metres, and the photo-interpreters had required to exaggerate the contrast in order to gain an impression of the topography. As a result, unlike their predecessors, Dave Scott and Jim Irwin really were about to descend into *terra incognita*.

For his knowledge of the LM, and because of his experience of the radar problems suffered by Apollo 14 during its landing, Ed Mitchell was CapCom for the descent. More than any other active-duty astronaut, he would be able to relate to the problems that Scott might face. "Falcon, you're 'Go' for PDI," he advised.

Jim Irwin, Apollo 15 lunar module pilot.

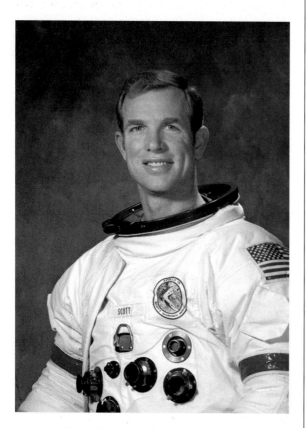

Dave Scott, Apollo 15 commander.

Al Worden, Apollo 15 command module pilot.

"Roger," Scott replied matter-of-factly.

With everything checked out in the hectic minutes following AOS, a calm had set in as Scott and Irwin waited for the clock to run down to the moment that Falcon's engine was to fire.

"Auto-ignition," Scott announced.

The DPS ran at 11 per cent of its rated thrust for a few seconds, to enable the gimbal to precisely align the thrust vector through the vehicle's centre of mass, then it throttled up to 100 per cent. During this early phase, Irwin verified that the AGS agreed with the more-capable PGNS, to be sure that the AGS could be relied upon in the event of an abort.

"You're 'Go' at 3 minutes," Mitchell called.

"Roger, 'Go' at 3," Scott acknowledged. "Altitude light is out; velocity light is out." The radar had come up on schedule. They were a little high, but it was nothing that the computer wouldn't be able to correct.

The PGNS navigated with respect to an idealised model of the Moon that did not include topographical detail. The radar measured the spacecraft's height above the surface. The altitude data enabled the computer to work out the height of the surface above its global reference level, and Irwin rapidly compared this with a chart in the flight plan to check that the radar data made sense; it did; Houston confirmed it; and the computer was told to use the data. Falcon was still a little high as the 4-minute mark came around, but it was almost on course 3 minutes later when it passed 3 km over the Apennines. The computer throttled the DPS back to 57 per cent. Scott input commands to check that his hand-controller was functioning. In Houston, the flight dynamics officer reported that the spacecraft was heading for a point 1 km south of the target. Although the flight director's inclination was not to tell Scott, because he would realise the error as soon as the LM pitched upright and enabled him to see the terrain ahead, Mitchell pointed out that this would not occur until the LM was 6 km from the target, and Scott would prefer to correct his path sooner rather than later; the flight director acceded. "Falcon, Houston," Mitchell called. "We expect you to be a little south."

In fact, Scott had been straining against his harness for some time in an effort to check the ground track through the lower corner of his triangular window. Although the peak of Mount Hadley Delta was visible off to the left, he could not see the rille slicing across the plain straight ahead. In the simulator, in which his window had presented the view from a TV camera that 'flew' over a plaster-of-paris model of the landing site, he had been able to sneak a view of the rille just before pitch-over. The fact that he could not see it led him to suspect that, irrespective of what the computer said, he was coming in 'long'. He was also astounded by the sight of Hadley Delta projecting half-a-kilometre *above* him. Intellectually, of course, he had known that he would be in the mountains once Falcon began its steep descent, but the simulator had offered only a view ahead, and the clarity of the texture of the summit was stark. Irwin, who was stealing occasional glances out of his window, was presented with the even more imposing Mount Hadley; he later likened this phase of the descent to flying a high-performance jet through a mountain pass.

"Okay," acknowledged Scott, accepting the news that Falcon was running a little south of track. This lateral error paled into insignificance compared to the prospect of landing long. He was to set down 2 km east of the rille. If he was just a little long he would land slightly closer to the rille, which would be acceptable, but if he was heading for the rille itself, which was over 1 km wide, he was in serious trouble, and since a pilot's instinct is to 'extend', his recovery would lead either to an abort or to a landing on the far side of the rille, which would be disastrous for the science plan. No matter how he stretched in his harness, he could not see the rille. The computer reckoned it was on course, so Houston was unaware of the mission commander's concern. But a few minutes later Scott's fears were relieved. As the LM pitched over at an altitude of 7,000 feet, he was presented with a view of the entire plain with the rille where it was supposed to be – the simulator had been misleading.

Irwin read off a succession of altitude, rate of descent and LPD angles, the latter indicating where the computer intended to land. It was difficult for Scott to verify the computer's aim, because the plain didn't look right. The simulator's model had been sculpted in accordance with the interpretation of the Lunar Orbiter imagery, but boosting the contrast to reveal topography had enhanced the crater rims. As a result, many of the landmarks that Scott had expected to see were so ill-defined that he could barely discern them. The fact that most of the craters were actually *rimless*, and so didn't cast significant shadows, did not help. They were indistinct, but he was able to identify a line of four large craters dubbed Matthew, Mark, Luke and Index. He was to land near Index. He began to look for a level site on which to set down.

"You're 'Go' for landing," Mitchell confirmed as Falcon passed through 1,000 feet.

At 400 feet, when Scott took full-control, Irwin deleted the LPD angle from his cycle and added propellant-remaining. Irwin served as Scott's eyes within the spacecraft while Scott focused his attention outside the window. "I've got some dust," Scott noted as he passed through 120 feet. In fact, this was an understatement, since the DPS plume was blowing so much dust that the gently undulating surface was completely obscured. As there were no rocks poking through the dust to provide a direct indication of his motion, he reluctantly reduced his rate of descent to 3 fps and 'came back in' so as to fly the rest of the descent on instruments, with only an occasional glance outside to verify his senses. Having the luxury of being 'fat' on propellant, he cut his rate of descent to 1 fps and maintained a slight forward creep while eliminating the lateral motions.

"8 feet," Irwin called, to start another cycle, but before he could read the rate of descent the light came on to indicate that one of the long probes projecting from the footpads had hit the surface. "Contact!" The DPS had been uprated to accommodate the mass of the Rover and the extra consumables. To prevent the lengthened engine bell from embedding itself in the regolith and causing an explosive blow-back, the engine had to be shut off as soon as the probes made contact. Scott eagerly obliged. Shattering though the free-fall drop was, he and Irwin were more concerned by the fact that as the LM settled it tilted backwards through 10 degrees.

"The Falcon is on The Plain at Hadley," Scott announced triumphantly a few seconds later. He meant more than that they were safely on the embayed mare plain; this was also an elliptical reference to the parade ground at the West Point military academy.

"Roger, Falcon," Mitchell acknowledged, barely audible over the applause in the control room. As soon as the LM had been powered down, he signed off. "You did a real good job on that descent."

"We appreciated all your help," Scott acknowledged.

SITE SURVEY

The 'H'-class crews had ventured out immediately after landing because they had a single night on the surface and their outings lasted only 4 hours. But a 'J'-class LM was capable of three nights and the improved PLSS could sustain a 7-hour period outside, and rather than make an immediate EVA Scott had decided to end the day with a 'Site Survey' to report the geological context.

Yet-to-fly astronaut Joe Allen, the mission scientist, had been assigned as primary CapCom for the surface operations. When Allen took over, Scott described the view out of his window. "We're sure in a *fine* place," he assured. "We can see St George, it looks like it is right over a little rise." This was a 2-km-wide crater on the northern flank of Mount Hadley Delta, near where the rille swung north after running around the western side of the mountain. "And we can see Bennett Hill." This was a peak way off on the far side of the rille. It had been named for Floyd Bennett, the member of the flight dynamics team who devised the steep descent that had enabled Falcon to set down at the confined Hadley-Apennine site.

"Did you see the rille on the way down?" Allen asked excitedly.

"Sure, Joe!!" Scott laughed. He did not mention his concern that the view had not matched the simulator.

Some 2 hours after touchdown, while Scott and Irwin ran through the final checks for depressurising the cabin, Al Worden made his first orbital pass. Viewing through his telescopic sextant, he readily identified the LM at the western tip of the bright streak where its plume had disturbed the regolith. He reported them "almost directly north of Index, near November Crater". November was within a couple of hundred metres of the target point. Although it had not been the most accurate landing of the program, this time it really didn't matter. Or at least it wouldn't matter so long as their vehicular transport functioned properly. If the Rover failed to deploy, then they would be reduced to walking traverses, in which case it would be crucial that they know precisely where they had landed.

As soon as the cabin air had been vented, Scott sat on the drum-shaped cover of the Ascent Propulsion System (APS), lowered the hatch in the roof, and withdrew the docking system. He then stood on the engine and put his head out of the hatch. "What a view!" Even though he didn't really need it, he accepted the map that Irwin provided, and then he conducted an initial circuit of the horizon. "I can see Pluton, Icarus, Chain and Side." These were craters in the North Complex that they planned to visit. He swung around to the south. "St George, Window, Spur." These were their targets on Mount Hadley Delta.

The clarity of the horizon on the airless Moon was fabulous. It had been so for previous crews, of course, but in this case there was so much more to look at.

Irwin handed Scott a Hasselblad with a 60-mm lens. "You want 22 frames." Like everything else on the mission, this panoramic sequence had been carefully planned.

Before Scott began running overlapping frames around the horizon, Allen turned to another point. "Dave, does the trafficability look pretty good?" Terrestrial radar data had suggested that the site would be so littered with rocks that the Rover would be seriously inhibited. The resolution of the Lunar Orbiter photographs had not been sufficient to enable the photo-interpreters to confirm or deny this.

"It sure does, Joe," Scott replied enthusiastically. In the LM's immediate vicinity there were no rocks larger than about 20 cm in size, and the coverage was no more than a few per cent.

"Do you see November?"

"I don't, Joe." Although he had evidently set down very close to this crater, it was difficult to locate. Once the pan was complete, Irwin took the camera back and then handed Scott a Hasselblad with a 500-mm lens so that he could take telephoto shots of prominent features. Finally, having exchanged the film of the first camera, Irwin passed this up to Scott, who reshot the pan in colour.

"We're coming up on 15 minutes," Allen noted.

"Okay, Joe, we've got all the photos." Scott's next task was to give a commentary on the site. In the Backroom the geologists were listening with eager anticipation.

"On the distant horizon to the northwest,

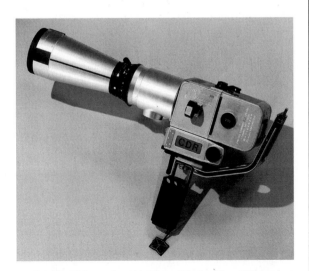

Apollo 15 introduced a Hasselblad with a 500-mm focal-length telephoto lens.

apparently across the rille, I can see a very large mountain." This was Hill 305 on the map. "All the features around here are very smooth. The tops of the mountains are round; there are no sharp jagged peaks." Nor were there any boulders on the slopes. "The largest fragments I can see are around Pluton. There are no boulders at all on St George, Hill 305, Bennett Hill or, as far as I can tell, up at Hadley." One of their objectives was to sample boulders that looked as if they may have rolled down mountains. "The horizon is really the North Complex. Chain, Icarus and Pluton are all very rounded subdued craters. The southern rim of Pluton is on the same level as the plain, but the northern rim is somewhat higher. The inside walls are fairly well covered with fragments up to – I'd estimate – 3 metres. They're just sort of scattered around and maybe the walls have 5 per cent coverage." The mound beyond Pluton had been named Schaber Hill, for Gerry Schaber, the geologist who had helped them to plan the traverse to the North Complex.

"Mount Hadley is in shadow," Scott continued, "although I *can* see the ridge line on top. It too, is smooth." He could see the summit of a hill beyond (Hill 22) and it was also rounded. So much for the science fiction 'moonscapes' with their jagged terrain. He swung around, through east, where the Sun made it impossible to resolve detail in the Swann Range, to the southeast. Mount Hadley Delta was awesome, but there was an even more remarkable sight to its left, and some 20 km away. They had named this Silver Spur after Lee Silver, another of their mentors. "I can see lineaments dipping northeast. They appear to be uniformly spaced, each maybe 3 to 4 per cent of the total elevation of the mountain." It looked for all the world like a tilted massif with heavily eroded outcrops of some depositional bedding. "I can't tell whether it's internal stratigraphy, but there are definite linear features, dipping at about 30 degrees." From his viewpoint, it appeared to be a veritable cliff, but he knew from overhead imagery that it was not. Nevertheless, it was clear that there was a steep drop-off from the summit where the western flank had been excavated by a crater that was hidden from view.

"As I look up to Hadley Delta itself, I can see what appears to be a sweep of linear features that curve around from the western side, dipping to the east at about 20 degrees. These are much thinner; each less than 1 per cent of the total elevation." He was amazed. Lineaments were one of the many things that they had hoped to see, and they were everywhere.

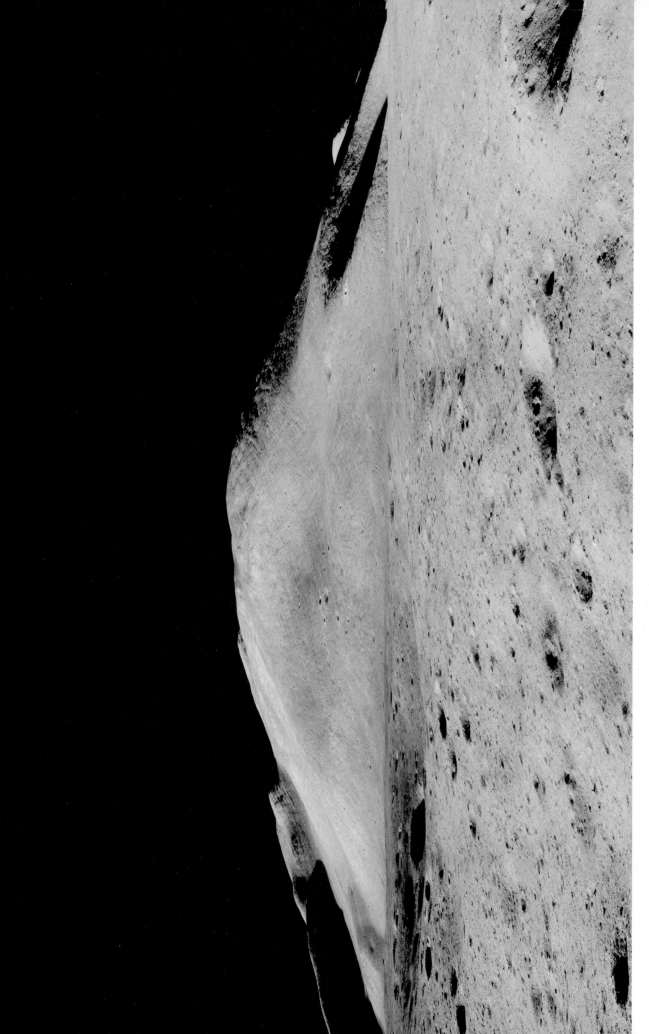

This part of Dave Scott's 'site survey' panoramic sequence shows Mount Hadley Delta with the crater St George in its flank (right), a chain of small craters running up its flank (centre), and the seemingly cliff-like Silver Spur (left). The mountain rises 3,000 metres from the plain.

Scott looked at the craters Spur, Window and Front that he and Irwin hoped to visit on their second EVA. Although spaced out several kilometres along The Front, as the mare-facing side of the Apennines was called, these craters were no more than 100 metres in elevation off the plain, so the traverse would be confined to the base of the mountain. Compared to the cratering density of the plain, the slopes seemed remarkably clear. "The craters on the side of Hadley Delta are rather few," Scott reported. But there was a chain of craters running from the plain to one-third of the way up the slope. "I might associate those with the South Cluster." It appeared that these craters had been made by debris which had flown in from the north to excavate the cluster of large craters on the plain at the foot of the mountain. In addition to lineaments, he had been asked to look for any sign of slumping, but there were no landslips. "I see nothing to indicate any flow downhill, only some subtle changes in topography." He turned to the other side of the mountain. "There's a bright fresh crater right next to St George, on the eastern side." It was strikingly bright. By now he was back around to the rille, so he paused to catch his breath; his horizon report had taken fully 6 minutes.

"Superb description, Dave!" Allen congratulated. He passed on a request from the Backroom for an assessment of the local area.

"I'll just take a quick look at the near field for you," Scott said obligingly. "It's all generally the same. The crater density is somewhat higher than I'd expected. Their sizes are mostly less than about 15 metres. The only *large* crater that I see is to the southeast. It has a very subtle rim, and almost no shadow in the bottom of it." As he looked around, he tried again to relate the craters to the map prepared by enhancing the contrast of the Lunar Orbiter pictures, but it was difficult. "I think that's one of the things that was deceiving on the descent; there are very few deep craters." On the other hand, the absence of deep pits would surely make it easier to drive at high speed, which was excellent news. "Trafficability looks pretty good. It's hummocky but I think we can manipulate the Rover in a straight line." In fact, the more he saw of it, the more he liked Hadley-Apennine. "It looks like we'll be able to get around pretty good."

"Sounds like we're in business, old friend."

"We are indeed," Scott confirmed.

"Dave, you're coming up on 30 minutes," Allen gently prompted.

"Okay, Joe. There's just *so much* out there, I could talk to you for hours. Do you have any specific questions before we call it quits?"

"We'll talk to you again," Allen promised, "once you button up."

There was no imperative to end the survey, it was just that 30 minutes had been allocated in the timeline. "Coming down," Scott informed Irwin, who, having been too busy to look out of the window during the descent, and who was now restricted to the view out over the rille, had listened to Scott's report in silence.

A GOOD NIGHT'S SLEEP

One goal of the Site Survey had been to help to relieve the emotional high of what had been a long and exciting day.

"We'll say a pleasant 'goodnight' to the two of you," Allen hinted gently. It was important that the astronauts get a good night's sleep in order to be fresh for a full day's exploration.

"Okay, Joe," Scott laughed.

"It's been an outstanding day," Allen said in summary.

"We're enjoying it," Scott assured.

Allen was synchronised with the crew's sleep cycle to enable him to support their external activities, so Bob Parker, who had a reputation for dozing, took over for the 'night shift'.

The LM was full of apparatus. It was surprisingly roomy in weightlessness, but it was cramped in a gravity field. The only place to stand was at the controls. After Scott and Irwin had stripped off their suits (the first crew to do so) and stored them on the APS, they installed their hammocks. During a night in a high-fidelity trainer they had found the hammocks awkward, but in the low lunar gravity they proved to be comfortable. Resisting the temptation to chat, they made a deliberate effort to fall asleep, and it worked.

The LM cabin was slowly leaking air, so Houston's wake-up call was made an hour earlier than scheduled. The flight controllers had spotted the leak soon after the astronauts had retired. As the power-saving low-rate telemetry did not provide full visibility of the systems, the flight controllers had debated whether to awaken Scott to switch the telemetry to high-rate in order to enable the leak to be identified, but because the leak rate was slow the flight director decided to let the astronauts sleep while the engineers monitored the problem as best they could. With high-rate telemetry, it was evident that the vent valve of the urine dump system had not sealed properly. When the valve was recycled, the leak ceased.

"It looks like you were getting pretty good sleep," Parker observed. The flight surgeon had been monitoring their biomedical telemetry.

"Yessir," Irwin said enthusiastically. "That's the best sleep I've had on the flight." He had found it easier to sleep in the hammock than he had when weightless.

"How was Dave?"

"Just fine, Bob," Scott assured. "I was way down in sleep when you gave us a call."

"We lost a little sleep on the ground," Parker pointed out, alluding to the ongoing discussion about the leak. "I couldn't even fall asleep at my console!" he joked.

"That *is* amazing," Irwin laughed.

Given that they were up, Scott and Irwin opted to make an early start on the EVA preparations. It wasn't just a matter of donning the suit, adding the PLSS and opening the hatch. The lengthy procedure was listed on cards that were hung on the control panel, and the men verified each other's actions. Even though he had suited-up in training many times, and knew it all by heart, Irwin was meticulous. Conrad and Bean had worked from memory, and had sometimes neglected a step, which had resulted in wasted time backtracking because most of the steps had to be undertaken in sequence.

VENTURING OUT

"As I stand out here in the wonder of the unknown at Hadley," Scott began, as he took his first step onto the lunar surface, "I sort of realise there's a fundamental truth to our nature: Man *must* explore." He paused, awed that this was not another simulation. "And *this* is exploration at its greatest." This was almost as profound as Armstrong's commentary, but history favoured those who pioneered rather than those who consolidated.

"I see why we're on a tilt," Scott laughed as he stepped away from Falcon. The surface was hummocky with heavily eroded craters. "Tell the Program Manager, the rear leg's in a crater and the rim is right underneath the engine bell." This was a reference to Jim McDivitt, who had insisted that they shut down the DPS as soon as the 'contact light' lit. "Sorry about that Jim," Scott said directly to McDivitt, whom he knew would be sitting in the back row of the control room, where the managers monitored proceedings, "but IFR,

you know!" He peered into the shadow under the LM to see if the engine had hit the ground. Although it seemed to be undamaged, the pictures that they later took to document the state of the vehicle would show that the bell was slightly buckled.

The first task on previous missions had been for the commander to take a scoop of soil and a few pebbles, but this time this sample had been assigned to the LMP, so once Scott had finished his walkaround inspection he made a start on unpacking the MESA.

"I'm going to come on out," Irwin announced. Although this was a statement, he was actually giving his commander an opportunity to suggest otherwise.

"Come on out," Scott agreed. "It's nice!"

Instead of dropping straight down off the bottom rung of the ladder, Irwin pushed himself away and landed on the rim of the footpad. With the LM tilted, the footpad was not flat on the ground. As it flipped flat on its universal joint under his weight, Irwin was momentarily disorientated. Although he still had a grip of the ladder with one hand, his PLSS tugged him back and he spun around and almost fell. As he recovered, he took in the view. "It's *beautiful* out here!" To convey something of the scene to Allen, whose TV field of view was restricted to the ladder and a wildly tilted horizon, he drew on their skiing trips in Idaho. "It reminds me of Sun Valley."

Irwin went to an undisturbed patch of ground near the tip of Falcon's shadow for the contingency sample. He selected a site that was littered with tiny fragments, set between two small subdued craters. A sample taken close to the LM was inherently unrepresentative, as the DPS plume tended to blow away the finer material. Taking the sample out west minimised this effect. As he scuffed the regolith with his boot, Irwin observed that his overshoes were already stained by the fine powdery material. "That's really soft dirt."

"Sure is," Scott agreed.

"That makes for easy trench-digging," Allen chipped in.

"Always thinking, eh, Joe," Irwin observed sardonically.

"Looking ahead," Allen countered. As CapCom, one of his tasks was to plan ahead, and the fact that the trenching might prove to be easier than expected might mean that it would free up a little pressure on the timeline. In fact, of course, he was simply provoking Irwin about his dislike of the soil-mechanics experiments.

WHEELS

Once Scott had dismounted the TV camera from the MESA, he set off to stand it on a tripod so that Houston could observe the deployment of the Rover. He had been out for about half an hour, and was on schedule. Irwin was actually several minutes ahead. This was an excellent start, but if time was going to be lost, it would be in releasing the surface transportation, whose deployment mechanism was about to be exercised for the first time. "Let's take a look at our Rover," Scott announced.

While Irwin ascended the ladder to access the lanyard which would slip the pins that held the tightly folded vehicle against the side of the descent stage, Scott inspected the 'walking hinge' that was to support the package during its deployment. To his amazement, he found that the hinge had popped. It seemed likely that the hinge had been released by the shock of the landing. It was simple to reset, however. As Scott continued his inspection, he began to wonder about how the LM's tilt would affect the Rover's deployment, because the tests by the supplier had been made with the rig perfectly upright. He unstowed the lanyard attached to the deployment mechanism, and then stepped well back. Irwin pulled the pins and a spring-loaded rod eased the top of the package several degrees out from the LM. Back on the ground, Irwin took another lanyard, this one attached to the top of the package, and moved out beside Scott.

"Here we go," Scott said optimistically.

Deploying the Rover was most definitely a two-man operation. Scott pulled the lanyard which controlled the mechanism that unreeled cables connected to the top of the package, and Irwin pulled on his lanyard to ensure that the package did not snag. Nevertheless, it was a jerky, decidedly comical process. It was not entirely manual. It had been deemed insufficient simply to have the astronauts lower the package to the ground and then assemble it, like a kit. The Boeing and Marshall Space Flight Center engineers had devised a folded configuration that incorporated spring-loaded actuators so that by the time the vehicle reached the ground it was almost ready to drive. The first of these systems was triggered when the Rover was half-way down. On Earth, the designers watched eagerly. The Rover had been built with its chassis in three sections, with the two end pieces, with the wheel assemblies, folded over the upper deck. The underside of the chassis was facing out from the LM, so that when the vehicle reached the ground it would be the right way up. The first part to unfold was the rear of the vehicle. The chassis plate hinged out and locked into position and then the wheels, free of their confinement, sprang out to the sides. The front section was identical, but wouldn't trigger until the Rover was almost horizontal.

"She's coming down okay," Scott said. Evidently, the LM's tilt was not impeding the deployment system. But then it seemed to stick and the cables went slack. "Can you pull it out a little bit?" Scott prompted. Irwin tugged. The front-end sprang into place. In rapid succession, this pushed the chassis clear of the LM, articulated the walking hinge arm, and lowered the rear wheels to the ground. Scott released his lanyard and went to detach the cables. Another lanyard had been fitted to lift the Rover off the hinge, so he pulled it and the front end was lowered to the ground. "It looks like it's loose to me." Because it was askew, he took hold of one side of the rear chassis and lifted it. "This thing's nice and light." He lined the vehicle up, then gave it a walkaround inspection. One of the pins that locked the chassis segments together had not fully engaged, but this was easily fixed.

Because everything else seemed to be in order, Scott moved to the next step in the plan. "Let's turn it around." The two men stood, one each side of the centre of the chassis and hoisted it between them, but could not draw it clear of the LM. "Wait a minute," Scott moaned, fearing that their luck had run out. "It's not disconnected." He manoeuvred as far into the gap between the Rover and one of the LM's legs as he could, and inspected the deployment system. Lowering the vehicle to the surface should have released the pin-connectors of the walking hinge on either side of the central section of the chassis. It was supposed to be a foolproof mechanism because gravity would pull the connectors from their sockets – nevertheless, for some reason it had become fouled.

"I think we can lift the front up," Irwin suggested.

"We can try," agreed Scott. While Irwin worked into position to reach down to lift the front of the Rover, Scott went to the rear to drag it clear once it was free. "Forget it," he countered, on realising that Irwin's PLSS was in danger of becoming caught in the framework of struts of Falcon's leg. "Hey, Houston, any suggestions?"

"We're working it," Allen assured. It wasn't long before he was back with a plan. "Pull the Rover out as far out as you can, away from the

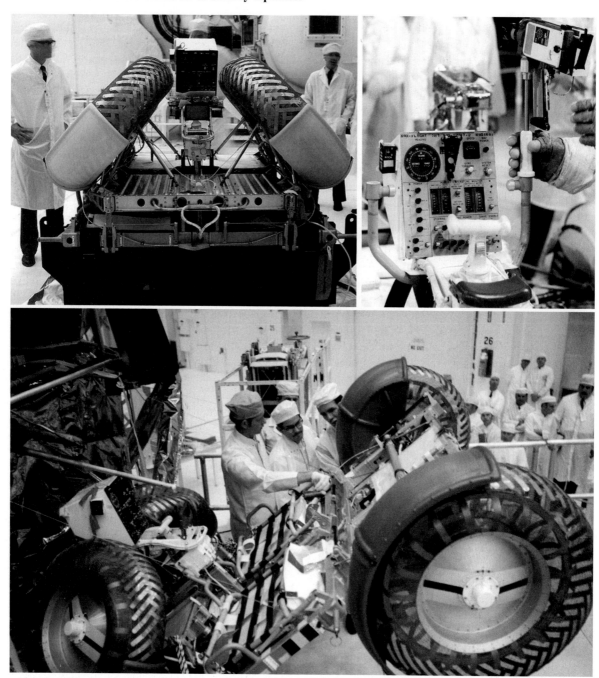

The Lunar Roving Vehicle folded into a wedge-shaped compartment in the side of Falcon for carriage. On the lunar surface, pulling a lanyard would cause the chassis to hinge down and the wheels to pop into position. It was driven using a T-shaped stick set between the seats, in front of which was a panel of status and navigational instruments (upper right).

LM, and *then* lift the front end." Drawing the Rover out first might give Irwin sufficient clearance to squeeze in to lift the front end of the vehicle, and in so doing hopefully release the arms of the hinge.

"We'll give it a try," Scott replied. He drew the Rover so far from the LM that the hinge rotated and lifted the front wheels off the ground. He would have to be careful not to let it slip, otherwise the hinge would draw the vehicle back in and trap Irwin. Although Scott warned Irwin that his PLSS was again becoming entangled with the leg struts, Irwin remained in place. As he lifted the front end further off the ground, the hinge arms fell out of their sockets. It was later realised that the plan for gravity to release the hinge had been foiled by the fact that Falcon was tilted at a 10-degree angle, because this caused the Rover to come down askew, twisting the arms of the hinge sufficiently for friction to prevent them from disconnecting from their sockets. The problem had cost them 5 minutes. After Irwin had lowered the front end of the Rover to the ground, the two men resumed their positions at its sides, hoisted it and walked it through 90 degrees so that it could be driven either forwards or backwards away from the LM. Leaving Scott to configure the Rover, Irwin resumed unpacking apparatus from the MESA.

The ingenious design of the Rover meant that it required only a few minutes to set up. Because the wheels had been stowed in such a tight space, their fenders had been designed to extend fore and aft, and they just required to be drawn out. A box with the instruments and the steering control was mounted between the seats, and it was a simple matter to raise and lock it. To save on mass, folding seats had been installed, and their backs were rotated into position. Finished, Scott climbed aboard. The crew stations matched the LM. The Rover was to be steered by the commander, in the left seat. It was a fly-by-wire system using a T-handle. For convenience, this was angled slightly towards him. Irwin's role was to navigate. As he shuffled to get comfortable in his seat, Scott realised that the training had been misleading. The 'J'-mission suit provided a higher degree of mobility, including the ability to bend at the waist to sit on the Rover, but it was still inflexible. On Earth, the combined mass of his body, the suit and the PLSS had forced him right down into the seat, but in the weak lunar gravity his mass did not yield the same weight, and he found himself reclining in the seat as if he were in a low-slung sports car. Unfortunately, the pre-fitted seatbelt was unadjustable. Because his field of view through the helmet was so limited, the catch of the seatbelt had to be eased into the lock by feel. It was something of a struggle, but he finally managed to belt himself in.

In working through the pre-start checklist, Scott reported a 'zero-volts' reading on one of the two batteries, but because it gave an acceptable 'amp-hours' reading he dismissed this as an instrument problem. Each battery was designed to supply 120 amp-hours of power, which would be sufficient for the entire surface traverse with a significant margin in reserve. A second battery was redundancy against failure. Each wheel had its own drive motor and the vehicle had both front and rear steering that could be used independently or in concert. It had a sustainable top speed of only 12 kph, but with a turning radius matching its length it had manoeuvrability, which was more important on Moon. After configuring the circuit breakers to feed power from the suspect battery to the rear steering, Scott was ready for a once-around-the-LM test drive. The T-handle was an intuitive system: push to go forward; tilt left to turn left; tilt right to turn right; pull back to brake; and all the way back into the detent to apply the parking lock. At the throw of a switch, pushing the handle forward meant reverse. There were no foot pedals. He eased the handle out of its detent, and moved off at a stately 3 kph. "We're moving!" he announced delightedly.

Irwin, working at the MESA, grabbed the 16-mm movie camera to record the historic first drive.

"Tell me if I've got any rear steering," Scott asked. He couldn't turn in his suit to see whether the rear wheels were steering.

"Yeah, you have."

"I don't have any *front* steering!" Scott reset the system to feed power to the front steering, but it had no effect.

"Just rear steering, Dave," Irwin confirmed.

"Press on," Allen advised.

"That's a good idea," Scott agreed; in this terrain he should be able to drive with just the rear steering.

The overhead in time of deploying the Rover had to be set against the benefit that it would deliver. However, there was an inviolable constraint relating to what would happen in the event of the Rover breaking down. Apollo 14's traverse to Cone had provided solid data on the rate at which a man consumed oxygen and coolant water in harsh conditions. Although the Rover offered a wide radius of operations, it had been decided that for safety the astronauts must not venture further from the LM than they could walk back using the consumables available to them at that time. As this radius decreased with every passing minute, it was vital that they set off immediately and reach the most distant point as soon as possible. They would then collect samples at locations consistent with the diminishing operating radius. As they were to venture so far from the LM, they had the Buddy Secondary Life Support System, which was a hose to enable coolant water to be pumped between the suits in the event that one PLSS failed. Each backpack had an Oxygen Purge System mounted on top, and this could provide 30 minutes of oxygen under strenuous activity, or an hour at a more relaxed pace. In the event of a suit failure, the sampling would be curtailed and the Rover driven straight home.

On an 'H'-class mission, the first EVA had been devoted to setting up the ALSEP and the second to the traverse. The old suit had been capable of 7 hours of life support, but this included 1 hour prior to egressing and 2 hours in case of problems re-entering the LM. The new suit's endurance was 9 hours, of which 7 hours could be spent outside. This made it possible to assign *both* the ALSEP deployment and a short traverse to the first EVA. The walkback constraint demanded that the traverse be done first, and the consumables margin meant that once the astronauts were back at the LM it would be possible to extend the EVA by half an hour or so in order to overcome a difficulty in deploying the scientific instruments.

"Jim, let's get on with it," Scott decided. He parked alongside the MESA to ease the task of loading the Rover.

When Irwin checked to see how much 16-mm film he had used, he saw that the counter hadn't moved! It transpired that a piece of tape had been applied to protect the winding mechanism of the flight-rated camera. This had not been on the training camera. Unaware that he was to have removed the tape before loading the magazine, Irwin had not done so. Its presence prevented the film from engaging the sprocket. Unfortunately, although there was nothing wrong with the camera, this 'procedures' failure undermined the plan to record the traverses. Nevertheless, Irwin followed the plan and mounted the camera on a pedestal on the Rover, just forward of the control station.

As Scott repositioned the TV camera to enable Houston to monitor the loading of the Rover, it showed the shadowed face of Mount Hadley. For the geologists, until now limited to an overhead perspective of the mountain, this was a truly astounding sight. Returning to the MESA, Scott unpacked the Lunar Communications Relay Unit, which he installed on a rotating joint at the front of the Rover. He then added the Television Control Unit to enable the camera to be controlled from Earth. After unpacking the High-Gain Antenna, he mounted the pole on the deck in front of his seat, with the dish antenna placed high enough not block his view while driving. He aligned the dish by peering through an optical sight which projected out at an angle from its base. Since it had to be within 2 degrees of the Earth to carry a television signal, the camera could be used only when the vehicle was stationary. However, as long as the stubby Low-Gain Antenna in the centre of the deck was pointing within 30 degrees of the Earth it could send telemetry and provide bidirectional voice contact. With the Earth almost

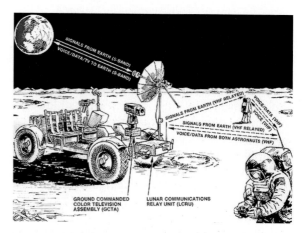

The Rover was independent of the LM's communications systems. On the front of the vehicle was a TV camera that could be remotely operated from Earth, its signal being transmitted using a high-gain antenna. Owing to the narrowness of the beam's width it was not possible to provide TV while driving, and, upon halting, one of the astronauts had to aim the antenna at Earth using an optical sight.

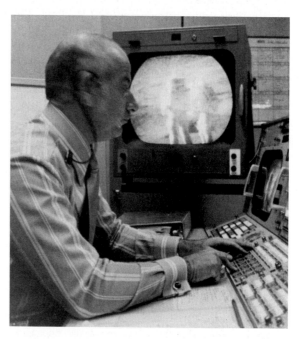

In Mission Control, Ed Fendell operates the TV camera mounted on the Rover.

directly overhead, it would need little maintenance.

"I'm going to put the TV on the Rover now," Scott said. He ran across, picked up the camera and its tripod, took it back to the MESA and then unplugged its umbilical so that the camera could be remounted on the mobile system. "She's all yours," he reported a minute later.

In Houston, Ed Fendell, head of the Communications System Section of the Flight Control Division, began to manipulate the camera by remote control.

"It moves!" Scott exclaimed delightedly.

"There's the TV!" Allen announced.

Scott left Fendell to run through the pan and zoom settings to test the camera. If the remote control unit hadn't worked, this would have been a serious blow to the geologists, who were looking forward to *participating* in the exploration and seeing for themselves rather than simply listening to the astronauts' reports and trying to visualise the scene second-hand. The television camera on Apollo 11 had provided the world with a sense of participation in the exploration of a new frontier. Its loss on Apollo 12 had convinced the flight controllers of the need to have some oversight of surface operations. When Shepard and Mitchell had walked out of shot to climb Cone Ridge, the limits of a fixed camera had been demonstrated. A mobile camera that was controlled from Earth therefore represented a major advance. It gave NASA an opportunity to recapture its popular audience, it enhanced the flight controllers' oversight, and it facilitated the Backroom's participation. Considering that it was an adjunct to the Rover, it was a considerable return on a relatively small investment.

Meanwhile, Irwin had carried the geology pallet from the MESA and installed it on the rear of the Rover. Once this tool carrier had been mounted upright behind the seats, a flap was folded down onto the rear deck to provide a work surface. He then loaded the rack with sampling tools and bags. Whenever possible, they were to put samples in individual bags, each of which was numbered to enable the samples to be correlated with the photographic record. On reviewing Apollo 14, Scott had decided that he and Irwin would use formal geological sampling procedures. Jim Lovell had chosen the motto 'Knowledge from the Moon' for Apollo 13, but had not made it to the lunar surface. Scott had taken this motto to heart, and wanted his mission to be remembered as a triumph for scientific exploration.

For a while, Fendell followed the loading activity with the TV camera, but then he swung around to the south and zoomed in on St George, the 2-km crater on the lower flank of Mount Hadley Delta. The low-definition TV image was sufficient to show a hint of the lineaments running across the base of the mountain, so even as the crew prepared their vehicle the Backroom was doing real geology.

Of the three traverses, the first was to be southwest to the rille and then around to the eastern rim of St George. The second would be south to Mount Hadley Delta via the South Cluster, and then east along the mountain's flank. The third would run due west to the rille, swing right and run along its rim for a kilometre, and then cut northeast across the plain to the North Complex. The geological priorities were to sample Mount Hadley Delta, the rim of the rille, and the mare plain. Mount Hadley itself, some 20 km to the northeast, was not on the agenda other than as a target for photography. An 'H'-class mission with a single walking traverse could never have done this site justice.

"I'm going to turn you off," Scott announced suddenly. It was time to initialise the Rover's navigational system, and to do so he had to drive to a suitable reference site. The 'nav' used a strapdown single-axis inertial reference to track the bearing of the landing site, and it derived range and total distance travelled from odometers on the wheels. It had to be initialised with the Rover on a known heading, so he set off due west and parked on a flat spot. The process took several minutes, so while he waited he munched on the food stick contained within his helmet.

When Irwin joined him, they donned their personal kit. Each fitted a Hasselblad with a standard 60-mm lens onto a frame mounted on his chest. Scott strapped a pair of sampling tongs to his left leg, and Irwin strapped a scoop to his right; these were attached to 'yo-yo' reels so that they could be drawn out when required and released to snap back into place for convenient carriage; it was a simple device intended to increase their sampling efficiency. Conrad and Bean had lugged the HTC around on their brief traverse. This frame carried tools around its periphery and a sample bin at the centre. The HTC was mounted on the MET for Apollo 14's traverse, and enabled Shepard and Mitchell to collect much larger rocks than would have been feasible if they have been restricted to what they could carry (as Lovell and Haise would have been, in fact, if they had made it to Fra Mauro). To enable Scott and Irwin to work away from the Rover at the geology stations, it had been decided that each should strap a large Sample Collection Bag (SCB) to the side of his PLSS that would face out when he was on the Rover; so Scott's was on the left and Irwin's was on the right. By the time they finished preparing for the traverse, they were exactly 2 hours into the EVA. Although they were 35 minutes down in the timeline, their prospects were looking pretty good.

A depiction of the Hadley-Apennine landing site, showing the planned traverses: the first to the craters Elbow, St George and Flow; the next to Dune in the South Cluster and along the eastern flank of Mount Hadley Delta to the craters Spur and Window; and the third along the rim of the rille, returning via the North Complex.

"Let's do a little geology!" Allen urged.

"Okay, 'Mister Navigator'!" Scott prompted Irwin.

As Irwin climbed aboard the Rover for the first time, Scott was surprised at how much this made the vehicle rock. "Easy, Jim!"

"Huh?" Like Scott when he had been alone, Irwin had no idea that he was having such an effect on the vehicle. As he attempted to settle, Irwin found that no matter how hard he tried, he couldn't fasten his seat belt. "I'll just hang on," he suggested.

"Let's start out right," Scott insisted. At the speed that he hoped to drive – which was flat out – the Rover was sure to bounce around. He went around and tugged on Irwin's belt until it would link up, then clambered aboard and strapped himself in.

ACROSS THE PLAIN

"We're moving, Joe," Irwin called.

"Roger," Allen acknowledged. He started a log

of the Rover's progress, so that a time and motion analysis could be conducted on its performance.

This first traverse was complicated by the fact that nobody knew *precisely* where the LM was. Until the Rover's 'nav' could be checked against a known feature, and the location of the LM worked out by backtracking, precise navigation would be impossible. But their ultimate objective, St George, was in view, and Station 1, at Elbow Crater (so-called because it was on the bend of the rille where it swung left to cut across the mouth of the bay) was nearby. A small crater slightly north of Elbow had been selected to serve as a checkpoint. This had been called Canyon because it was on the lip of the rille. If they had landed on target, a 2-km drive on a heading of 203 degrees would have taken them to Canyon. In fact, the LM was 600 metres north and 175 metres west of the target point, so this heading would meet the rille a kilometre north of Canyon. This didn't matter though, because on reaching the rille they would need only to swing left and go south to Elbow. Irwin was to track their progress on a map.

A view from orbit of Hadley Rille and the craters Autolycus (near) and Aristillus (beyond) on Mare Imbrium, one or other of which seems to have 'rayed' the plain at Hadley-Apennine.

In addition, he was to look for any variation in surface texture that might confirm the presence of a ray from either Aristillus or Autolycus, a pair of large craters a few hundred kilometres to the north, and as they approached the rille he was to report whether it had a raised bank, as subdued topography was difficult to identify in overhead imagery.

Scott promptly ran up to 9 kph and made a zigzag around a small crater to assess the Rover's handling. The low-slung chassis enhanced the sense of speed. The way in which it bounced in the low gravity made the ride distinctly "sporty". When the Rover met the rise of a large hummock, it barely slowed and there was no sudden drain on the battery; it just shot up the slope and over the crest. At full speed, the wheels threw up a "rooster tail" of dust. As this settled, it stained the track with lighter albedo material, just as had the engine plume of the LM in the final phase of the descent. Although this was not noticeable at ground level, the route across the dark plain was discernible in photographs later taken from orbit. It had been feared that the wheels would stir up so much dust that the astronauts would become coated, but the fenders proved to be excellent. "No dust, Joe," Scott reported.

"We suggest that you go straight towards St George Crater," Allen called, several minutes into the drive, "and you'll find Elbow okay." With St George so prominent, there was no real need for a checkpoint. Indeed, given the uncertainty of the location of Falcon, the act of *finding* Canyon simply to use it as a reference would have been a waste of valuable time.

"There's a large depression off to the left," Irwin reported a few seconds later. He didn't know it, but this was Last Crater. His navigational problem was exacerbated by the fact that, contrary to what he believed, they had not yet driven *onto* the map segment that he had chosen.

"It's really rolling hills, Joe," Scott observed. It resembled Fra Mauro in this respect. "Up and down we go. We're going to have to do some fancy manoeuvring." In fact, he couldn't even take his eyes off 'the road' long enough to make a regular cycle of the instruments. "The Rover handles quite well. It's got very low damping compared to the trainer but the stability is about the same and there's a lot of roll." Driving onto a relatively flat stretch, he slammed the T-handle forward to run up to maximum speed, and then shot off the end into a patch of hummocks. "Whoa! Hang on."

"Bucking bronco," Irwin agreed.

"When you back off on the power, it keeps right on going," Scott pointed out. To slow down, he would require to apply the brakes. Faced with a large crater ahead, he swung west to detour around it, and was momentarily dazzled by the backscattered light. "The zero-phase lighting is pretty tough, Joe. It all looks flat." It wasn't flat, however; it was just that all the shadows were behind the objects that cast them. As soon as he could, he turned back onto a southerly heading. Clearly, if they were ever faced with a long drive directly down-Sun, it would be better to zigzag either side of true in order to remain out of the worst of the glare.

ON THE RILLE

"You may want to turn the 16-mm camera on," Allen reminded them as soon as the line that he was tracing on his map indicated they should be approaching the rille.

But the rille was not yet visible, and in any case Irwin was preoccupied. "I'm still looking for Rhysling!" The Rover's 'nav' placed them 1.1 km from Falcon, and if they were where he thought they were, then this relatively sharp-looking 150-metre crater ought to be visible; but there was no sign of it. There *was* a subdued 60-metre crater straight ahead that had a distinctive 10-metre crater within it, so he scrutinised the map for this pattern. But it was no easy task to *read* the map, owing to the rough ride. "It really bounces," he complained. Looking up, he spotted a likely candidate ahead. "That large one coming up could be Rhysling." The 'nav' indicated 1.3 km to the LM, which was consistent with Rhysling, but they were really running on a track that was essentially parallel to, and a kilometre north of the one that he believed, and this was the 125-metre crater they had dubbed Nameless.

"We've got a ridge in front of us," Scott reported.

As they ran up the incline, a distant mountain poked over the horizon. "I can see Bennett," Irwin noted.

"There's the rille!" Scott announced as they reached the crest of the shallow ridge that ran parallel to the rille.

"And a lot of blocks," Irwin said. The coverage of rocks was dramatically greater on the gentle incline from the ridge to the rille than it had been on the plain. "We're right on the edge of the rille," he added for Allen's benefit. He had not done much of a job of navigating *to* the rille, but now that they had reached it he tried to figure out

The actual Rover traverses and sampling stops made by Dave Scott and Jim Irwin.

precisely where they were. They were evidently further north than expected, so he looked south for Canyon or, better, Elbow. There was an indistinct ellipse way over to the left. Such a big crater on the rim of the rille could only be their first sampling target. "I see Elbow!" In fact, the ridge ascended to the south, and this placed Elbow somewhat above them.

"It's subtle," Scott agreed as he slowed down for a moment in order to inspect the rille. The slope down from the ridge appeared blocky not because there were more rocks as such, but because the ongoing process of cratering had thinned the regolith. Impacts at the rim had tossed part of their ejecta over the edge of the rille, thereby preventing the build-up of fragments which, on the plain, had evidently accumulated to a depth of about 5 metres. Along the rim itself, where there was a sharp transition in the slope from about 3 degrees to 30 degrees, there was an exposure of the mare substrate. This was excellent news, as no bedrock had been sampled *in situ* before. Indeed, some people were still insistent that there was no bedrock, as such, beneath the regolith. Scott stared across to the far wall of the rille, but only the upper part of it was visible above the local horizon of the near rim, and he couldn't make out very much. "I don't see anything that would suggest layering, but there's a lot of big angular blocks." Taking Irwin's advice, Scott swung left in order to run south along the ridgeline about 75 metres back from the rille's rim, well clear of the rocks.

"We were on a heading that was a little too far west," Irwin reflected, having finally gained his bearings. This explained why during their approach he hadn't been able to correlate the map with reality. Now that he knew where he was, navigating should be much simpler. Driving south, he studied the route that they were to take along the flank of Mount Hadley Delta the next day, but from this vantage point his view was so limited that he could see no further than Spur Crater. One of the things that the geologists wanted them to locate was a boulder that had rolled downhill, as this would give a sample from much higher on the massif than they could possibly drive. He saw one boulder near Spur that should be accessible.

As they ran up the shallow incline towards Elbow, Irwin observed a field of large boulders marking the near rim of the rille. "It's the first good concentration that I've seen." These were indeed rich pickings. "This field is very similar to that Apollo 14 saw at the top of Cone," he added to convey the image to Allen and the Backroom.

Now that Irwin had a line of sight down into the rille, he saw that the wall below St George was comparatively free of debris. It was so clean, in fact, that he could see horizontal banding on it. As they progressed south, he gained a view around the bend in the rille. "I can see into Head Valley." This east–west section of the rille had been named after Jim Head, one of their geology tutors. "I can see the *bottom* of the rille, now," he added a few moments later. "It's flat. I'd estimate some 200 metres at the base." Rim to rim, the rille was in

excess of 1 km wide. It was 350 metres deep, and its walls were quite steep, which gave it a flattened floor. "It's very smooth," he added. Although there was debris clinging to the slope, there did not seem to be any large-scale slumping. Nevertheless, there were some tracks where large boulders had rolled down, some of which were sitting on the floor, fully exposed. As they continued south, he got a foreshortened view of the wall opposite St George, which was more blocky than the bare slope directly below the crater. It was as if there were outcrops everywhere except below St George. Because this crater predated the mare, and thus the rille, it was essentially irrelevant: the significance of the slope beneath the crater was that it was a buried part of the massif, whereas the rest of the rille was a section through the mare plain.

"I can see what we thought was Bridge Crater," Irwin continued his commentary. This feature had been so-named because Lunar Orbiter imagery had suggested that it might be practicable to use the crater's rim to cross from one side of the rille to the other without descending into the trench. There were no plans to do this, of course, but the possibility had been studied as an option in the event that they landed on the far side of the rille, in order to gain access to St George and The Front to achieve the primary sampling objective. But Bridge was simply a depression in the far wall. "It definitely would *not* have been a place to cross." If they'd landed over there, they would have been confined to the west bank.

Scott had been reflecting on Irwin's observations of the wall below St George. It seemed to be a somewhat shallower slope than the rest of the rille. "It almost looks like we could drive down in, on this side," he mused.

"Standby on that, Dave," Allen warned. There was barely time to achieve the objectives on the near-side of the rille, without expanding the mission.

"We could drive down," Irwin agreed, "but I don't think we could drive back *out*." The slope below St George may well have been shallower than the rest of the wall, but it was still too steep for the Rover.

SAMPLING ELBOW

"We're on the high point," Irwin informed Allen, "east of Elbow."

The western rim of the 350-metre crater abutted the rille, so they were to establish Station 1 on its eastern rim.

"Stupendous!" Allen replied with typical enthusiasm.

With the Rover now at a known reference, the range and bearing would refine the landing site. They were 3.3 km from Falcon rather than the predicted 2.7 km, and had driven a 4.5 km dog-leg. The traverse had not exceeded the allotted 26 minutes, however, so at least the navigational difficulties had not imposed a time penalty.

Scott gazed up at Mount Hadley Delta, then turned and looked along the length of the rille. "This ought to give the folks back home something to look at!"

Elbow itself was not impressive. Although it was the same diameter as Cone at Fra Mauro, it was not such a pit and there were few blocks on its rim.

Following the plan developed for establishing a geological sampling station, Irwin took a pan while Scott aligned the TV antenna, then Fendell made a pan so that the geologists could familiarise themselves with the site.

Gordon Swann was the Apollo Field Geology Principal Investigator and, as such, he was in charge of the Backroom. A large-scale map was updated to keep track of the crew's position. A chart of the planned and actual timeline, and any adjustments that would have to be made to overcome delays, was sent to a screen in the main control room so that the flight director and CapCom could see it. The items on the checklist for each station were ticked off as they were accomplished. A card file was used to record the data pertinent to each sample. The cards were the main means of tracking which material was collected, when, and where; with annotations describing the orientation and type of a sample as gleaned from the astronauts' descriptions and from the TV. Another display in the Backroom presented a chronological record of events, so that anyone who lost track while occupied with something could catch up. A photogrammetry team pasted together Polaroid pictures taken from the TV while Fendell made his opening pan, and this was marked up with interesting features. It could be used to direct Fendell to specific features for follow-up viewing. It was also annotated with the locations of the samples. Stenographers made a transcript of what the astronauts said. Some geologists, away from the hubbub, watched and absorbed what was going on, reflected on its significance, and then wrote a summary once the EVA was complete. For Apollo 15, the science team was as highly coordinated as the main flight control operation. And Allen, as the interface between the Backroom and the astronauts on the lunar surface, was far more than simply the duty CapCom, he was the mission scientist, a post introduced to reflect the importance of scientific research in the 'J'-mission format. A physicist by training, Allen had accompanied Scott and Irwin on terrestrial field trips on which traverses were simulated, and so, in a very real sense, especially given the Rover's TV, he was 'on the Moon' too.

"Why don't we do a quick radial sample," Irwin suggested. In fact, this was the *only* item on the checklist.

A 'radial' involved sampling at different distances outside the rim of the crater in order to establish any difference in the material excavated from different depths. For a crater such as Elbow, the material on the rim could have come from as deep as 75 metres but, although a Copernican Era crater, the ejecta had been reprocessed by later impacts. In fact, there was a 4-metre crater on Elbow's eastern rim. In the division of labour, Scott selected the samples. There was a slabby rock in the small crater, but it was much too big. But just to the east there was a single rock among a scattering of pebbles. The three-legged gnomon provided a sense of scale and a vertical reference, and it was always placed west of the target so that the orientation was evident in the pictures that documented the sampling. They had rehearsed the sampling procedure so often that it was automatic, and they worked as a team without specific verbal coordination. Irwin stood due east and took a long-shot down-Sun that included his shadow, the gnomon on the far side of the sample and, often, his commander's feet; Scott was in the frame because his role was to take close-up pictures from either the north or the south in order to provide a stereoscopic cross-Sun view of the sample.

With the prior-to-sampling documentation complete, Scott extended the tongs on his yo-yo, and lifted the rock. The subangular fragment[1]* was caked with dust. It was so friable that he had to take care not to break it as he brushed off the dust. "I see a lot of 'sparklies' in there!" Observing a variation in albedo, he declared the rock to be a breccia. Time was precious on the Moon, and the sampling procedure allowed only a few seconds to make comments about a sample. This done, he put the rock into the numbered bag that Irwin had prepared as his part of the routine. In fact, the

---

* Information pertaining to each sample is given on page 369.

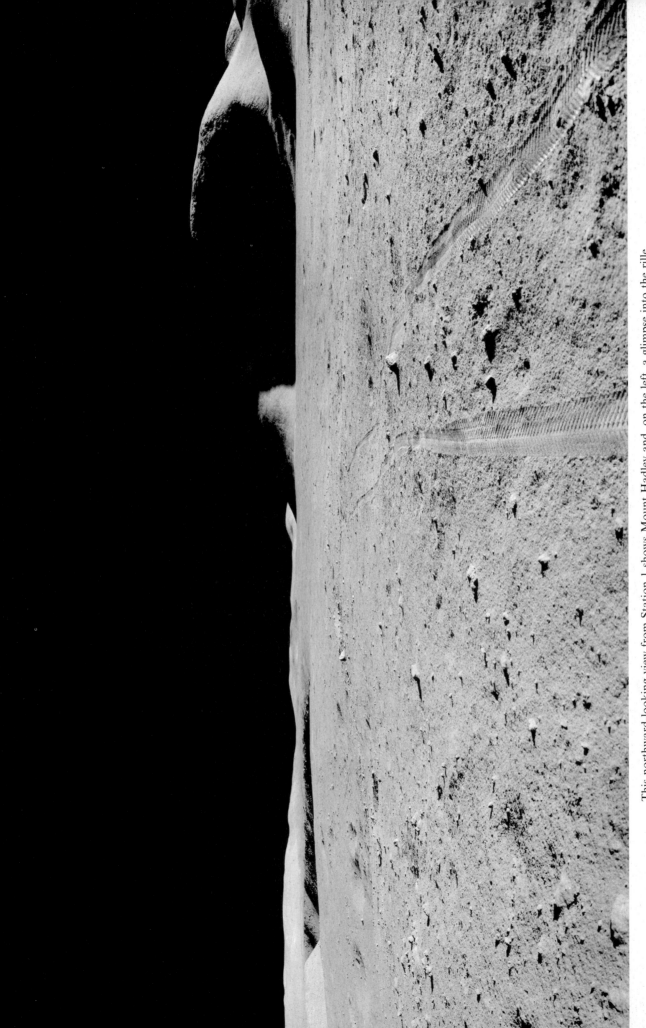

This northward-looking view from Station 1 shows Mount Hadley and, on the left, a glimpse into the rille.

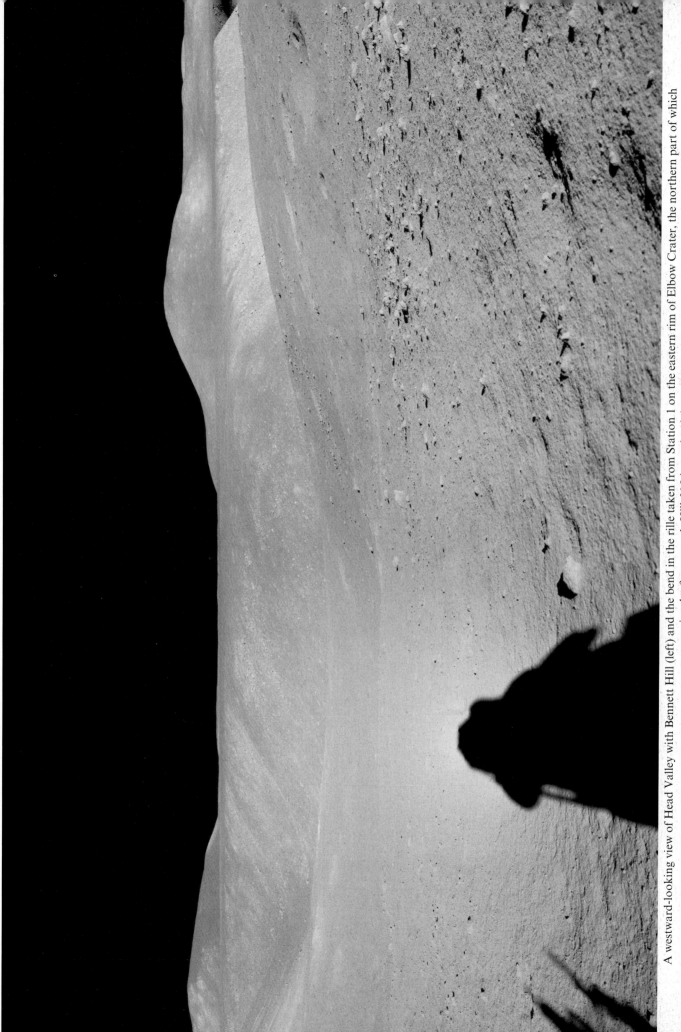

A westward-looking view of Head Valley with Bennett Hill (left) and the bend in the rille taken from Station 1 on the eastern rim of Elbow Crater, the northern part of which occupies the foreground. Hill 305 is on the right.

In conducting a radial sample east of Elbow, Dave Scott and Jim Irwin used formal sampling procedures, including photographing a rock prior to lifting it. While Scott took cross-Sun pictures from close in, Irwin added a picture looking down-Sun and a wider 'locator'. In each case the three-legged gnomon was placed by the sample, with the pendulum indicating local vertical and the photographic calibration chart illuminated. Here they are sampling a rock on the rim of a small crater just outside Elbow – the near rim and southern part of which occupies the foreground.

rock was not a breccia, it was a coarsely grained basalt. But one end was more mafic than the other, with the less mafic end being enriched in feldspar which, being a white mineral, had caused the difference in tone that had prompted Scott's misidentification.

With the first sample 'in the bag', Scott moved 20 metres further east to select the second from a cluster of small rocks on a flat patch near a group of subdued craters that were no more than a few centimetres across. Similarly textured, the rocks were evidently related to one another. They all had fillets, and a few were partially buried.

"Shall I get a little soil with this one?" Irwin enquired.

"Yeah."

Scott used his tongs to dig the rock[2] from the regolith, lifted it, and scraped off its cover of dust; it was basalt. "By golly! I see olivine. And there's a big lath of plage in there, about a centimetre long." He'd seen an oblong crystal of plagioclase ('plage') in the olivine basalt. "It's a light-grey matrix with millimetre-sized grains and 2-mm phenocrysts in it." Phenocrysts were crystals composed of other minerals. "Let me get another one." He dug a second rock out of the regolith, but it was too big for the bag so he lifted a third[3] and added it to the bag without bothering to inspect it. Irwin put some soil into the bag to complete the sample.

"Let's hop on out, and get one more," Scott said, delighted to be 'geologising' at last. He went another 35 metres east, to a site which was pocked with small craters. Having selected a partially buried rock that was free of dust, he drew out his tongs – and broke the yo-yo! While this would not prevent him from continuing to use his tongs, it was an inconvenience because it meant that he would have to carry the tool in his hand. He lifted the cobble-sized rock[4] and inspected it. It was coarsely grained basalt. He popped it into Irwin's bag. A second rock[5] was more interesting. It was a fragment of regolith breccia – the 'instant rock' that is formed by the compression wave that is transmitted through the ground during an impact. It included fragments of a pyroxene-rich basalt.

"Back to the Rover?" Irwin asked. It was unlikely that a sample taken any further from Elbow's rim would be related to the crater.

Scott answered indirectly. "Joe, do you want us to press on to St George?"

"Move on," Allen replied immediately. With the radial sample done, the objective at Elbow had been achieved. On a lunar geology field trip,

there was time only for a series of brief sampling stops. The difference between this and the previous missions was the rigour of the sampling process. As a result, the radial had taken 10 minutes. However, just as in the case of the drive, the on-station activity had been completed within the planned time. As a bonus on this mission, of course, many experts in the Backroom had been making observations at the same period, and they would be able to replay the tape as often as they wished. As a result, much more had been achieved during this brief stop than two astronauts picking up a couple of rocks on the rim of a crater that they never even ventured into.

ST GEORGE

A hundred metres beyond Elbow, as Scott swung southwest around the bend in the rille, they started up the flank of Mount Hadley Delta. The Rover barely slowed.

"What a beautiful view looking up that slope," Irwin enthused.

"You can see the lineaments cutting across," said Scott.

The flank of the mountain had a streaky appearance. The most obvious marks ran across the slope and dipped to the east, but there were also more subtle lines running directly up-slope just above St George. Intriguingly, the dipping lineaments ran right across the crater, the great age of which was indicated by its smooth rim.

Station 2 was to be half a kilometre south of Elbow, on St George's northeastern flank. As the primary objective of this traverse, it had been assigned 45 minutes. The checklist called for a lengthy radial of the ejecta blanket in order to investigate the vertical structure of the mountain, a trench to find any layering in the regolith, and a 'comprehensive' sample of the regolith. Contrary to expectation, however, there was no field of rocky debris. As a result, far from being able to use the crater as a 'drill hole' into the massif, it was evident that there would be little point in driving all the way to the rim. Scott was entranced by the really bright crater immediately east of St George. A sizeable fresh crater, it might have excavated something interesting. He was in two minds about switching the sampling objective, but the crater was at least another kilometre further up-slope. As he pondered the issue, they happened upon an unusually deep crater about 10 metres across. Should they sample it? The absence of ejecta meant that the impact had simply stirred up the regolith, which was evidently pretty thick on the lower flank of the mountain.

There was an isolated rock sitting on the surface further up-slope. "Why don't we just go straight over to that?" Irwin suggested.

"That's what we'll do," Scott decided. When they arrived at the rock, there was a fresh-looking 6-metre raised-rim crater nearby as a bonus.

"What a view back to the rille!" Irwin exclaimed as Scott swung right in order to park cross-slope.

Armstrong and Aldrin had had the "magnificent desolation" of their open plain. Conrad and Bean had had the excitement of visiting Surveyor 3 in its crater. Shepard and Mitchell had had the slog up Cone Ridge. But Hadley-Apennine was majestic.

"This is spec-tac-ular!" Scott confirmed.

"I can hardly wait," Allen chipped in, to prompt Scott to align the TV antenna.

As soon as he had done this, Scott went to inspect the boulder. As Irwin completed his pan, he caught his commander bending over the rock.

At a metre and a half in size, this was the largest rock Scott had yet come across. "It's very angular, with a very rough surface texture, and it's got glass on one side of it with a lot of bubbles of about a centimetre across. There's a linear fracture through one side; it almost looks like it might be a contact. It looks like we have a breccia on top of a crystalline rock." The fact that the glass spanned the contact indicated that it was a coating. Fendell zoomed in so that the Backroom could join in the assessment. Scott looked for any sign that the rock had rolled down the hillside but there were no tracks. It had evidently been tossed *up* onto the mountain by an impact down on the plain. In fact, there was a slight depression immediately down-slope of the rock, and this seemed to be where it had hit, displacing the regolith. As his yo-yo was broken, Scott stuck his tongs into the soil nearby in ski-stick fashion, so that he would have both hands free.

"It's probably 'fresh'," Allen mused. "Not more than three and a half *billion* years old!"

"Can you imagine that!" Scott replied, in awe of the implications. "It's been here since before creatures roamed the sea."

They opted to sample the displaced regolith first. Even in perfect conditions, soil sampling was a two-man task, but it was difficult to stand still on the slope. Shepard and Mitchell had traversed a slope on foot, but Scott and Irwin were the first to try to *work* on one. It was only 8 degrees, but with their feet slipping in the soft soil it felt much steeper. To scoop efficiently, Irwin stood down-slope. If he had been up-slope, his PLSS would have tipped him over when he bent forward. Scott stood to one side with the bag, and Irwin turned to him to pour from the scoop. It took several scoops to fill the bag.

"You know what we're going to do once we're through," Scott chuckled. "We're going to roll it over and sample the soil beneath."

"Atta boy, Dave," Allen said approvingly. The regolith beneath the rock would have been shaded from cosmic rays and the solar wind, and a comparison of the regolith from alongside and beneath the rock would provide a measure of how long the rock had been present.

"I want to get a close-up of that contact," Scott decided. In order to have the best possible focus, he set out to rest one end of his tongs against the rock and shoot down the handle, whose length he knew. As he reached for the tongs, he realised that the fact that they had remained upright was an interesting soil-mechanics observation, and so he photographed the tongs standing in the regolith and then, on noting that the hole did not collapse as he retrieved the tool, he photographed the hole too. The distraction over, he shot the contact in the rock.

"Okay, let's try the hammer." Smashing rocks was one of Scott's tasks, so for convenience the hammer was carried on the side of Irwin's PLSS. He struck the top of the rock a heavy blow. "Man!"

"Got one?" Irwin asked, thinking that he must have missed seeing the chip fly off.

"Naw, it's hard!" Scott delivered several more blows, but to no effect. "And after all that instruction!" Evidently, his training in how to sample a rock with a hammer had been in vain.

Irwin had been studying the surface of the rock, and he thought that he could see a fracture line. "Dave, if you hit it up here, it'll break."

"Right here?"

"Right there."

Scott bashed the line of weakness and a chip flew off. As Irwin went to fetch it, Scott knocked off a second chip, this time from the up-slope side of the rock in order to sample the other side of the contact. As he had surmised, the chip from the top[6] of the rock was a breccia. It was so angular that it was bounded by five joint-surfaces, and had split off from the rock along two of these surfaces. The glass coated two of the other surfaces. The light-grey matrix was almost 25 per cent clasts. Although the fragment from

A wide view of Dave Scott inspecting the angular boulder selected for Station 2 on the flank of Mount Hadley Delta.

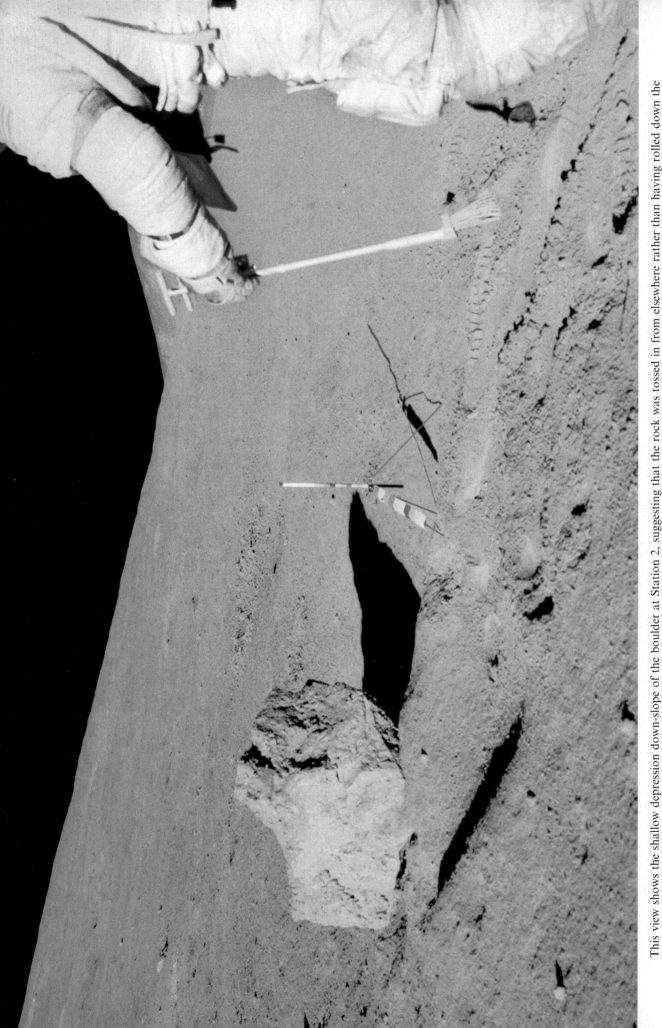

This view shows the shallow depression down-slope of the boulder at Station 2, suggesting that the rock was tossed in from elsewhere rather than having rolled down the flank of the mountain.

A close view of the Station 2 boulder from up-slope.

The Station 2 boulder from the same perspective as the previous illustration, but after Dave Scott had tipped it onto its side to examine the regolith beneath. Note Jim Irwin's scoop and rake resting on top of the rock.

the up-slope corner[7] of the rock proved to be a breccia also, its dark matrix was vuggy with cavities up to 4 mm across, the presence of which meant that it had once been very hot and had cooled slowly. When the samples were studied on Earth, it was realised that the entire rock was essentially a large regolith breccia of a contact that separated a pyroxene basalt like that which they had sampled at Elbow from a KREEPy 'non-mare' basalt. Not only was this the first evidence of the local mare overlying non-mare basalt, but the ages of the two types of rock made the contact an *unconformity* spanning some 500 million years. However, contrary to the astronauts' banter about the boulder having been on the mountainside since before life emerged from the Earth's ocean, the exposure age revealed that it was excavated only 1 million years ago. In all probability, it was ejecta from an impact right on the rim of the rille.

"We'd like you to finish this rock, and press on with the comprehensive sample," Allen said. The Backroom was multidisciplinary, and the soil-mechanics specialists were eager for their part of the Station 2 checklist to be performed.

"Okay," Scott agreed. He stowed the hammer back on Irwin's PLSS. "Why don't you go get the rake; I want to roll this over."

As soon as he saw Irwin start back to the Rover, Allen pointed out that the TV camera had snagged its cable. Fendell had been attempting to untangle it but, despite this frustration, Allen had refrained from interrupting the sampling to seek assistance.

After tipping the rock over to the west so that it would not cast its shadow on the exposed soil, Scott looked at its underside: much of it was covered by glass, and in one place this seemed to have slickensides etched into it, suggesting that it had once suffered abrasion. There was rather a lot of glass for a pick-up coating. Maybe it was *in situ* melt?

Irwin returned and sampled the newly exposed soil, then, finished with the scoop, detached this from his universal handle, rested it on the rock, attached the rake to the handle, and moved 10 metres east to take the comprehensive sample. The rake had been designed by Lee Silver to trawl the regolith. Comprehensively sampling all the fragments contained in a swath of soil would permit the size distribution and relative populations to be determined. The 1-cm-spaced tines were intended to retain pebbles and to reject fines. This was the tool's first use, and Irwin was eager to see how well it worked.

"Anything?" Scott asked as Irwin finished his first swath.

"Nothing!"

"Dig deeper," Scott suggested.

Irwin tried again, scraping a second swath alongside the first. Fendell zoomed in to enable Silver to watch the result. "Two little frags," Irwin reported. It was hardly a statistically valid sample. The paucity of fragments was because the regolith had been thoroughly gardened.

In fact, now that he thought about it, Scott had not seen any variation in the regolith as they had driven up off the plain onto the mountain's flank. "Let's call it quits," he decided. "We'll pick up a double core."

"Right on!" Allen approved.

"Do you want soil with the comprehensive?" Irwin asked, the increased level of his voice indicating that he was addressing Allen rather than Scott.

"One bag of soil," Allen said. Instead of retrieving the scoop from the top of the boulder, Irwin used the solid curved edge of the rake to scrape up some soil, and this tool proved to be somewhat easier to pour from than the scoop.

"Hey, Joe," Scott said. "We've got a crater up here with a fairly fresh rim. Would you like the core on its rim?"

"Standby." Allen bounced the question through to the Backroom.

"There's a change in albedo on the rim," Scott added encouragingly. "It's much lighter."

"Roger, Dave. Drive the core through the rim."

"I thought you'd say that!" Scott laughed.

Irwin turned and faced the crater that they had parked alongside.

"No," said Scott, "I was thinking of the bright one." He indicated towards another crater, about 100 metres further up-slope.

"It'll probably take 5 minutes to get up there," Irwin cautioned. It would be a fair climb in the soft soil. Their radius of action on foot on the slope was extremely limited.

"Joe, what do you think?"

Allen had no idea which crater Scott had in mind, but he was tracking the passage of time, and suggested that they remain near the Rover.

It was Irwin's job to take the core, so the tubes were carried on Scott's PLSS. While Scott took the prior-to-sampling pictures of the selected site, Irwin screwed the tubes together. This time Irwin stood up-slope in order to gain the advantage of the higher elevation in driving the tube into the ground. The first section went into the ground easily, but then the fine-grained soil compacted and it took several

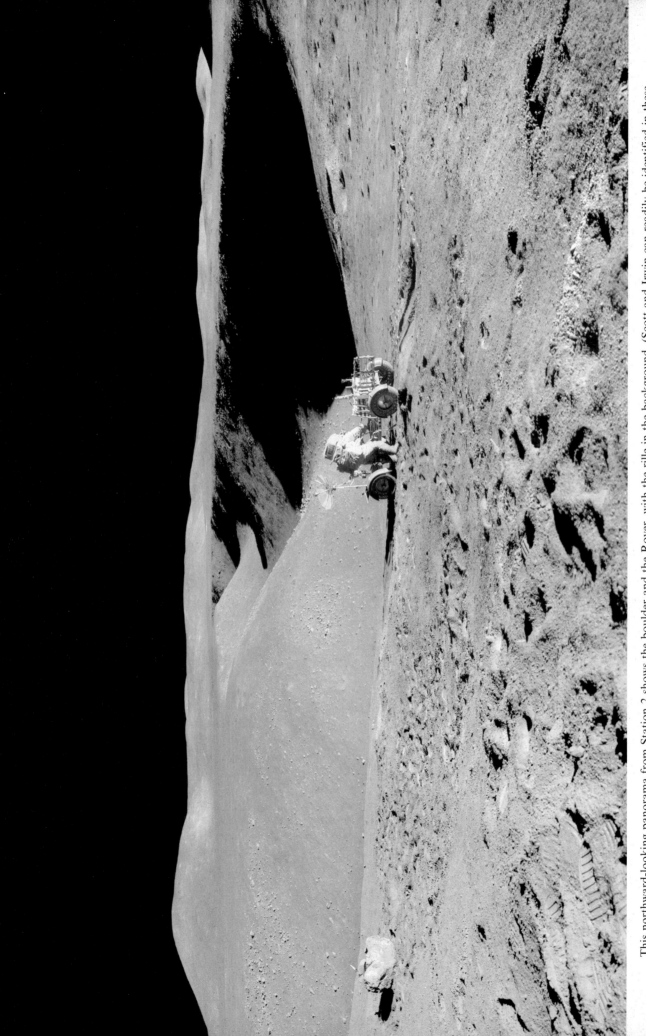

This northward-looking panorama from Station 2 shows the boulder and the Rover, with the rille in the background. (Scott and Irwin can readily be identified in these pictures by the fact that Irwin strapped his sample bag to the right-hand side of his backpack, and Scott had his on the left so that when they were seated on the Rover their bags faced outward.)

This telephoto view from Station 2 shows boulders littering a section of the floor of the rille. (This area appears in the previous illustration, just above the antenna of the Rover.)

heavy blows of the hammer to finish the process. In an effort to improve core sampling, Apollo 15 had been provided with a new, wider tube, and great care had to be taken to ensure that the bottom of the sample was not lost during extraction.

As Irwin took a second pan from a slightly different position in order to facilitate stereoscopic analysis, Scott retrieved the Hasselblad with the 500-mm lens and took telephoto shots of the north–south section of the rille that cut across the mouth of the embayed valley.

HEADING HOME

"We're going to eliminate Station 3," Allen announced as Scott and Irwin prepared to board the Rover.

The plan had been to drive off the mountain and then run 1 km east to sample the small crater that they had named Flow, in order to attempt to determine whether (as overhead imagery suggested) material had slumped onto the plain, but Scott's observation that there had been no change in the regolith during the drive onto the mountain had led to this being deleted. The cancellation of Flow regained most of the time that they had lost prior to starting the traverse.

As the front-steering was inoperable, Scott ran down off the mountain at an angle and at a modest 5 kph. This proved a wise decision, because one of the front wheels dug into a soft patch of regolith and the Rover spun through 180 degrees, adding a touch of excitement to the ride. At Elbow, they departed from their previous tracks and started out across the plain, relying on the Rover's 'nav' to get them home. They may well suffer the occasional difficulty pinning down a specific site, but with such a distinctive horizon there was no way that they could become disorientated.

Irwin stared off to the right. "Could that be Dune?" He meant the most westerly of the large craters in the South Cluster. He consulted his map. "Oh, it's Fifty-Four."

"Fifty-Four's on the Moon!" Scott chanted with delight. He had named the crater to honour his *Class of '54* at West Point.

Looking ahead to Mount Hadley, Irwin saw something intriguing on the western foothills that had just been exposed by the mountain's receding shadow. There was a thin stripe running along the base of one hill just above the mare 'shoreline' that was strikingly reminiscent of a flood-mark, but with magma rather than water. There was clearly an intriguing geological history to this little bay in

the Apennines.

About 125 metres southwest of Rhysling, just after crossing the septum between a pair of extremely eroded 30-metre craters that were completely free of debris, Scott spotted a rock sitting on top of the surface all by itself. "There's a vesicular basalt!" he exclaimed. He drew to a halt just beyond the rock.

"We're stopping," Irwin told Allen as a matter of routine, to assist him in tracking their progress.

"Let me get my seatbelt," Scott said innocently He disembarked, fetched his tongs and ran back to get the rock. As he did not have the gnomon, he poked his tongs into the ground to provide scale and shadow references for the photographs. Meanwhile, Irwin distracted Allen with a masterful description of how many of the small craters on the plain were full of regolith breccia. "Okay, I've got it," Scott announced as soon as he was back aboard, giving Allen the impression that he had simply attended to his seatbelt. This 'target of opportunity' remained a secret from the scientists until it was found in a rock box by the LRL 'receivers', whereupon Scott confessed. This rounded scoriaceous basalt[8] was riddled with smooth-walled vesicles. In view of the manner in which it was sampled, it became known as the 'Seatbelt Rock'.

"I can't believe we came in over those mountains," Irwin said as Scott fairly tore across the plain to make up time. He was entranced by the Swann Range to the east that formed the backbone of the Apennines. "It's just a *beautiful* little valley."

Upon cresting a shallow rise, they saw Falcon a kilometre ahead, precisely where the 'nav' indicated it would be. After a total run of 10 km, the inertial reference had barely drifted. Being accurate to within a hundred metres, it was more than adequate for navigating a traverse. Although St George's rim had not lived up to expectation, Scott and Irwin had the satisfaction of knowing that their first traverse had been a remarkable *operational* success.

ALSEP FRUSTRATIONS

The final task of the first EVA was to deploy the ALSEP. While Irwin set up all the other instruments, Scott was to emplace the Heat-Flow Experiment by drilling a trio of 3-metre holes, two to accept thermal probes and the third to give a core sample of the material in which the probes were installed. This experiment was first assigned to Apollo 13, which had not reached the Moon.

This northeast-looking view of Jim Irwin at the Rover after the first traverse shows the 4,000-metre-high Mount Hadley in the background, its western face in shadow.

Although Apollo 14 had inherited its predecessor's landing site, this experiment was not carried over. Gaining insight into the lunar 'heat engine' was considered a high priority, so Scott had high hopes that it would assure that his mission received a prominent mention in the annals of lunar science.

Unfortunately, the drilling became an ordeal. Progress on the first hole was rapid for the first 40 cm, then slowed dramatically. Thinking that he had hit a rock, Scott applied his one-sixth-gravity weight to assist the drill in boring through, but the pace remained slow. When progress halted completely at a depth of just 1.6 metres, Allen relayed the Backroom's recommendation that he give up and insert the first string of probes. Scott readily agreed, but then discovered that the extreme torque had locked the drill's chuck, and he had to use a wrench to release it. Making no better progress on the second hole, he ran out of time at a depth of 1 metre, and was obliged to retreat to the LM leaving the drill in place. It was clear that he would have to return later to finish the task.

THE LONG DRIVE SOUTH

The second excursion was to see Scott and Irwin return to Mount Hadley Delta, this time via the South Cluster. Since it had been possible to identify the position of the landing site from measurements made by the Rover's 'nav' on the first outing, Irwin expected the task of navigating to be much easier.

Upon starting up the Rover, Scott had a welcome surprise for Allen. By some mysterious process, the front steering was working. "You know," he joked, "I bet you let some of those Marshall guys come up and fix it!" After a short test drive, he reported that the vehicle was much easier to control than it had been using just the rear steering.

As before, Irwin described the view while Scott concentrated on driving. Within minutes of starting, they passed just to the west of a subdued 450-metre crater with a 25-metre crater half-way down its eastern wall. Had this exposed an outcrop of mare bedrock? The walkback constraint meant that they would break the outbound drive only if they were to find something truly exceptional. A kilometre out, Irwin saw the rim of Salyut ahead; this was a crater they had named in honour of a crew of Soviet cosmonauts who perished on a mission several months earlier. Scott veered around the western rim. As he drove on at a fairly racy 9 kph, he glanced down to check the vehicle's instruments, and was astonished to find upon looking up again that it was too late to avoid a small crater.

"Whoa, baby!" Irwin exclaimed as they bounced across the depression.

"We're all right," Scott reassured.

"Great machine," Irwin concluded.

Approaching the 2-km point, Irwin reported that he could see the South Cluster, the closest crater of which was Crescent. He also noted that the coverage of blocks was increasing. It was very patchy, however. Crossing the septum between a pair of 20-metre craters, he saw that the eastern crater was full of debris but the western one was clear. As they raced past Earthlight, the crater they had named after a novel by Arthur C. Clarke for the reason that it depicted a journey through the Apennines, he reported the presence of large blocks in its wall.

To pass through the much larger craters of the South Cluster would soak up time, so their route veered around the western fringe. The problem was that the craters had such subdued rims that they were far from obvious when viewed from the surface. However, the increase in the coverage of half-metre-sized blocks was an indication of the cluster's proximity, and Scott slowed down in order to manoeuvre through the debris. Irwin observed that the smaller craters, by which he meant those that ranged up to several metres in size, were deeper than their counterparts out on the plain, and some were distinctly asymmetrical with their ejecta preferentially positioned on their southern rims, both of which facts supported the case for the cluster being the result of a splash of debris which arrived from the north, possibly from either Autolycus or Aristillus, which were 150 km and 250 km away respectively. The 460-metre crater Dune, the westernmost of the large craters in the cluster, was assigned as Station 4. They were to have given it a 20-minute stop on the way south, but the problem with the drill had meant that time had to be reserved to resume this task at the end of the second EVA. But something had to be sacrificed to release that time, and sampling The Front was more important than Dune. Nevertheless, unless access to the crater was denied by dense litter, Scott was to run up onto the rim of the crater and Irwin was to make a drive-by assessment of its interior; if there was exposed bedrock, they would stop off on the return trip to sample it. Although the fringe of the cluster was considerably more heavily littered than the plain, Scott had

Dave Scott on the Rover just before beginning the second traverse. Note the map affixed to the control panel, the 16-mm movie film camera above it, the TV camera at the front, and the pallet of sampling tools in place at the rear.

no trouble approaching Dune, and as they drove around the western rim of the crater Irwin reported its wall to be laced with large blocks. On seeing a prominent ray of ejecta littering the plain beyond the crater's southern rim, Scott steered west in order avoid it.

After 26 minutes of concentrating 'on the road' Scott halted for a brief rest. Irwin retrieved his camera and took a south-looking pan of their destination from this 'pit stop' site.

The closer they got to their destination, the more imposing Mount Hadley Delta appeared. On Earth, it is rare to find a massif rising from a flat plain. Mountain belts on Earth result from extremely slow crustal deformation driven by plate tectonics. In contrast, the ranges of mountains peripheral to the Imbrium Basin were formed in an instant. It was only later that upwelling lava filled the cavity to create the mare plain. To the geologists, the benefit of the Hadley-Apennine site was that it combined the mare plain on which to land with ready access to the mountains forming the rim of the Imbrium Basin. The objective of this traverse was to sample the ejecta of some of the craters along The Front in the hope that these had penetrated the talus that had slumped off the mountain prior to the mare flooding, and in so doing had excavated the massif. Ideally, sampling such craters would yield anorthosite, the material that was believed to have formed the ancient crust as the magma ocean solidified.

The contact between the plain and the mountain was marked by a shallow trench, evidently the result of isostatic adjustment – either the massif settling or the mare withdrawing.

After using Spur Crater to check the Rover's 'nav', the plan was to swing east and drive along the flank of the mountain for 3 km to two craters called Dandelion and Front, the latter of which marked the maximum walkback limit and therefore was to be Station 5. Scott and Irwin were to select the most appropriate spot for Station 6 as they drove back to Spur, which was to be Station 7. But as they ran diagonally up in search of their checkpoint, the up-hill view made Spur difficult to spot. When Allen provided the estimated 'nav' coordinates, Irwin noted that that these matched their current position.

"I don't see it," Scott complained. A moment later, however, he saw a small ridge ahead that looked as if it might be the rim of the crater. "Maybe to the right there, huh, Jim?"

"I'll buy that," Irwin acceded.

With the 'nav' verified, Scott swung left and, after cresting a shallow rise, saw the boulder that Irwin had pointed out the previous day. It was a bit above the elevation of the planned drive, but appeared to be accessible. "We'll pick up that big rock on the way back!" he promised.

Now that they were driving east, Irwin could see Mount Hadley. The western face was now illuminated, and he could see "really remarkable" lineaments that dipped to the west. "The whole mountain has the same linear pattern, very closely spaced," he enthused. The effect was spectacular, but only Scott and Irwin would appreciate it because the low resolution of the TV only hinted at it. Scott's insistence on having a 500-mm lens was vindicated, because when he later documented the mountain using this lens it captured a level of detail that could not have been extracted by enlarging photographs taken using the standard 60-mm lens. Although the pattern hinted at the exposure of a tilted depositional structure, post-flight analysis concluded that it was a lighting effect caused by the shadows of the mottled surface combining under the low-angle illumination, because on that part of the mountain the Sun was only a few degrees above the horizon.

As they drove along The Front, it dawned on Scott that there was little purpose in sticking to the plan, so he swung upslope and halted. "I think this ought to do it," he announced.

Allen, surprised, reminded him of the need to sample a crater which had a blocky rim.

"There aren't any like that, Joe!" Scott countered. "They're all very subtle." There were no large rocks, just an abundance of small fragments.

"I guess we want to continue towards the east," Allen prompted gently.

But Scott was reluctant. There was little variety of rock types. They had spent 42 minutes reaching this point which, although 5 km from base, had involved a 1.5 km detour around the South Cluster. If they continued to Front Crater and, as seemed likely, it had nothing dramatically different to offer, then that part of the traverse would have been a waste of time. Since their current position appeared to be typical of The Front, he felt they should curtail the drive, delete Station 5 and call their current position Station 6. After sampling this site, they should go and examine the boulder above Spur.

Realising that the commander in the field had made his decision, Allen concurred, "Okay, Dave. That sounds good."

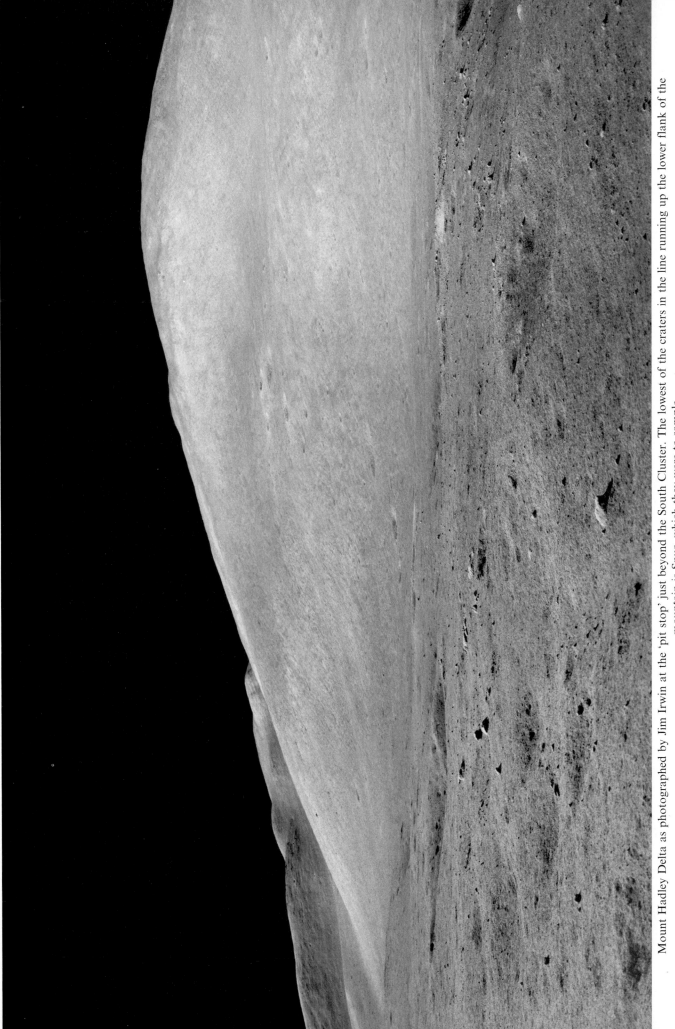

Mount Hadley Delta as photographed by Jim Irwin at the 'pit stop' just beyond the South Cluster. The lowest of the craters in the line running up the lower flank of the mountain is Spur, which they were to sample.

WORKING ON THE SLOPE

The 'typical' site at which Scott had decided to stop was about half-way between Spur and Window, which put it about 3 km from St George, where they had been the day before, but at an elevation of 100 metres above the plain they were much higher. It was not until Scott disembarked and found just how much he had to lean into the hillside that he realised how steep it was. Also, standing still was difficult because the thick soft soil slumped away. "We're on a *steep* slope," he warned Irwin. "By golly. This Rover is remarkable!" In the absence of a vertical reference, the slope had been deceptive, and their vehicle was clearly more capable than he had been led to believe. On turning around, he saw the rille to the left, Mount Hadley ahead and the rolling peaks of the Swann Range to the right. "*Look at that!*" he exclaimed in amazement.

After setting up the TV antenna to enable Allen to join in the fun, Scott surveyed their site. "Let's go up," he decided. Retrieving his tongs and the gnomon, he set off 20 metres up-slope to a fresh-looking 1-metre crater on the rim of a subdued 3-metre crater. It wasn't until he reached the crater that he got his first view into it. "It's got glass in the bottom." This implied that it was a high-energy strike that had vaporised the impactor and shock-melted the regolith. The general area was clear, but there was a concentration of small clods around the pit.

"Shall I sample the glass?" Irwin asked.

"Start with the middle," Scott prompted. "And then we'll pick up the rim."

The sample[9] was a regolith breccia that was lightly coated by glass. "See how it's all welded," Irwin observed

"It's an intricate pattern," Scott agreed.

Irwin sampled the regolith from the eastern rim, and then Scott snapped the after-sampling documentation. The well-rehearsed procedure had taken just 2 minutes.

"Let's find ourselves a couple of fragments," Scott urged. Moving cross-slope, he went over to what, on this bare slope, passed for a big rock: fully 15-cm long, with a fillet.

"Have you noticed a variety of rock types?" Allen asked.

"We can't tell any differences as they sit on the surface," Scott replied. "They're all covered with dust." After lifting the rock,[10] he scraped it clean. "It's a fine-grained breccia with white clasts in it; the matrix is black. And it has glass within a fracture on the side." In fact, the glass had

penetrated the rock through the crack and lined its internal surfaces.

Continuing the sweep above the Rover, Scott came across a partially buried rock near a cluster of small fresh-looking craters that were full of clods. It was a light-grey fine-grained breccia[11] with tiny white clasts contributing no more than about 10 per cent of the rock. He turned it over to inspect the face that had been in the ground. "Ah-ha! The bottom has slickensides." He moved to allow the Sun to illuminate the rock, and turned it end over end. "I see some glass spattered on one side. And I also see a little orange crystal in there." It was a good find. "This is definitely a *different* kind of breccia, Joe." It was too large for an individual bag, so he dumped it straight into Irwin's SCB. He found another "interesting" one several metres further on. "I think that little rock hit there," he explained once Irwin joined him, "and left a mark about a foot from its present position." The geometry implied that the fairly rounded rock had come in from the east, and rolled to its final position. He leant on his tongs with one hand and reached down to grab the rock with his other hand.[12] The albedo variation suggested clasts in a moderately grey mass. "I guess I'll just have to call it a breccia."

Reflecting on their progress so far, Scott noted that the size range of the rocks on the slope was bimodal: there were small fragments and there were cobbles, but there were no intermediate sizes. The variety was also rather limited: apart from the glassy regolith breccias, thus far they had found only fragmental breccias. The purpose of sampling on the mountain was to find massif material, which was expected to be crystalline, and in all probability anorthositic. If the rocks had been clean, it would have been easy to search for anorthosite, which is strikingly white. But with all the rocks caked with dust, it was impossible to tell what one was without picking it up and scraping it clean. If they were to stick rigorously to the time-consuming routine, they might jeopardise their scientific mission. It was a dilemma: to document a small number of rocks, or simply to grab as many as possible in the hope of achieving a greater variety.

"There's a high-albedo crater to the west, Dave," Irwin pointed out helpfully. It might have dug something up.

Scott decided the crater was too far away to visit on foot. "We'll head that way with the Rover, when we get going." He moved a little way down-slope and picked up another rock.[13] "Another microbreccia," he sighed. He lifted another.[14] "Oh boy! Look at the bottom of that, Jim."

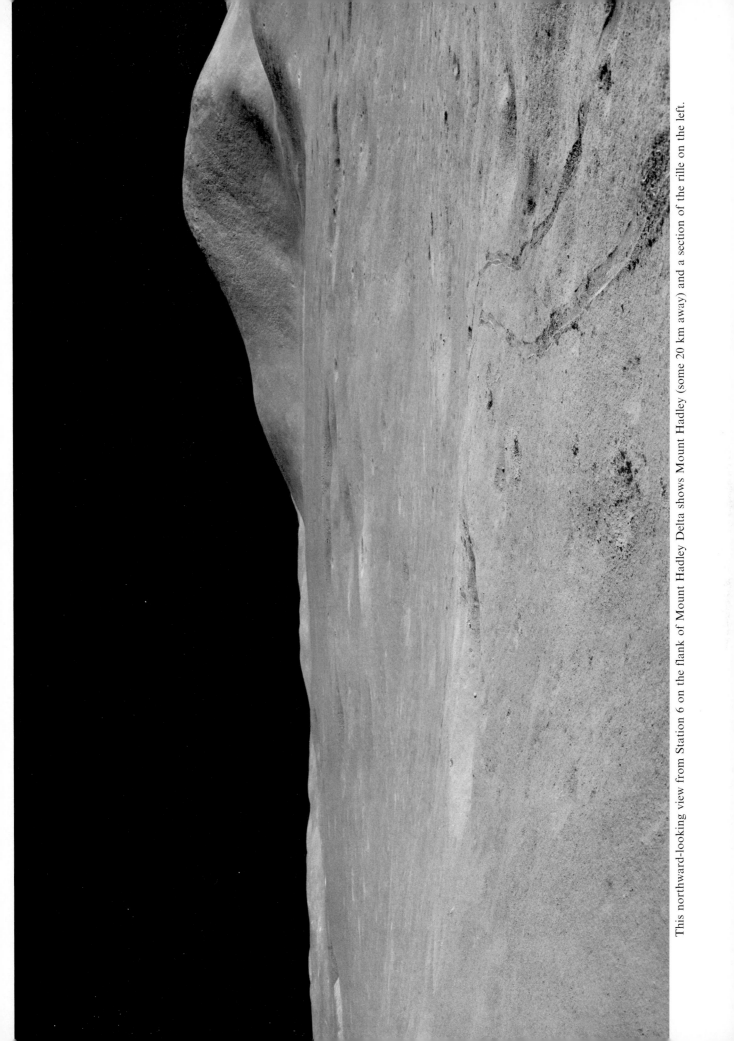

This northward-looking view from Station 6 on the flank of Mount Hadley Delta shows Mount Hadley (some 20 km away) and a section of the rille on the left.

"All glassy," Irwin observed.

Although Scott dismissed it as another breccia, this was a porphyritic basalt with one nearly-flat face and a parallel internal fracture plane. It would turn out to be the only basalt they found at Station 6 that could be said to be indigenous. Irwin started to scoop up a little soil for context, but Scott told him that they had plenty, and it all seemed to be the same anyway.

After spending almost 25 minutes methodically sweeping above the Rover, Scott decided to examine the 12-metre crater that was about 15 metres below where they had parked. It was by far the largest crater in the immediate vicinity. It had a slightly raised rim and, to his surprise, there was a bench in the northern wall. He decided to take a look at the bench because it might contain rocks from beneath the regolith. As the inner wall on the northern side would cancel out the slope of the mountain and offer a horizontal surface, he side-stepped down the eastern rim and then entered the crater.

Irwin took up position at the eastern rim, ready to take the down-Sun picture once his commander selected their target. As Scott surveyed the rocks, most of which were too large to sample, Irwin pointed out that Scott's boots were exposing a light-toned material beneath. He proposed that they should trench this. Scott initially agreed, but then had second thoughts – a trench was an *in-situ* soil mechanics experiment, a core would give the geologists an opportunity to study the depositional structure in detail. They decided to get the core from the flat spot on the upper rim on their way up to the Rover. In the meantime, Scott had documented a cluster of rocks in the centre of the crater. There was one main rock surrounded by a litter of fragments which looked as if they had broken off on impact. He sampled one of the fragments. "Another breccia!" He bagged it, then added several more fragments without bothering to look at them. As the fragments were frangible, he decided to try to chip the main rock. "I'll take a whack, and see if it'll come open." After pulling the hammer from Irwin's PLSS, he momentarily dipped down on one knee and delivered a blow to the top of the rock, shattering it into several pieces that scattered around.

"Not bad at all!" Allen congratulated.

Scott retrieved one of the fist-sized fragments.[15] It was a breccia whose large light clasts were themselves brecciated, indicating that the rock had been through several significant impact events. The largest clast (15 mm wide) proved to be a pyroxene-enriched feldspathic material. For good measure, Scott popped a second fragment[16] straight into the bag. No sooner had he done so than Allen slipped in a reminder for the trench. Scott thought they had settled on a core as the way to sample the layering in the regolith. "Would you like a trench, or a core, Joe?" he asked.

"We'd like one of each, if we could, Dave," Allen replied.

"A trench *and* a core?" Scott echoed wryly. Clearly some in the Backroom were calling for a trench and others for a core. "We'll go up and trench it first, to see if it's worth coring," he decided.

Upon reaching the horizontal patch beyond the up-slope rim, Irwin started on the trench. He aligned it north–south so that the Sun would illuminate its western wall and made 20 scrapes using the scoop. Although the powdery soil at Station 6 would readily slump when subjected to a load, the near-vertical walls of the trench retained their integrity. To take the picture, Scott went around to the north and stood on the lip of the rim, then carefully took one step down into the crater. It was a precarious stance, since the slope of the inner wall added to that of the mountainside, making a 20 degree incline. As he tried to climb back up again, the soil slumped beneath his feet, he lost his balance and fell onto his hands and knees just inside the rim of the crater. "How about a hand, old buddy?" he chuckled. Irwin, still on the level spot above the crater, yanked him up without difficulty.

At this point, Allen passed on another request from the Backroom, this time for a SESC: the can would preserve the sample in pristine condition. Irwin scraped out the trench to ensure that it had not been contaminated by their antics, then took a scoop from the floor for the can. As soon as this sample was stowed, Allen reminded them of the core. When Irwin pointed out that his trench had not shown any depositional structure, Scott decided that they should return to the lower side of the crater, and so they once again side-stepped down beside the crater's rim.

"The soil is more granular here, too," Scott noticed, upon reaching the bench. He chose an undisturbed site a few metres away from the shattered rock. Having rehearsed the various sampling routines in training, the two men were able to work as a team without indulging in verbal coordination. Scott turned his back to Irwin so that Irwin could take a core tube from his PLSS. Irwin then turned to enable Scott to collect the hammer. In the event, Irwin was able to push it all the way in by hand. "Oh! Easy," Scott applauded.

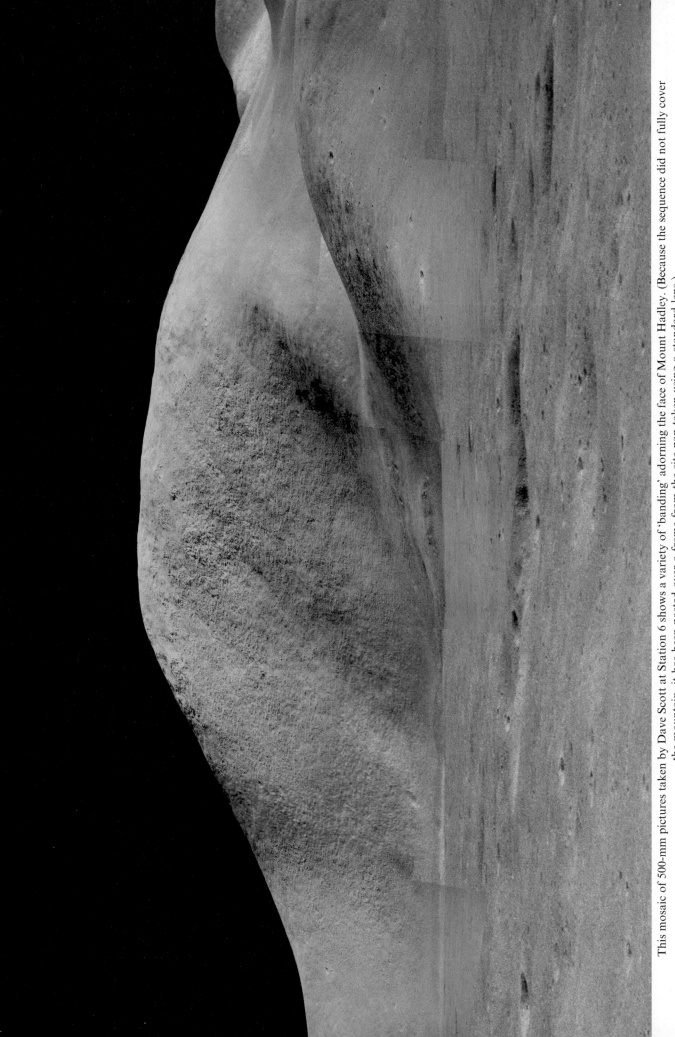

This mosaic of 500-mm pictures taken by Dave Scott at Station 6 shows a variety of 'banding' adorning the face of Mount Hadley. (Because the sequence did not fully cover the mountain, it has been pasted over a frame from the site pan taken using a standard lens.)

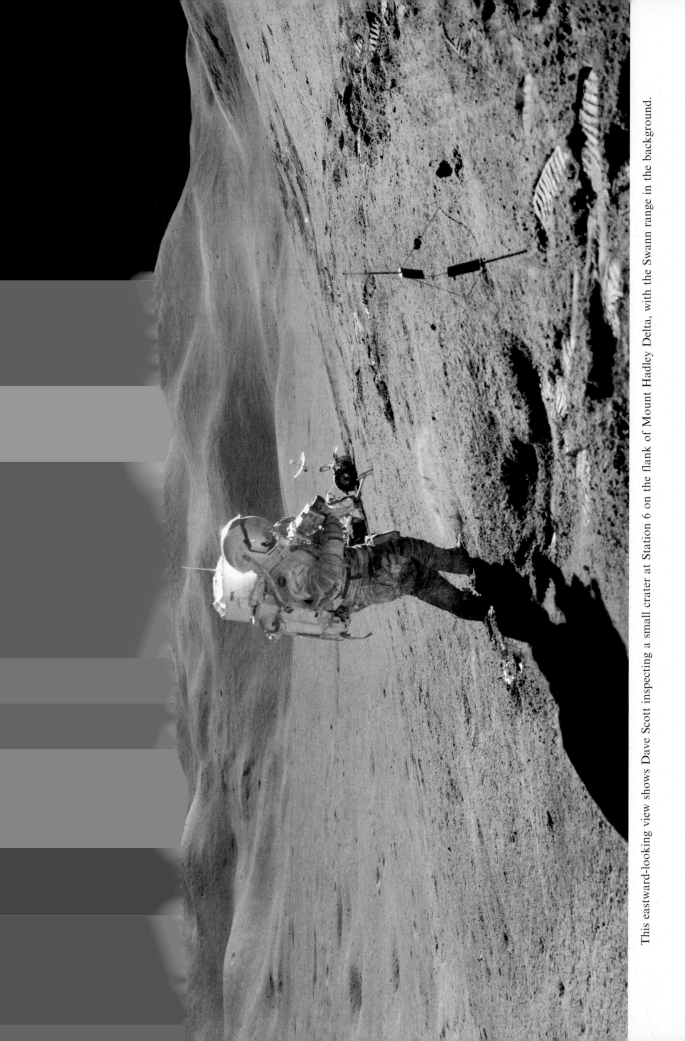

This eastward-looking view shows Dave Scott inspecting a small crater at Station 6 on the flank of Mount Hadley Delta, with the Swann range in the background.

Dave Scott is in position to take stereoscopic cross-Sun pictures in the 12-metre-diameter crater at Station 6. At left is one such picture taken prior to sampling (top), after breaking the rock apart (centre), and after retrieving a fragment (bottom). This kind of documentation enabled geologists to reconstruct the context of each sample. The sample bags are on a dispenser attached to his chest-mounted Hasselblad. Note that Scott's boots have exposed a layer of lighter subsurface material.

At Station 6, Dave Scott took this telephoto picture of Falcon on the plain, in line between the craters Dune in the South Cluster and Pluton in the North Complex.

"It's all the way in, Joe," he added for Houston's benefit.

"Anything there?" Allen enquired.

Scott inspected the end of the up-turned tube, and reported it to be full of dirt. It turned out to be a very deep layer of finely grained soil laced with angular fragments up to 2 cm in size. "It's a good core," he confirmed.

"I like cores like that," Irwin agreed, delighted that it had been obtained so easily.

"That might even be a *great* core," Allen added playfully.

"Let's work *above* the Rover from now on!" Irwin suggested, as he trudged back up the hillside. Surely, if they *must* walk up-slope, it would be better to do it in short spurts between individual samples rather than one long trek afterwards. Scott, some distance behind, contrasted the deep indentations made by his colleague's boots with the shallow imprints of their vehicle's wire-framed wheels. It was so striking that he paused to take a picture for the soil-mechanics people.

When they finally reached the Rover, Allen forwarded *another* request from the Backroom, this time for a soil sample. "And we're after a large volume, so shovel it in." Irwin obliged, and as Scott shook the bag to settle its contents he noted that it was so much easier to seal a bag when there was no air – in contrast to in training on Earth. Before they could leave, there were chores to perform. While Irwin put a new film magazine onto his camera, Scott retrieved the Hasselblad with the 500-mm lens and documented Mount Hadley, the Swann Range and outcrops near Mount Hadley Delta's summit. He was particularly impressed by the sight of Falcon on the plain, in line between the South Cluster's Dune and the North Complex's Pluton. The clarity of the detail in the airless environment was astounding.

Station 6 had investigated three types of terrain: the fresh glassy-pits; the 12-metre crater; and the open mountainside. Most of the samples were lithified breccias rich in feldspathic clasts. A few contained clasts of mare basalt. There was only one clearly non-mare basalt. However, there was nothing that was definitively part of the massif. The only remaining chance to find anorthositic material was Spur, but as this was much smaller than St George, whose ejecta was reduced to fines, it was starting to look as if they would leave the mountain empty handed.

"We're moving out," announced Scott as he restarted the Rover.

Allen said that they should spend 15 minutes at the boulder to the west, and then investigate the

fresh-looking crater that Irwin had noted. Although the boulder was only 200 metres diagonally up-slope, it was above an inflection in the mountain, where the slope was steeper and distinctly hummocky. When the wheels began to slip and he had to offset the steering to 'crab' across the slope, Scott praised the Rover as the "best tractor" he had driven. On approaching the boulder, Irwin was able to look down into Spur, 30 metres below them and 250 metres further west, and reported a massive block on its northern rim, right where the ground ought to be flat. Despite Irwin's recommendation that they should work above the Rover, Scott parked above the boulder, which was unfortunate as the slope was now about 15 degrees. In order to save time on what was to be a brief stop, Scott opted not to align the TV antenna.

"Whatever you say," Allen acknowledged.

"This is *steep*!" Irwin exclaimed, as he eased off the Rover's down-slope side. It would be near-impossible to get back on. With no chores to tend to, Scott retrieved his tools and set off "to attack" the boulder which, at 3 metres long, was the largest rock they had yet sampled. Struggling to maintain his footing, Irwin felt obliged to criticise his commander. "It's gonna be a bear to get back up, you know."

Allen picked up on the lack of mobility. "Hey troops," he called, "I'm not sure you should go down-slope very far – if at all."

"It's not far, let me try it," Scott decided. "Jim, you just stay there."

"I think we can side-step back up," Irwin offered.

"It's not that difficult," Scott reported.

Allen had to concede to the commander in the field: "Proceed carefully."

"Okay, I'm half-way," Scott noted for Allen's benefit, but actually he was having second thoughts. "I'll come back up."

"We should've parked right beside it," Irwin reiterated.

"I think I will." Scott had just decided to move the Rover.

"How's the footing?" Allen persisted.

"The *footing* is alright," Scott insisted, stretching the truth a little. "It's just that you have to work pretty hard to get back up."

Allen wanted to see what was going on. If Irwin was standing beside the Rover, then surely he could aim the antenna. Even a fuzzy TV picture would be better than staring at a blank screen. "Jim, are you still up near the Rover?"

"Wait a minute, Jim," Scott interjected before

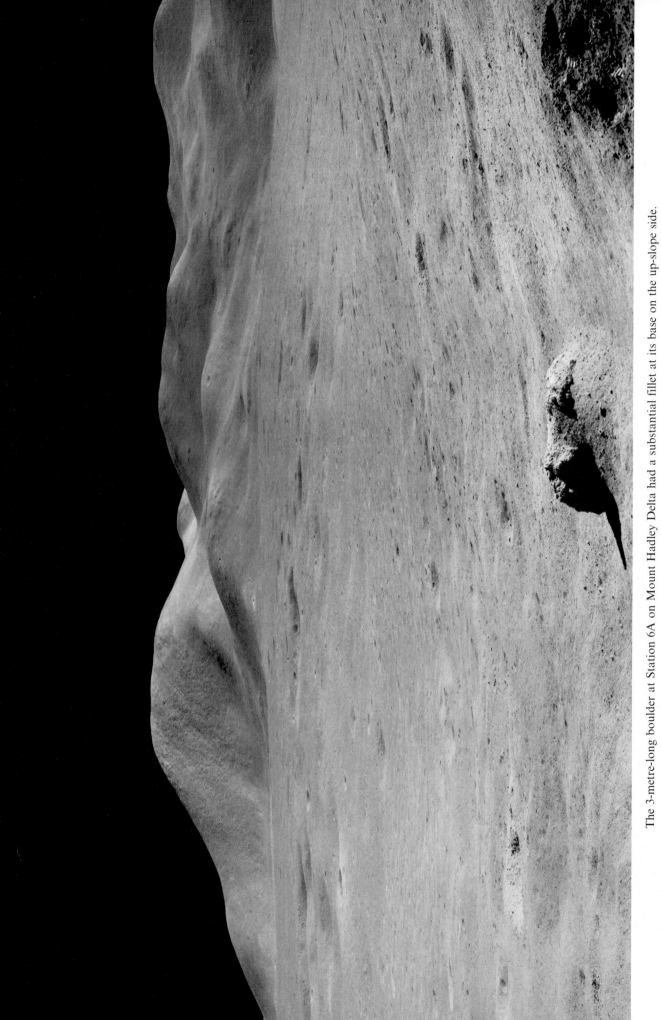

The 3-metre-long boulder at Station 6A on Mount Hadley Delta had a substantial fillet at its base on the up-slope side.

Irwin could reply to Allen. "Let me drive the Rover down there."

"Jim," Allen prompted. "Just take a rough guess as to where the Earth is. We're going to give the TV a try."

"I'm going to drive down!" Scott repeated. He clambered aboard from the up-slope side of the vehicle, reversed, and turned to face directly down-slope.

"Do you want me to come over there and get on?" Irwin enquired.

"Stay there," Scott ordered.

"You're proceeding down towards that large block now?" Allen asked.

"Very gently," Scott chuckled. He parked just east of and below his objective. "It sure is a big rock."

"We'd like you to take a rough guess with the antenna," Allen persisted, hoping to cajole Scott into providing TV coverage.

Feeling the Rover slip as he eased off it, Scott grabbed hold of it and dug his feet into the slope to stabilise the vehicle, but he was up-slope and poorly positioned. "I tell you what, Jim, we'd better abandon this one."

"Afraid we might lose the Rover?" Irwin asked

– a remark hardly likely to calm the mounting concern in Houston.

"You come down," Scott ordered.

As Irwin made his way past the boulder, he thoughtfully paused due east of it to take a down-Sun prior-to-sampling picture, and as he lined up the camera he noticed something strange. "This rock has got *green* in it; a light-green." Green was usually an indication of olivine. The picture taken, he continued down and joined Scott at the Rover, which was so precariously situated that it actually had one wheel off the ground. "Where do you want me to hold it?"

"Stand on your side," Scott suggested. By standing down-slope, Irwin would be able to stabilise the Rover by leaning against it, which would be easier than trying to hold it in place from up-slope.

"Use your best judgement," Allen broke in. "The block's not *all that* important."

"I'd just as soon not spend too much time here," Scott assured. It was now only Irwin's report of a green tint to the rock that was sustaining his determination. It had taken over 5 minutes from his decision to move the Rover, to actually parking it by the boulder, which was twice

It was as he took this down-Sun picture of the boulder at Station 6A that Jim Irwin observed that its mottled surface was encrusted by a light-green material.

As Dave Scott took cross-Sun pictures of the boulder at Station 6A, his line of sight also recorded (out of focus) Jim Irwin holding the Rover – its left-rear wheel off the ground – in order to prevent it sliding down the steep slope.

as long as it had taken to drive over from Station 6. If the time spent testing the slope on foot was added, then 7 minutes had elapsed, and this was an unacceptable overhead on what was to have been a 15-minute stop. And it was not even as if he had been able to start work at that time, as he had then been obliged to hold the Rover in place. As a result, as Scott finally moved in to take his first close look at the rock, almost all of the allocated time had expired. "It's a big breccia," he said dejectedly. "That's all it is." And then, almost as an afterthought, "I don't see any green."

"Maybe you have to look down-Sun to see it," Irwin prompted. In the harsh light, the tone of a rock was sensitive to the angle at which it was viewed. "It looks like a light-green layer; a fairly thick layer."

"You mean on the surface?"

"Yeah, on the surface."

You're right!" exclaimed Scott. The tint was very subtle. Irwin had been fortunate to spot it.

"It would be great if we could get some of that green material," Irwin prompted tactfully.

"I'll get it!" assured Scott. After all the time that they had invested in this boulder, there was no way he was going to leave without chipping off a piece. However, he didn't have the hammer; it was on Irwin's PLSS. "I think I can get a sample with my tongs," he ventured. The green material was either a very frangible half-metre-wide clast, or a coating that the breccia had picked up. He easily prised off a chunk.[17] The hue of the newly exposed surface was much brighter. Realising that the breccia itself was loosely consolidated, for good measure he broke off a fist-sized fragment[18] that contained a large white clast.

It transpired that the green material owed its hue to magnesium oxide, which was present in the form of glassy spherules believed to have been spewed out by a fire fountain. Because the site was on the rim of the Imbrium Basin, it seemed likely that the fountain was associated with deep faults that were made by that impact. The host breccia fragment was slightly recrystallised, indicating that the rock had once been hot. The glass formed a thick crust on the breccia. Although not 'black-and-white' in the Fra Mauro sense, the breccia was strikingly mottled. And, significantly, none of the clasts (in the sample, at least) were mare-basalt, they were of the KREEPy 'non-mare' variety. But where was it from? Although there was no track to indicate that it had rolled down the mountain, neither was there an imprint nearby to mark where it had hit after being thrown in. However, the up-slope fillet meant that it had been in place for a

long time, and the 1.2-billion-year age measurement was consistent with it being ejecta from either Autolycus or Aristillus.

"I think we'll call it quits on this one," Scott decided. Their struggle to sample this rock had not been in vain: the green material was a genuine *discovery*, and that was the main point of the mission.

"Sounds great, Dave," Allen agreed. However, he was concerned at the timeline. Although they had been on-site at what would be called Station 6A for 18 minutes, only the final few minutes had been spent working on the rock; the rest of the time had been spent recovering from the mistake of parking up-slope. He relayed the news that the Backroom had deleted the fresh-looking crater.

SPUR'S TREASURE

Faced with the difficulty of trying to reboard the Rover from down-slope, Irwin said he would rather walk down to Spur, but Scott refused point-blank; instead, he drove down to a depression, and Irwin scrambled aboard. But because Irwin was unable to engage his seatbelt he had to hold on as Scott proceeded diagonally down-slope to Spur. Having learned his lesson, Scott went for the northern rim of the crater, which cancelled the slope of the mountain's flank. With the luxury of a level site, they would be able to work efficiently. As Irwin had noted, there was a boulder, clearly a breccia with a dark matrix mottled by white clasts, right on the rim; a location which suggested that it had come from the deepest excavation. Scott parked 60 metres up-Sun of it, so that Houston would be able to monitor their sweep along the rim crest towards the boulder. Since the 100-metre crater was 20 metres deep, there was every prospect that they would find material excavated from the massif.

Allen announced that they had only 30 minutes. For this time to be significantly extended, they would need to have discovered something more important than was likely to be offered by the deferred sampling of Dune.

Spur's rim was strewn with fragments, and Scott's attention was drawn to one that had "a white vein on top".

"Get it!" urged Allen.

"It's a breccia," Scott opined, describing the rock as they photographed it. "It's a dark grey rock that looks like a pinnacle, with a small breccia on top of it with a light-to-medium grey matrix and about 20 per cent white clast." He

Having just parked on the northeastern rim of the crater Spur, Dave Scott aligns the Rover's high-gain antenna.

At Station 7, Dave Scott stands by with tongs to sample the rim of the crater Spur. The 'before' and 'after' pictures (left, top and bottom respectively) show the small piece of glass that was found in the hole where the rock with the "white vein" had been. Notice that virtually all of Scott's suit is stained with dust.

picked it up and studied it more closely. "There are sparklies!" Looking back down, he saw that the regolith was "caked", so he suggested they sample it. As Irwin took a scoop of soil, Scott saw something shining in the hole where the rock had been, so he dug this out and put it into the bag with the soil. Analysis of the rock[19] confirmed that the clast was crystalline; it was a shocked fine-grained plagioclase-rich gabbro. The presence of vesicles in the host matrix indicated that this had once been hot, but not so hot as to melt the crystalline clasts. The shiny fragment[20] retrieved from the hole was just a chip of glass.

As Scott moved a few steps further west around the rim, Irwin announced that he could see some more green material. Believing Irwin to mean that Scott's boots had exposed something, he looked at his tracks but couldn't see anything. But Irwin was insistent. Then Scott saw the rocks Irwin was referring to. "It *is* green," he exclaimed. To check, he raised his gold-plated visor. "No, it's grey; the visor makes it green." Nevertheless, they took a sample. "It's light-grey, very fine-grained," Scott reported. "It looks like a basalt, with some less-than-millimetre-sized vesicles." As he scraped it, the material flaked. "Maybe it's not a basalt!" he chuckled. He wasn't sure what it was. For good measure, he lifted another fragment from the same cluster, and added it to the bag. The loosely consolidated regolith breccias[21,22] proved to be welded by the same type of green glass as had coated the Station 6A boulder.

As Scott pondered whether to go straight to the big breccia on the rim, which was now only 15 metres away, a rock with "a white corner" caught his attention. It was almost as if a brilliant white rock was sitting on a pedestal.

"What do you think the best way to sample this would be?" he asked Irwin.

"You could probably lift that top fragment right off."

"Let me try." As soon as Scott grasped the white clast[23] using his tongs, it broke free of its base. "Oh, man!" he exclaimed as he held the rock up to the Sun. "Look at that *glint*!"

"I can almost see twinning in there!" Irwin prompted excitedly. He could hardly believe his eyes; the highly regular matrix – like the spines of books on a shelf – was as clear an indicator as they could have hoped for.

"I think we've found what we came for," Scott announced.

"Crystalline rock, huh," said Irwin.

"You better believe it," Scott laughed. "Look at that plagioclase." As he studied the crystal more closely, he realised that it was almost *pure* plagioclase. "I think we might have ourselves something close to anorthosite." This was the term for a rock that was at least 90 per cent plagioclase. "What a beaut!"

"Bag it!" Allen urged.

Tiny fragments of plagioclase-rich material were in the mare regolith returned by earlier missions, but finding anorthosite on the Apennine Front strongly supported the proposal that the first lunar crust was a scum of buoyant aluminous minerals that floated to the surface of the magma ocean.

Apollo 15 had been portrayed by the Press as a search for the oldest rock on the Moon, and when a geologist seeking to explain how this sample would significantly advance the study of the origin of the lunar crust used the term petrogenesis during an 'instant science' press conference, this led to the journalistic contraction that gave the rock its popular moniker of the 'Genesis Rock'. The 0.27-kg fragment proved to be 98 per cent plagioclase, so it truly was anorthosite. Although a crystallisation age could not be given since it contained no mafic minerals, the strontium data indicated that it was *extremely* primitive, and it almost certainly was part of the original crust. At some point during its long history, its crystalline structure had undergone intense shock, so it was not in pristine condition. Nevertheless, Scott and Irwin really had achieved their objective.

The next task was to get a bit of the host. It looked to be loosely consolidated. "Maybe you could stick your scoop in and break off a chip," Scott suggested. Irwin did so, and a fist-sized chunk[24] split from the pedestal. After retrieving his prize, Irwin carefully raised it on his scoop and Scott eased it into a bag. "It's a clod," he confirmed.

Resuming the sweep, Scott saw a fragment displaying a contact between two types of rock, and he set about documenting it.

"I think we ought to get over to that big rock," Irwin suggested.

"We're getting there," Scott assured. The boulder was now only 10 metres away, and he had every intention of sampling it.

"I think the big rock is probably more important," Irwin persisted. The Backroom was divided over whether to seek many small rocks for variety, or to concentrate on boulders whose context was more apparent. Scott was trying to strike a balance, but Irwin had a keener sense of the ticking clock.

"This one we have *got* to pick up," Scott

At Station 7, Dave Scott stands by as Jim Irwin takes the down-Sun 'locator' of the "rock with the white corner". In the background is the large boulder to which they were progressing. Left, top to bottom: before sampling; after removing the "white corner"; using the scoop to smash the host; and after having sampled the host. The white fragment would later be nicknamed the 'Genesis rock'.

insisted, and he promptly did so. As he scraped away the dust he revealed a black-and-white breccia.[25] Its white part proved to be norite, which is plagioclase enriched with pyroxene. It was very shocked, and was dated at 3.86 ($\pm$0.04) billion years.

At this point, Allen reminded them that the checklist called for a comprehensive sample: "We think you ought to give some thought, pretty shortly now, to getting us a rake sample – if you can find a good area."

"Okay," replied Scott rather neutrally.

"That seems a shame," Irwin said, well aware that Allen was trying to tempt them back to the Rover. "We ought to sample that big one."

"We'll do that," Scott promised.

Although the big breccia was just 3 metres away, their progress towards it halted when Allen pointed out that Irwin's SCB was working loose. Scott went to attend to it.

"We'd prefer just a very quick sample of the block," Allen began ominously. The walkback constraint was looming. "Perhaps just photographic documentation."

"Okay, Joe," Scott replied, as he struggled to refasten the strap of Irwin's SCB's.

Half a minute later, Allen called again. "We want you to forget that large block entirely. We want as large a collection of small fragments as you can get." Given the straight choice, the Backroom had opted for a rake sample. "You'll probably be working near the Rover for that," Allen added, reinforcing the point that it was time to head back.

"We understand, Joe," Scott assured. But he had been saving the breccia as their 'dessert', and was not simply going to turn and walk away. It was time to split their efforts. "Why don't you get the rake," he said to Irwin. "I'll photograph this thing." The 1.5-metre-long block was actually resting just inside the crest of the rim, and in circling it Scott had to venture into the crater. It was sheared along several fractures parallel to the main axis, and there were some perpendicular lineations. It was also partially coated in dark glass. Although Scott did not have the hammer, he was able to sample it indirectly by collecting a sizeable fragment[26] that had evidently broken off. "It looks very much like the Apollo 14 rocks," he concluded.

"I'm set up," Irwin called, having configured his rake and selected a site right on the crest of the rim. He raked a swath about a metre long, shook out the fines, and inspected the result. "I've collected about fifteen rocks."

"A jackpot!" Allen exclaimed. It was certainly an improvement over the trawl at St George the previous day.

Irwin made a succession of parallel swaths, finding fewer pebbles each time, and after speculating that this was because he was moving away from the rim, he moved closer in and got another full load. Of the total of 78 fragments,[27] a few were mare basalts tossed up onto the mountain, but most were shattered breccias and only a few of these contained mare basalt clasts. The site was clearly dominated by 'frontal' material, and because Spur, a Copernican Era crater, was not as old as St George, its regolith was not as 'mature'.

Given the rich pickings on the rim, Scott proposed a core, even although this was not a checklist item. "All we really need is soil," Allen replied. "And maybe some grapefruit to football-sized rocks," he offered as a consolation. But this was not a 'hunting licence'; he wanted them to remain near the Rover. "We'll want to move in about 3 minutes."

"That doesn't give us time to do much," Scott pointed out. Looking around for a candidate, he spotted a rock right on the rim that showed a linear pattern. It was too large even for a football-sized sample, so Irwin struck it with his scoop, splitting it open along a fracture line.

"Good boy!" Scott said delightedly. Although merely a chip, the 4.83-kg sample[28] turned out to be the second-largest rock of the entire mission. It was a breccia with a very dark matrix hosting clasts several centimetres in size.

As Scott and Irwin hastily packed up, they had the satisfaction of knowing that the rich variety at Spur had made up for the blandness of the open mountainside, and they had found their anorthosite.

TANTALISING DUNE

While Scott concentrated on controlling the Rover as they drove directly down-slope, Irwin studied the South Cluster. Although his view was foreshortened, the area was evidently not the obstacle to trafficability they had been led to believe. The overlapping rims would be very hummocky, it was true, but the ejecta did not appear too bad. "It almost looks like you could drive around the eastern rim of Dune," Irwin mused, not so much to tempt Scott into making a diversion as to correct the record. Encountering their tracks at the foot of the mountain, Scott decided to follow them back to Dune.

The 1.5-metre-long breccia on Spur's rim was aligned radially with its 'snout' poking over the crest into the pit (lower left).

As they drove north, Irwin admired Mount Hadley at the other side of the valley. The lineaments on the 4,000-metre peak were spectacular. "That's really beautiful."

"That's the most *organised* mountain that I've ever seen," Scott admitted.

All the evidence suggested that the South Cluster marked the fall of ejecta from a major impact to the north. Although only a low-energy secondary impact, there was an excellent chance that Dune had dug deep into the lava flow that had embayed the valley. As they made their approach, Scott left their tracks to penetrate the fringe of the ejecta that was concentrated south of the crater, in order to reach the point where a 100-metre crater had cut a 'notch' in Dune's rim; his objective being a cluster of boulders that had undoubtedly been excavated by the impact that formed the notch.

Although the floor of Dune was fairly clear, the presence of clusters of large boulders running around its wall confirmed a study of the crater's morphology that had estimated the regolith on the surrounding plain to be about 5 metres thick.

When Scott announced that he had parked 30 metres short of Dune, Allen warned that they

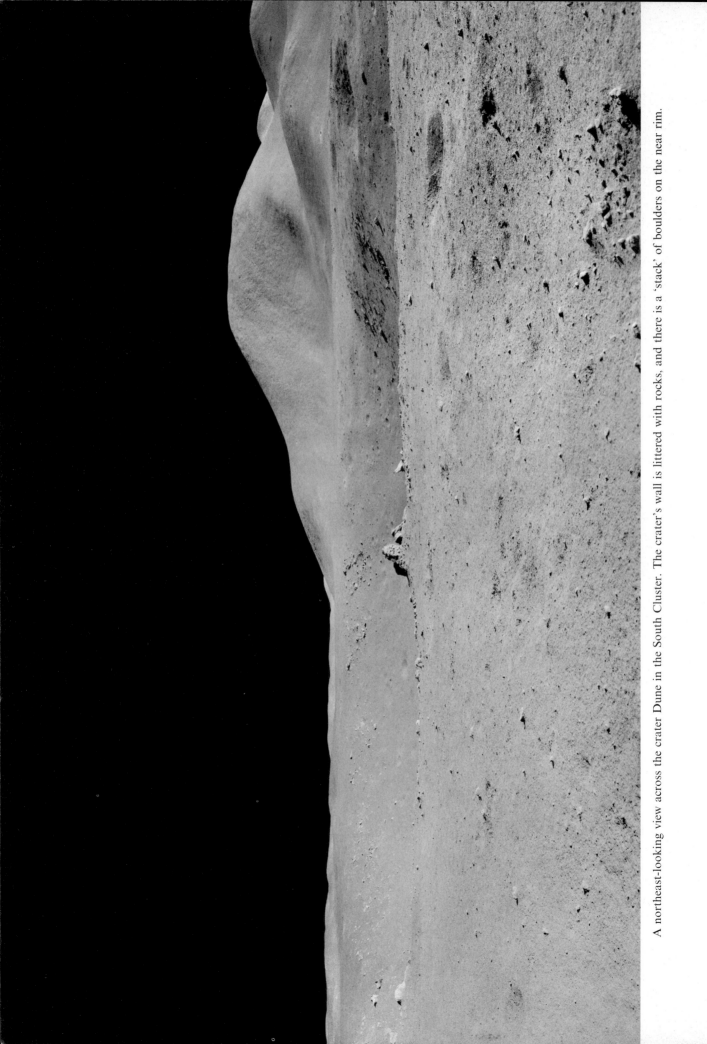

A northeast-looking view across the crater Dune in the South Cluster. The crater's wall is littered with rocks, and there is a 'stack' of boulders on the near rim.

These boulders on the rim of Dune represented material from the deepest point in the excavation.

would have to be off again in *10 minutes*, which allowed them only half the time that had been assigned to Station 4. The intention had been to perform a radial, but this was now impossible. They were limited to taking a few documented samples and raking a comprehensive sample. Although Allen offered to do without the TV in order to save time aligning the antenna, they got off to a poor start when Irwin tried to take the pan and found that his camera had jammed – the motor was running, but the screw which gripped the shaft had worked loose and was not advancing the film. Scott shot the pan, then looked around to see what they might sample in the vicinity of the Rover prior to going to the crater's rim. He selected a cluster of small cobbles and, with Irwin's camera out of action, took all the documentation himself. In order to save vital seconds, he bagged the samples without inspecting them. One[29] proved to be a coarsely grained gabbro with vugs lined by pyroxene, and the others[30,31] were porphyritic basalts, also vuggy, with a variety of mafic phenocrysts.

Having cursorily sampled the periphery of Dune, Scott ran over to take a look at the cluster of boulders perched on the rim crest. The largest was 'upright', evidently jammed in by the rocks at its base. The amazing thing about it, however, was that it was riddled with cavities up to 100 mm in diameter. It was clearly a magma that had solidified near the surface, yet its presence on the rim implied that it had excavated from the deepest point. Perhaps Dune had drilled 100 metres through a *succession* of lava flows. As he studied the surface of the rock more closely, Scott noticed that the lenticular plagioclase laths were randomly orientated. To supplement the standard documentation, he took a series of close-ups shooting down his tongs to measure the focus.

"Are you going to try to knock a piece off?" Irwin asked innocently.

"Oh, yeah!" Scott chuckled. He took the hammer from Irwin's PLSS, struck the edge of one of the cavities, and gained a small chip.[32] As with the samples by the Rover, it proved to be a porphyritic basalt that was thick with pyroxene phenocrysts, but in this case a glassy matrix made it a vitrophyre. Upon realising that the vesicles were graded in size and abundance running across the face of the rock, Scott decided to sample the opposite side of the rock to document this zoning. In these chips[33,34] the plagioclase laths were flow-aligned.

"Let's head back to the Rover," Scott announced, but this was a ruse to placate Allen's persistent calls to move out. Although Irwin set off back to the Rover, Scott, knowing that this time Allen could not see him, lingered to study the other rocks in the group. There was another vesicular rock beneath the one he had sampled. Being a lighter shade of grey with much smaller, more uniform vesicles, it was evidently a different lava. He was tempted to sample it to record the chemical variety, but there simply wasn't time. The fact that he was being hustled to resume drilling the heat-flow hole meant that this geophysics experiment had a higher priority.

Dune was a magnificent crater with a rim fully a kilometre around. Even in ideal conditions, they would not have had the time to do more than make a radial sample. Of necessity, an Apollo traverse was a series of dashes from one site to another in order to collect whatever could be sampled in the limited time available, in the hope that this would provide sufficient 'ground truth' to test theories. Scott's primary concern was to do this with geological rigour. By the time he finally returned to the Rover he had stretched Station 4 to 17 minutes, and thus almost restored it to the originally allotted time, so, in truth, they could not have been expected to achieve much more.

JOURNEY'S END

"Gee, it's nice to sit down, isn't it?"

"Oh, it is," Irwin agreed.

One of the advantages of the Rover was that the astronauts could rest between stations. It was not simply a matter of taking a break, though. While they were seated they drew less oxygen and coolant, and so the total time available to the traverse was substantially greater than would have been possible on foot. Thus the Rover not only expanded their radius of operations, it also extended the effective duration and made more efficient use of it. This was in addition to facilitating navigation and enabling them to carry more tools and collect a heavier load of rocks. All in all, therefore, the Rover was vital to the success of the J-mission format.

"Dave," Allen called cheerfully, "the seismometer is picking up the 'rumble' of the Rover rolling across the plain."

"Is that right!?" Scott cheered.

As they drove by a large crater, Scott and Irwin debated its identity. "It's probably Arbeit," suggested Allen.

"Arbeit!" Scott laughed. On the nominal plan they were to have done Station 8 at Arbeit. Since

Irwin considered the soil-mechanics experiments that he had been assigned to be manual labour, Scott had named the crater after the German word for 'work'. Although this station had been deleted, Irwin doubted that he had escaped the activity, because it was sure to be rescheduled for later.

"This is making me seasick," Irwin complained lightheartedly as they drove through a very hummocky patch at full speed.

"What do you expect, travelling on the *mare*," quipped Allen, with a wit as dry as the lunar 'seas'. Then he asked more seriously: "I'm wondering if you caught sight of the crater that you described as having bedrock in the bottom?"

"Are you contemplating a stop there?" Irwin asked. He had been watching for the crater that he had remarked upon early in the southbound drive, but had not seen it.

"Negative; just curious."

"Tracks upon tracks, Jim," Scott said jovially as they approached the landing site.

"It looks like a thoroughfare!" Irwin agreed.

"Home, sweet home," Scott laughed.

THE SAGA OF THE DRILL

After parking the Rover alongside the LM, Scott and Irwin dismounted to unload the trove from The Front. Unfortunately, because Allen had attempted to second-guess the stowage, and tried to micromanage the transfer, this took an inordinately long time. When Irwin reported that he was ready to seal one of the Sample Return Containers, Allen sprang his surprise: they would need the box later, for the tubes from the deep core sample.

"Well, Joe," Scott sighed. "You didn't say anything about us getting a deep core." He had accepted the need to finish drilling the second heat-flow hole to emplace the thermal probes, but had presumed the associated core sample to have been cancelled. So! This was why Allen had been reluctant to extend the sampling at either Spur or Dune. But there was no point making a fuss. Once the Rover had been off-loaded, he left Irwin by the LM and drove to the ALSEP site, parking so as to enable the TV to watch him at work.

"Try to get the drill in at least another section," Allen urged. "Recycle the chuck several times, then start the drill and put only a few pounds of force." When he had encountered resistance the previous day, Scott had leaned on the drill in an effort to assist it to penetrate, but the Backroom had since decided that this could have been

counterproductive, as it may have encouraged the external helical flutes of the stem to clog with chips of what was clearly an extremely compacted substrate. It had been decided instead to let the drill determine its own rate of penetration, and whenever it locked he was to hoist it up a little and twist it in the hope that this would clear the flutes sufficiently so that when the drill was released it would resume burrowing.

"Jim," Allen called. "How're you doing?"

Irwin was trying to coax his Hasselblad to advance film. "I'm about finished."

"Okay, Jimmy. Sounds good." Allen then asked him where he had put the as-yet-unused manual core tubes. When Irwin replied that they were on the Rover, Allen pointed out that this was just perfect. "I don't know how to break this news to you," he teased, "but you're going to do Station 8 out at the ALSEP site."

"Thanks, Joe!" Irwin laughed good naturedly.

"I've got the drill on," Scott interjected. He had fitted another section to the stem, and was ready to start.

"We've got lotsa power left," Allen assured.

"I'm putting very little force on it," Scott reported, "but it binds up."

"Any luck by trying to pull it back a bit?"

"No, it pulls me right on down with it!"

"Clear it by lifting, as you turn the power on," Allen urged. Fendell zoomed in the TV just as Scott yanked the still-rotating stem about 15 cm out of the hole and then twisted it in place. "Hold it right there. Let it run a while." The hope was that this action would strip the hole of the debris that was presumed to be impeding progress.

After a moment, Scott let the drill drop back into the hole, and it promptly drew itself in, then stopped just below its previous point. Keeping the drill turning, Scott repeatedly yanked it up to free it and then let it claw its way back, each time gaining a few centimetres.

"We've got power to burn," Allen repeated. "Jim," he continued, "we would like you to do a pan out at the ALSEP." Scott and Irwin had trained to work together on Station 8, with Irwin digging and Scott taking the documentation and helping with the bagging. As things stood, Irwin was effectively *waiting* for Scott to finish the heat-flow work, and Allen was keeping him busy.

"It's tightening up again, Joe," Scott warned. "I'm not putting any force on it at all; it pulls itself in and then binds up."

"We're going to ask for about two more minutes on the drill, then call it quits," Allen said. But seeing that Scott was making some progress,

This view of Falcon from the southeast shows its rear leg resting on the inner slope of a small crater. Note all the discarded packaging. The Rover is at the ALSEP site.

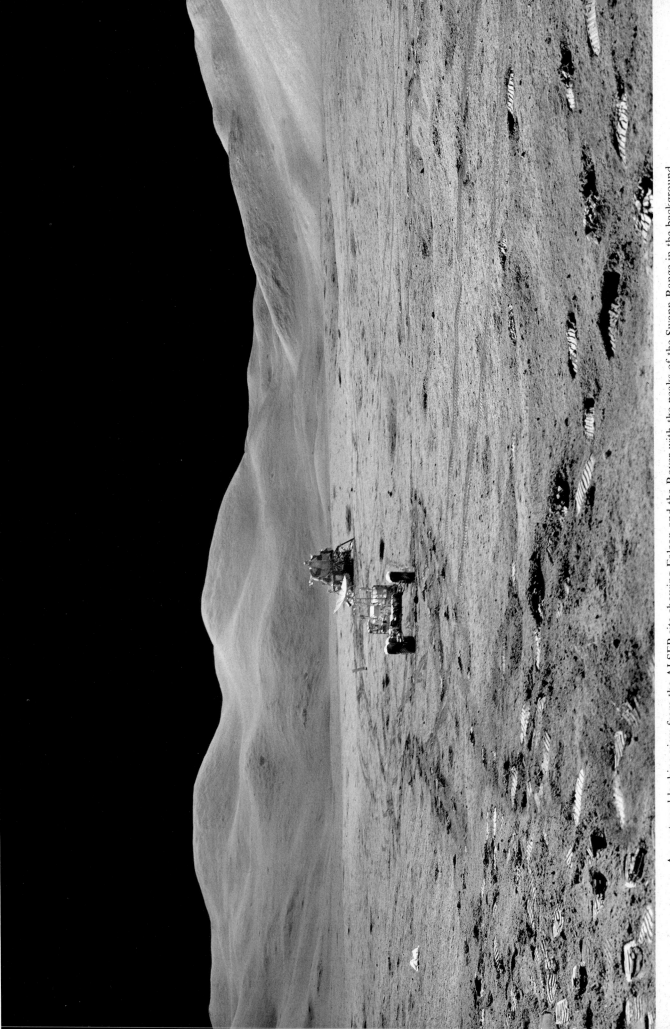

An eastward-looking view from the ALSEP site showing Falcon and the Rover with the peaks of the Swann Range in the background.

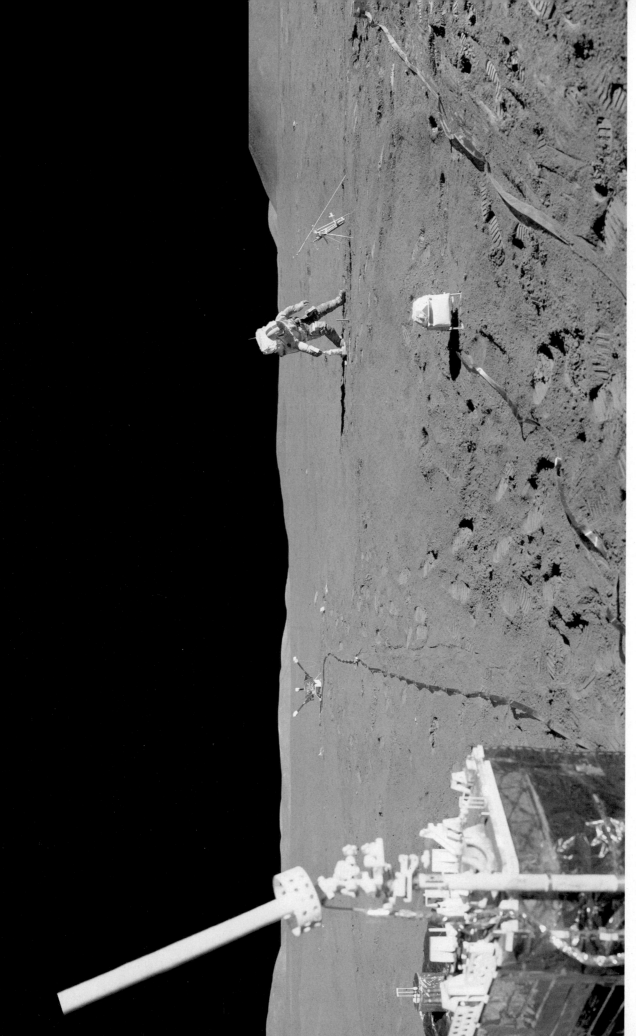

The Central Station of the Apollo Lunar Surface Experiment Package (foreground), the triaxial Lunar Surface Magnetometer and the boxy Solar Wind Spectrometer. The cable snaking right leads to the Heat-Flow Experiment. Dave Scott is bending to lift the drill. The rod inclined on the tool rack to his left is the 'rammer', used to emplace the thermal probes in the holes.

Allen deferred, with the result that it was five minutes beyond the 'two minute' mark before Allen told Scott: "One minute of drilling left."

"Okay, Joe!"

Post-flight analysis revealed a design flaw. At the joint between stem sections, the flutes were of reduced size, and it transpired that this interrupted the upward flow of debris, creating a compacted zone that impeded the drill's penetration.

When it was evident that Scott was making no further progress, Allen announced: "We're satisfied with this hole. Go ahead and emplace the heat-flow probe." As both holes were about 1.5 metres deep, the experiment ought to produce usable data.

"Okay!" said Scott enthusiastically. At last! He disengaged the drill and started to insert the string of sensors into the tube, but the rammer met an obstacle. "I think that one of the stems may have been bent," he speculated.

"Standby," Allen acknowledged.

While waiting for Allen to seek the advice of the experimenter, Scott extracted the probes and then he eased the rammer into the tube to investigate the problem. To his amazement, the rod did not penetrate more than about a metre below ground level.

"The ALSEP pictures are complete," Irwin reported. Allen told him to prepare for the Station 8 activities.

"It won't go in any further," Scott repeated. Maybe it was time for the application of measured force? The tube could not be too badly bent. The rod was so slim that it should curve around a slight arc. On the other hand, if the rod broke, it would block the tube. "I could try and push it in."

"Standby," repeated Allen.

When the Backroom gave its approval, Scott tried to drive the rod deeper into the tube. "No way."

"We'll take what we've got," Allen conceded.

It was later realised that lifting and rotating the drill in an effort to clear the flutes had caused the stem possessing the bit to disengage, and when it seemed that he had broken through a layer and was making progress, what had really happened was that the open upper stem had begun to *core* beside its companion, and when he inserted the rammer this met the material at the bottom of the tube.

For Scott, the experiment had become an exercise in frustration. The extra drilling had already consumed 25 minutes, and he had yet to start the third hole. What might they have achieved with another half-hour at Spur or Dune? Was the heat-flow data worth it?

"Jim," Allen called his other charge. "How're you doing?"

"Oh, I picked up a couple of rocks," Irwin replied. On spying a fragment of high-albedo rock[35] some 30 metres northeast of the Central Station, he had taken it without interrupting the discussion about the drill. This piece of non-mare basalt, the only one found at the landing site, was likely part of the ray of ejecta that was surmised to run north–south across the plain. He had also lifted an extremely dusty dark rock[36] that proved to be a glassy breccia.

"What's next on the agenda?" demanded Scott after he had emplaced the thermal sensors.

"Get Jim started," Allen replied, "then do the deep core." Fendell framed the TV to monitor their progress.

"Do we want the whole Station 8 activity?" Irwin asked, wondering whether to do a comprehensive sample as well as dig a trench.

"Comprehensive sample first, I reckon." Scott suggested.

"Start with the trench, Dave," Irwin countered.

"We'll depart this site in about 30 minutes," Allen called, "so you're looking real good on time – there's no rush." With the walkback constraint no longer a factor, Houston's attitude to time had changed. In effect, the landing site was in a different 'time zone' to a traverse route.

"Joe," Irwin called, "do you only want the trench 12 inches deep?"

"Whatever you think is reasonable."

"I'm down that far already!" He had made rapid progress, but the experience with the drill suggested that he would soon encounter consolidated material. Although he had excavated a variety of fragments, some of which were white while the others were black, there was no evidence of layering. But then he came upon a flat surface of consolidated material that was littered with fragments of black glass. "I think I hit bedrock!" As there was no way that he would be able to dig through this "hard pan", he began to extend the trench to the south in order to accommodate the penetrometer with which he was to measure the strength of the material at the base of the trench.

Although this experiment was intended to be a two-man task, Irwin volunteered to do it alone in order to enable Scott to take the core. At the drill site, Scott placed the treadle plate of the coring apparatus on the ground, put his boot on it to hold it still, and began to drill. With the tube holding the material rather than forcing it aside, he was able to sustain a rate of about a few centimetres per second through Irwin's hard pan –

Jim Irwin uses his scoop to dig a trench by repeatedly scraping loose material back between his legs.

which was not solid rock because Scott could feel the vibrations as the drill encountered layers of different densities.

"Coming up on 5 minutes," Allen warned.

By this point, Irwin had taken the requisite readings on the floor of the trench. For the final part of the test, he dug the penetrometer into the ground 10 cm from the edge and pushed it to measure the lateral strength of the regolith. It took more force than he expected, but it was satisfying to see the wall collapse. As time was running out, the rake sample was cancelled.

Once the core had penetrated 2.4 metres, Scott called a halt. The sample was to document the regolith within which the heat-flow probes were embedded, and it was already far below that level.

"Just leave the drill on the stem," Allen prompted. It was time to return to the LM.

Scott was aware that if the planners stole time from the final excursion to retrieve the core, they would take more than they really needed in order to build in a margin, and in so doing would jeopardise the traverse to the North Complex. "I'll just try it now," he insisted. He took up position with the drill between his legs, and pulled the stem up almost 20 cm on the first attempt. "I'll get it!" he promised. But no matter how hard he tugged, the stem remained fast. "I'm not sure I'll *ever* get it out!" he exclaimed in frustration. Yielding to the inevitable, he left the tube in the ground and went to the Rover. After picking up Irwin, he drove to Falcon, where they raised the flag, took the standard 'tourist' shots, and finally re-entered the LM for supper after what had been a long day.

STARTING THE FINAL DAY

When Scott woke up the next day, he discovered that not only had time been taken from the third traverse, they were to extract the core *prior to* setting off, which was terrible news since any delay would be magnified by the walkback constraint. The exploration of the North Complex was being put at risk to gain *a single sample*. It would be the deepest lunar core yet, but was it *really* worth it? This was a significant dilemma. Scott was in command, but he was a member of a team. In effect, he was working for the scientists. If the tube refused to budge, should he assert his authority and abandon it? At what point would the futility of the operation become evident? It was therefore with a keen sense of frustration that he and Irwin drove out to the drill site.

"What should I do?" Irwin asked.

"The object is to pull the drill out of the ground," Scott replied sarcastically.

"Well, one of us get on one side..." Irwin mused, his meaning obvious. Perhaps their combined efforts would break the tube free.

"Let's try it," Scott agreed.

The two men stood either side of the drill, and each slipped one arm under one of the handles in order to lift it in the crook of his elbow.

"You say when," Irwin prompted.

"When!!!"

"A little bit," Irwin estimated, rather optimistically.

"It's got a long way to go." They tried again, but to little effect, and Scott decided that they were wasting time. "I suspected as much." The stem was jammed. "Joe, it looks to me like the only answer is going to be to back it off." He wanted to run the drill to clear the flutes, but without it penetrating any deeper – and hopefully without the vibration causing the far end of the sample to fall out of the tube.

"Let's do that," Allen agreed. In fact, he had only a vague idea of what was going on because, although the TV was operating, Fendell could not raise the camera from its 'parked' position.

Scott switched on the drill, which immediately drew itself back into the ground and locked tight, and he laughed despite the futility of his predicament. "It was a good idea, but it didn't work."

"Let me get an elbow under it," said Irwin, reverting to the previous plan.

"I don't think it's worth doing, Jim," Scott said. "We're not going to get it out."

Irwin persisted, however. "I could put a lot of pressure on it, this way."

Scott relented. "Let me try, too."

"Here it comes!"

"Now we're making a little progress," Scott announced more happily.

"How far did it move up?" Allen enquired.

"We got it up about 3 feet. I think maybe we can do it piecemeal." By this, Scott meant that instead of pulling the stem up in one piece, perhaps they would be able to draw it out one section at a time, dismount the drill, disconnect the exposed half-metre section of the stem, and then remount the drill to regain the necessary leverage. But this would clearly take a long time. He made the point of asking Houston to confirm that it was worth the effort. "Are you guys *that* interested in this thing?" There was no reply. In fact, there was a debate raging in the Backroom over the relative merits of the deep core and the final traverse. "We sure have invested an awful lot in this thing," he

Dave Scott salutes the flag on the Plain at Hadley, with Falcon and Mount Hadley Delta as a backdrop.

Jim Irwin salutes the flag on the Plain at Hadley, with Falcon and Mount Hadley Delta as a backdrop. Note that the Rover's TV is watching him do so.

noted in an effort to prompt a response from the ground, but to no effect.

Irwin sensed his commander's anguish. "You don't think there's any chance of us pulling it all out that way, do you, Dave?"

"Let's try again," Scott relented.

Irwin slid his shoulder under the drill, and they made such progress that the drill was soon too high for him, and so he grasped hold of the stem and tugged on that instead.

"Let's take a break," Scott sighed. He didn't say anything, but he had strained a ligament in his shoulder.

Looking around, Irwin noticed that the TV camera was aimed down. Surely with all of this effort being put into the drill, Houston couldn't be fascinated by the dirt under the Rover? "Joe, are you having trouble with the TV?"

Allen's response was immediate: "You better believe it!" Its drive mechanism had overheated in the Sun, and the clutch was slipping.

Scott went to the Rover and elevated the camera by hand.

When Scott rejoined Irwin, they gave the drill another tug and it began to move, and several seconds later it emerged from the hole. Scott swung it level to ensure that the material at the end did not trickle out. The next task was to detach the treadle, which had become jammed on the stem. Irwin held the drill while Scott twisted the treadle, but this didn't work. Scott went to use the vise mounted on the back of the Rover. The vise looked different to the one that he had trained with, but he inserted the stem and tried to twist the treadle free – and the jaws released. He examined the vise. "I hate to say this, but I swear it's on *backwards*!" The problem would later be traced to a diagram in the assembly manual that had been reversed. The jaws were to resist torque in one direction, and to collapse in the other direction for easy release. Set in reverse, the vise opened as soon as he placed *any* force on it! However, he could not simply flip the stem end for end, since there was no room for him to work on the opposite side of the Rover. Finally, he vented his anger. "How many *hours* do you want to spend on this drill, Joe?" And again, incredibly, there was no reply from Houston.

Irwin joined him at the Rover and offered to hold the other end of the stem, beyond the vise. This provided the extra grip required, and the joint loosened.

Meanwhile, it had not escaped Scott's notice that his repeated pleas had gone unanswered. "Joe," he began, "I haven't heard you say yet that you really want this *that* bad."

"It's hard for *me* to say, Dave." Allen didn't take the decisions: although mission scientist, in this situation he was merely the interface to the Backroom which directed operations. "How many sections to come?"

"Oh, standby, Joe," Scott snapped, his impatience showing once again. Couldn't they keep track? It was a six-section stem and he'd just removed the second section. He pointedly counted aloud the remaining sections: "One! Two! Three! Four!"

Allen backed off: "Thank you."

Scott continued to struggle with the stem.

"Dave and Jim," Allen began, this time rather gingerly, "this is Houston."

"Go ahead," Scott invited.

"What's you're best guess?" Allen asked.

Scott held the remaining stem up to the TV. "Joe, you can see what we've got. I just can't get them apart."

A few seconds later Allen was back, and his tone was noticeably different. "Put the stem on the ground, if you would, please. We'll pick it up on the way back."

Scott was not about to argue. "Good enough."

"We might be able to return it, just like that," Irwin said, wondering whether they might be able to return the stem to Earth without breaking it down into its individual sections.

The fact that it had taken so long to extract the core was preying on Scott's mind: "There's more time invested in it than anything we've done." They had been at the drill site for half an hour. And they had been on their PLSS resources for 45 minutes prior to that. By this point on the original plan they would have been at the rille. He sensed the wonders the North Complex slipping away. And it was not even as if they were now free to set off, they were to make a movie of the Rover being driven. "Get your camera," he told Irwin, but then he recalled that the 16-mm had been acting up. There was no point in him setting up for the test drive if the camera wasn't working. "Why don't you check it out first." Then he called Allen: "Joe, you never did tell me that the drill was that important. Just tell me that it's *that important*."

"It's that important, Dave," Allen came back immediately.

"Okay, good," Scott replied sardonically. "Because now I don't feel like I wasted all that time."

"Seriously, Dave," Allen continued, "that's undoubtedly the deepest sample we will have out of the Moon."

"Well, that sounds good," Scott conceded.

Having finally extracted the "deep core" from the ground, Dave Scott mounted it in the vise on the rear deck of the Rover in order to strip it down. Notice that the TV is watching Jim Irwin take the picture.

In fact, when the core was studied, it was found to have at least 42 distinct layers of material. This stratification greatly simplified the task of dating the levels of the sample. The uppermost 45 cm was heavily reworked by the gardening process, but beneath this the material was unsorted, and it turned out that the top of this section was 400 million years old. This was astounding, as it meant that activity on the plain in this period was confined to the upper half-metre of the regolith. On the Earth, this same interval corresponds to 80 per cent of the time since the start of the Precambrian Period during which continents split apart, drifted thousands of kilometres and formed mountain ranges upon colliding. Because the deeper material in the core was unsorted, the chemical variation represented true heterogen-eities, which made it a truly valuable sample. Doubtless Scott was silently thankful that he had not been expected to drill all the way to the bedrock!

Scott halted the Rover at the position that he had selected to start the Grand Prix.

"I'm ready for you," Irwin confirmed.

Scott was to perform a specific sequence of manoeuvres. He set off in a straight line at a steady 6 kph. "Tell me when to turn."

"Turn now."

"Okay, here comes the acceleration to 12 kph."

It was a serious effort to record engineering data, but it was clearly fun. "Ride 'em Bronco," Irwin laughed.

"Yeah, man," Scott agreed. While extracting the core sample he had been rather subdued, but now his mood was improving by the minute.

"You kicked up a very nice rooster tail," Irwin warned. Hopefully, it would not interfere with the ALSEP, whose instruments would be susceptible to overheating if this dust were to settle on their thermal shields.

"I'll come back across and give you a hard stop."

"How's that camera?" Allen interjected.

"It *feels* like it's working, Joe," Irwin replied. Once the test was over, he switched off the camera and examined the counter, which still indicated a full magazine. "Ah, shoot!"

"It didn't work!?" Scott asked in astonishment.

"It's not working."

"You're kidding!" Scott briefly considered switching the magazine and repeating the test, but rejected this as a waste of time.

"Let's press on and take a look at that rille," Allen announced.

"Joe, that's the best idea you've had all morning," Scott said.

As he dismounted the Rover in order to fasten Irwin's seatbelt, Scott reflected on the fact that the Grand Prix, which had been planned for the end of the day, had been brought forward – this was ominous, as it meant that engineering data had been given a higher priority than the geology of the North Complex.

RACING WEST

Scott drove west across the plain at full speed heading for Station 9, which was to be at Scarp, a small crater 100 metres back from the rille. After driving north along the rille for a kilometre or so, they were to turn right and go to the North Complex where, according to a study by Gerry Schaber, the fact that the older craters were flooded meant that the hills were once immersed in lava. The larger craters (such as Pluton) were excavated later. Mike Carr and Keith Howard were of the opinion that the dark mantling in this area was a pyroclastic deposit, which raised the tantalising prospect of sampling 'recent' volcanism.

"You should be passing Quark Triplet," Allen prompted, several minutes into the drive. These fresh craters had excavated a lot of light-toned litter. If there was time, they were to stop and sample it on the way back.

"I see them," Irwin confirmed. Having learned from the previous day's fruitless attempt to relocate the 'bedrock crater', he took note of the 'nav' coordinates.

Driving into the zero-phase, it was difficult to discern the nature of the route ahead, so Scott concentrated on dodging the obstacles that loomed almost without warning out of the glare.

"There's a very large depression ahead," Irwin reported a minute later. "We don't want to drive through that."

"Let's take a look," Scott decided.

"It's a very *large* depression," Irwin emphasised. As Scott drove around it, Irwin examined the interior. The slope was only 3 degrees, but it was 60 metres deep, with a fresh crater at its centre which, despite being confined, was surprising free of litter. As it was not listed on the map, he chose to name the depression 'Wolverine' after the University of Michigan, which all three members of the crew had attended at some time or other.

As they left this crater behind, Scott exclaimed, "Look at that rock." There was an angular fragment about 30 cm across sitting exposed on the surface just down-slope of a small crater. One part of the rock was coated in glass, as was the central part of the crater. It was clear that the crater had been made by the impact of the rock, with the geometry indicating the azimuth of the source. Since they had flown to the Moon to *geologise*, not just to dash about, Scott drew to a halt, released his seatbelt, turned to the side, snapped a picture of the rock and then strapped back in – barely a minute in all. One of the post-flight recommendations was that later crews should make better use of their traverses. Specifically, the LMP should take pictures every 100 metres or so in order to document the route. Furthermore, he should be provided with a pair of long-handled tongs to enable him to sample targets of opportunity without having to disembark.

"Another depression ahead," Irwin warned a minute later.

This time, Scott decided to drive straight through it. "It's pretty rough out here," he noted.

"It's like big sand dunes," Irwin mused. He suspected that the depressions in this area were actually extremely eroded rimless craters. "I thought we'd whip right over to the rille; I didn't think we'd have to contend with this type of terrain!"

"It was a lot easier driving yesterday," Scott agreed. But the rille was now visible cutting across the plain ahead, so they did not have much further to go.

A few minutes later, Irwin reported that there was a tight cluster of fresh-looking rocks directly ahead.

"You must be very near Scarp," Allen offered. He had been timing their traverse and trying to correlate the features that they mentioned with his map.

In fact, this 15-metre crater was not Scarp; that was 75 metres to the northwest. It didn't matter, however, because the objective at Station 9 was to assess whether the regolith thinned towards the rille, and they opted to sample this crater as a substitute for Scarp. Scott drove around the northern rim and parked due west, which was a shame because it meant that the seats blocked the TV's view of the crater.

With Irwin's Hasselblad out of action, Scott was obliged to take the pan. As he scrambled up onto the slightly raised northern rim of the crater, his boots sank into the loose material about 10 cm; it was like walking across a freshly ploughed field. "The centre of the crater is just *full* of very angular debris," he said in amazement. When he gripped a sample[37] using his tongs, it split apart.

He put the fragments into a bag and kneaded them between his fingers. "This stuff is really soft." In fact, it was probably the *freshest* patch of lunar surface ever trod by an astronaut. With the rim cursorily sampled, he moved one crater's diameter west to sample the outer fringe of the ejecta, and selected a dark fragment[38] that sparkled. "It's covered with dirt, but it looks just like a big piece of glass." Further out, he spotted a rock that was ribbed by slickensides implying that it had been subjected to abrasion, so he took a photograph for the record.

Having pronounced the crater to be "unique", Irwin remained on the rim to study its interior. The fact that it had created primarily regolith breccias indicated that the regolith had to be at least 3 metres deep. But there was a distinct bench running right around the centre of the pit. A bench was supposed to indicate bedrock outcropping. But if the crater had not reached the bedrock, what did the bench signify? It marked a clear boundary, with the material in the centre being considerably darker. It struck Irwin that this crater was a large counterpart of the small fresh glass-lined pits that they seen from time to time, only in this case the glass had coated the debris in the centre instead of forming a pool. As it turned out, this observation was the key to the bench. It marked the zone of shock-melting where the glass welded the compressed regolith.

ON THE RIM

"I get the feeling we're coming up to the real ridge line," Scott ventured after 2 minutes of further driving. The shallow ridge that they had found on the first excursion to the rille was evidently a general feature.

"There's some good blocks down there!" Irwin exclaimed as they crested the rise and he gained his first view of the slope down to the rille's rim.

"That's a *big* outcrop," Scott agreed.

"There are places all along here to sample large blocks," Irwin said. "It looks as if many are almost *in situ*."

This part of the rille's rim had been selected because the Lunar Orbiter imagery had suggested there might be an outcrop of bedrock. From the dense coverage of blocks, it looked as if this observation was correct. Furthermore, at the rim there was almost no regolith. Although the bedrock had been shattered by impacts, the blocks had not been tossed around. As they ran down towards where the rim fell away, they realised that

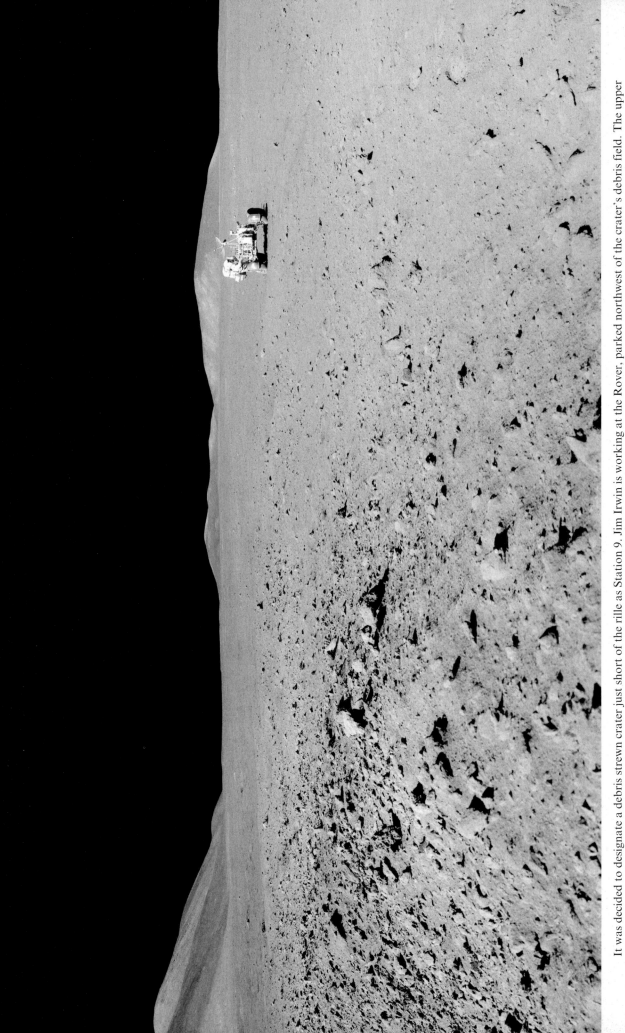

It was decided to designate a debris strewn crater just short of the rille as Station 9. Jim Irwin is working at the Rover, parked northwest of the crater's debris field. The upper part of the rille's opposite wall is visible in the middle distance, particularly to the right, and Bennett Hill lies far beyond.

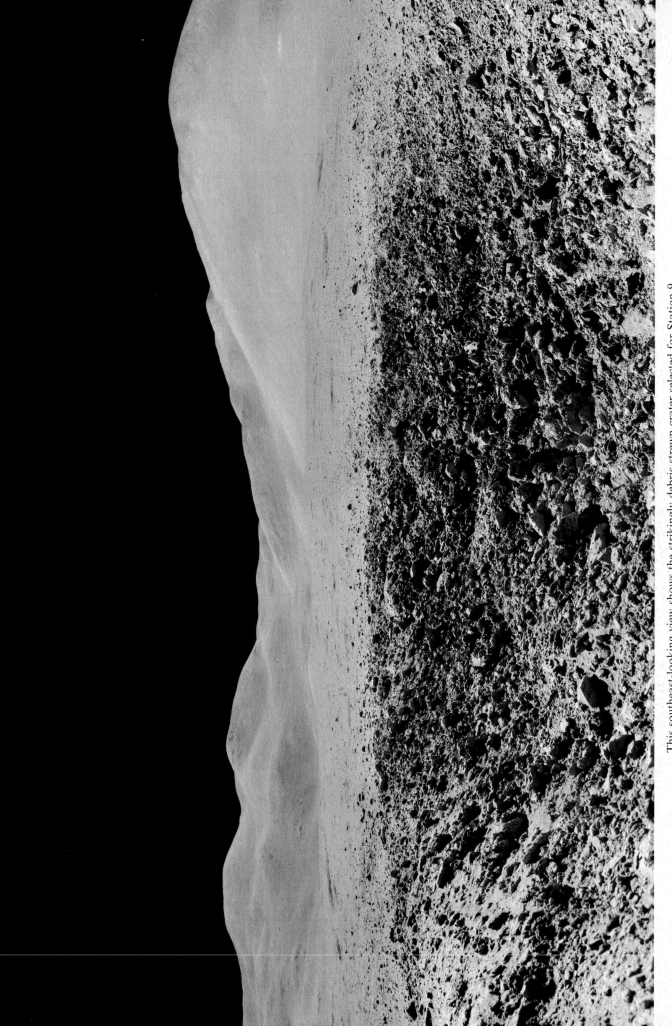

This southeast-looking view shows the strikingly debris strewn crater selected for Station 9.

As he made his way through the dense litter at Station 9, Dave Scott found a large angular fragment displaying a striated pattern.

the feature they had named the Terrace was a part of the rim that had slumped, and was flanked to either side by a shoulder.

After parking well back from the rim, Scott gave his chest-mounted Hasselblad to Irwin to enable him to take the pan prior to making a start on the sampling, while Scott took the telephoto camera and moved down-slope to document the far wall of the rille.

One of the reasons for landing at Hadley-Apennine was to exploit the rille to seek evidence of vertical structure that might indicate how the mare had been built up. Was Palus Putredinus a single massively thick lava flow? Or was it a succession of flows? And if it was layered, was there regolith sandwiched in between the layers that would give an indication of the timescale? This was likely to be Apollo's *only* opportunity to view such an exposure.

As Scott took his pictures he gave a commentary on what he could see on the wall opposite. Outcrops were evident at various points in the uppermost 60 metres, and it appeared that there were several layers. The most prominent was a light-grey layer about 20 metres thick situated just below the surface. It rested on a similar thickness of darker material, below which, in places, there was a 10-metre-thick outcrop that was flat-faced and displayed horizontal and vertical fracturing. The litter of rock on the slope was distributed in random patches. The near rim prevented him from seeing the floor directly beneath his position. The TV barely hinted at this detail, and when he saw his film upon returning to Earth Scott reported that even the telephoto lens did not capture fine detail that was apparent to the naked eye. As he stood on the rim, he realised that he faced a task as daunting as attempting to deduce the history of the Colorado Plateau by peering into the Grand Canyon for 10 minutes.

Meanwhile, having finished the pan, Irwin was undertaking a radial to sample the variation perpendicular to the rille. He went up-slope 60 metres and picked a highly vesicular fine-grained basalt[39] that appeared not to have been exposed for very long. On his way back, he spotted a rectangular half-metre rock with a distinctly layered structure; it was too large to sample, so he snapped a picture for the record. "Dave," he began, in the tone of voice that indicated he was about to make one of his eagle-eyed observations, "did you see the horizontal bedding south of us?"

"On the other side?"

"No, *this* side."

When Scott turned to look, Irwin pointed south. "I didn't even *look* on this side, to tell you the truth, Jim," Scott admitted. As he turned, he tripped over a rock that he had not realised was by his feet and, in full view of the TV, fell onto his hands and knees. "It's a pretty durable little fella," Scott chuckled as he wiped the dust off his camera. At Allen's request, he took a few telephoto shots of the outcrop on the near side that Irwin had pointed out, and then returned to the Rover.

Irwin proposed that they sample the blocks on the rim, and Scott agreed. Fendell caught them bounding down-slope.

"It's good firm ground here," Scott observed.

The largest crater in their vicinity was only 3 metres in diameter, but its rim was littered with metre-sized blocks whose planar surfaces implied that the bedrock had split along fractures. Scott selected an extremely angular vesicular rock and placed the gnomon down-Sun of it so that Irwin could undertake the documentation.

"This will be our last documented sample," Allen announced. They had been on-site for all of 25 minutes, and it was time to move on.

Scott dropped to one knee to deliver a series of heavy hammer blows to the rock. Although it was well indurated, he persisted and eventually gained a chip.[40] "It's a fine-grained crystalline rock. I've *got* to say that, because it has got millimetre-sized laths randomly oriented," he reported. In addition to small olivine phenocrysts, this slightly vuggy porphyritic basalt contained a variety of mafic silicates. As the rocks in this crater were indigenous, this sample was unambiguously from the uppermost lava flow.

"Are we going to have time to sample bedrock?" Irwin asked hopefully, meaning a rock that was still in the ground.

"Apparently not," Scott mused.

"Standby on that," Allen cut in. While they waited, Irwin bagged a small fragment of rock[41] from nearby for good measure. Allen came back once the Backroom had debated the priorities. "If you think you can reach *true* bedrock, then we are willing to give up the mare sampling." He was referring to the sampling of Quark Triplet on the way home.

This was all the leeway that Scott needed. There was a jumble of boulders directly down-slope. "I think that's true bedrock."

There was concern in Houston over the accessibility of the rim. "Out of sheer curiosity," Allen began lightheartedly, but fooling no one, "how far back from what you'd call the edge of the rille are the two of you standing now?" Scott answered elliptically. "How far back from the *lip*

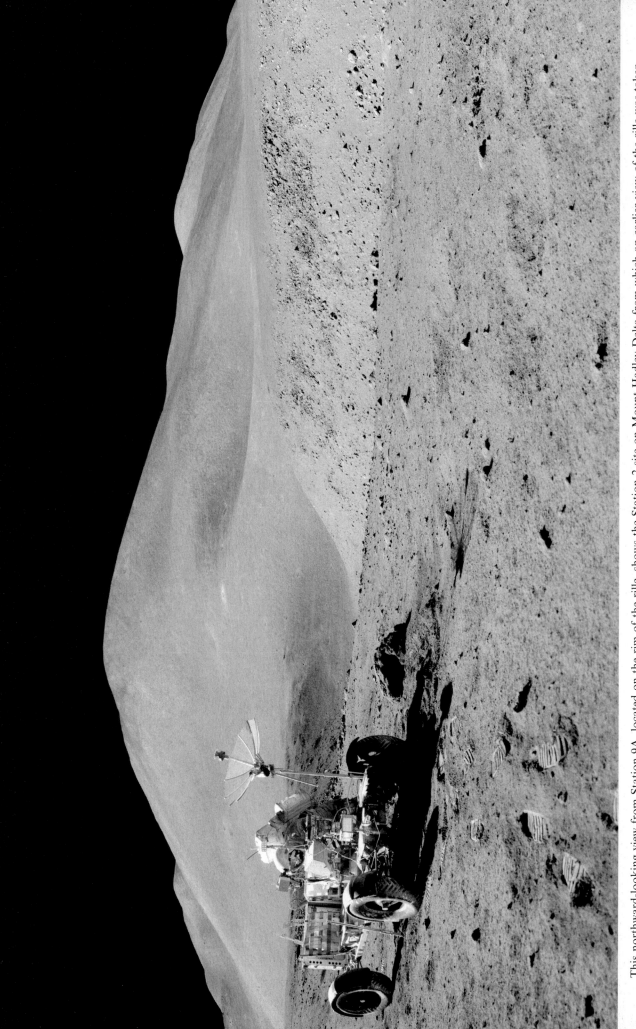

This northward-looking view from Station 9A, located on the rim of the rille, shows the Station 2 site on Mount Hadley Delta from which an earlier view of the rille was taken.

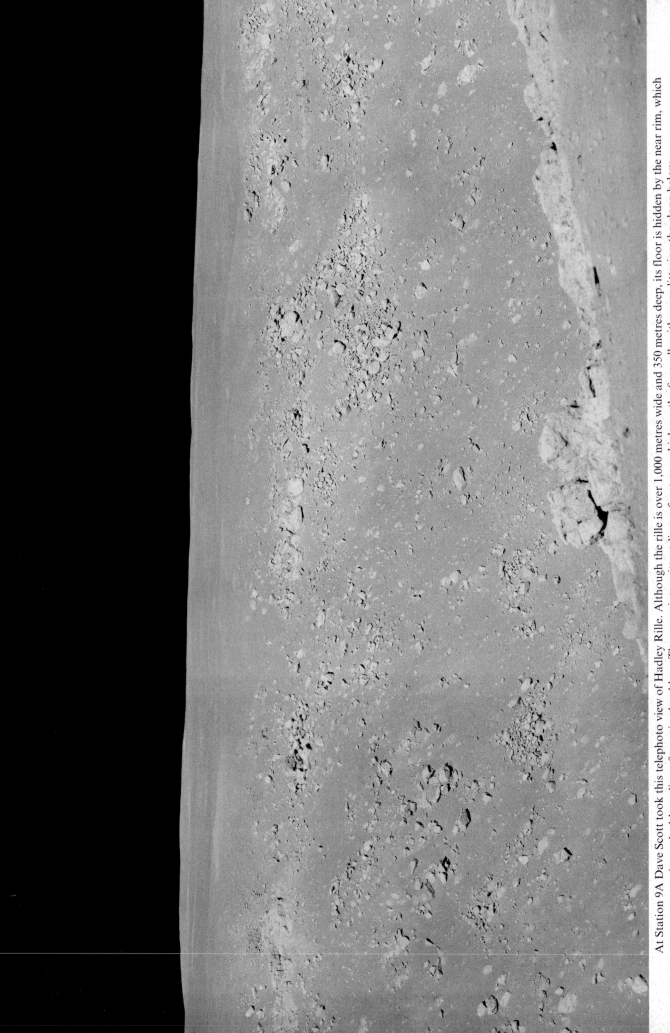

At Station 9A Dave Scott took this telephoto view of Hadley Rille. Although the rille is over 1,000 metres wide and 350 metres deep, its floor is hidden by the near rim, which is marked by a line of massive boulders. There are intermittent lines of outcrops high on the far wall, with scree littering the slope below.

Because of the way that the rille curved, from Station 9A it was possible to view a section of the eastern wall, where much of the rock was in outcrop.

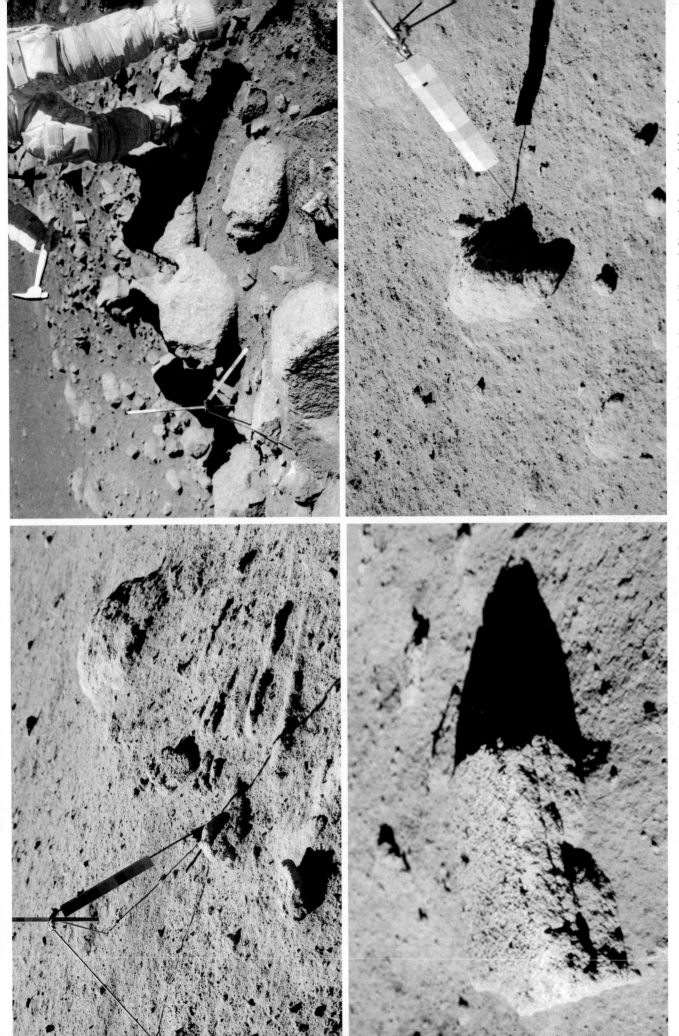

Station 9A: an almost completely embedded 'ribbed' rock (upper left); Dave Scott hammers a block (upper right); a tabular rock (lower left); and the rock which was later dubbed the 'Great Scott' (lower right).

Dave Scott samples a boulder at Station 9A. Large rocks situated so close to the lip of the rille (just 8 metres away in the background) were likely to be 'true bedrock'.

of the rille do you think you are?" Allen persisted.

"I can't *see* the lip," Scott pointed out. They were on the incline that ran down off the ridge towards the rim, where the slope increased sharply to 30 degrees to plunge down into the canyon.

From Allen's point of view, the rille was heavily foreshortened. "It looks to me like you're standing on the edge of a precipice."

"Gosh, no," Scott chuckled dismissively, and set off down-slope before the Allen could change his mind.

"Oh, you're getting it *here*?" Irwin said in surprise.

"*This* sure looks like bedrock," Scott replied, continuing towards the boulders that he had selected.

"Look to the north," Irwin prompted. There was a massive slab of rock on the rim, several hundred metres away.

Scott considered it, but decided they did not have time to venture that far. "Come on down here, and let's get a frag off one of these."

Upon reaching his target, just 8 metres from the lip, Scott selected a 1.5-metre flat rock that was resting on the surface. In making his assessment, he referred to the boulders as "fragments" because they were dwarfed by their context. It took a lot of hammering, but he gained several chips.[42,43] They were moderately vuggy porphyritic basalt with pyroxene phenocrysts. As a supplement, he bagged a few smaller rocks: one[44] was similar to the boulder, but was more finely grained and less vesicular; the other[45] was moderately vuggy and laced with olivine.

"It's just a mass of big boulders along the Terrace," Scott pointed out, describing the wider setting. He then returned to Allen's concern for their safety: "I guess we're just about at the lip."

"Amazing," Allen allowed. "When you finish, we'd like you to move back to the Rover."

"Yeah," Scott acknowledged automatically, but as usual he lingered to examine the nearby rocks.

"I was wondering," Irwin began, "if that outcrop to the north is the same." They were to drive north to take more pictures to facilitate a stereoscopic analysis of the rille. "Maybe we could stop there for the stereo pan."

"Okay," Scott agreed.

"We want a rake sample, a soil sample and a double core at the Rover," Allen reminded, as they made their way back.

While Irwin methodically raked, Scott scraped with his boots to see whether there was a subsur-

Apollo 15 mission scientist Joe Allen and back-up commander Dick Gordon in the Mission Operations Control Room, as Dave Scott and Jim Irwin take a soil sample on the rim of Hadley Rille.

face layer; he found it a few inches down. "Maybe we ought to make a trench," he mused. In Houston, a photographer snapped Joe Allen and Dick Gordon in discussion with Deke Slayton while, on the wall screen, Scott and Irwin worked on the soil sample.

The next task was the core, but Scott had his doubts about using a double-length tube. "We might find ourselves driving into bedrock if we're not careful."

"If we hit bottom, we hit bottom," Allen replied philosophically.

Irwin managed to push the tube in by hand about 30 cm before it stopped dead. "It feels as if it's hung up on a rock," he mused. He hit it with the hammer and it cleared the obstacle, so he continued to hammer it and penetrated a few centimetres with each stroke.

"I think you've got a good one," Scott complimented as the core came out easily. Irwin inverted the tube in order to inspect the penetrator and was amazed to find that there was a rock jammed in it.

While Irwin split the tube, Scott went to look at "a good vesicular rock" that had caught his attention. Irwin pointed out that he had already sampled such a rock, but Scott had been smitten. Although it was football-sized, he decided that it was simply too good to miss. He knelt to grab the 20-cm rock[46] and then rolled it up his leg until he could get a firm grip, then he dumped it on the Rover without an inspection. Later dubbed the 'Great Scott', this 9.6-kg coarsely-grained basalt was the largest rock of the mission. It was dated at

3.28 ($\pm$0.006) billion years, implying that it derived from one of the last flows to have come down the rille, and as such was an excellent find.

Although Scott and Irwin set off to do Station 10 at the outcrop that Irwin had noted, Allen had no intention of sanctioning any more sampling. "All we need is photography of the rille," he insisted. Accepting this without comment, Scott veered right to run along the transition zone where there were fewer large rocks and, upon encountering a fresh-looking 60-metre crater a few minutes later, he decided to stop. It was Rim Crater, but he believed it to be South Twin. Station 10 was originally to have been at North Twin, 100 metres further on, but the plan was history. He parked just beyond the crater, near an angular boulder that was sitting on the surface.

"All we need is photography," Allen repeated. Then he gave the bad news: "And we're looking to arrive back at the LM in about 45 minutes."

"Shoot!!" muttered Irwin. The drilling had taken its toll, and they had just lost the North Complex.

Concealing his disappointment, Scott remained silent. He retrieved the telephoto camera and went to take his pictures of the rille.

Irwin took his pan from the northern rim of the crater, which was about 10 metres deep. There were some very vesicular rocks, but most were non-vesicular. The debris was uniformly distributed but it was not as dense as at the Station 9 crater, and there was no bench. Once he had finished the pan, he went to take a look at the nearby boulder. At first sight, it seemed to be a composite of a dark extremely vesicular basalt and a light-toned rock, but when he looked for the contact he found a gradual transition.

"Leave it there, Jim," Allen joked when Fendell swung the TV around and showed Irwin at the rock. "The Rover's not stressed for it."

"I'd love to bring it back," Irwin laughed. There wasn't time to sample it, so he took a few close-up pictures and then went to snap some of the smaller rocks that were scattered around.

"Much as we hate to," Allen called, a minute later, "we're going to *have* to get you heading back east." He was well aware of the disappointment that Scott and Irwin felt at being denied even a flying visit to their final objective.

"We'd better leave something for the next guy," Scott said sardonically. "Joe, are you planning a mare stop on the way back?"

"We'd like you to go straight to the drill site," Allen replied.

Accepting that Houston wanted him to attempt to break down the stem, Scott said nothing.

Driving east, Scott and Irwin enjoyed the magnificence of the embayed mountain range.

"What a big mountain that Hadley is!" Scott observed.

"It's beautiful," Irwin agreed.

"I'd sure like to do some more geology up here someday," Scott said wistfully.

Although disappointed not to have been able to visit the North Complex, Scott and Irwin had the satisfaction of knowing that they had confirmed the potential of the 'J'-mission format.

As they drove back across the 'dunes' they encountered their tracks, and followed them east. They did not catch sight of Falcon until they rolled over the final crest, after which it was a 500-metre cruise across the plain. Scott drew up alongside "the trusty drill", hopped off to retrieve it, and then drove to the LM to unload the Rover.

After another frustrating session with the vise, Scott gripped the tube in his hands, one either side of a joint, gave a firm twist, and, to his amazement, released the joint. This left him with three single segments and a 1.5-metre triple that was deemed to be manageable, so he called it a day. It was later determined that the joints had been over-tightened by being forced into a denser regolith than expected.

FINAL ACT

Once the rock boxes had been packed, Scott rounded out the surface activity with a personal item. Having been cued in by Allen, Fendell already had the TV framing the mission commander.

"In my left hand," Scott began, "I have a feather; in my right, a hammer. And I guess one of the reasons we got here today was because of a gentleman named Galileo, a long time ago, who made a rather significant discovery about falling objects in gravity fields. And we thought: where would be a better place to confirm his findings than on the Moon?" Fendell zoomed in a little, but was careful not to crop off the ground at Scott's feet. "And so we thought we'd try it here for you. The feather happens to be, appropriately, a falcon feather for our Falcon. And I'll drop the two of them here and hopefully they'll hit the ground at the same time." He did, and they did.

"How about that!" Allen called delightedly over the applause in the control room.

"Mr Galileo was correct," Scott concluded. His historic demonstration over, he parked the Rover up-Sun of Falcon at a spot named the 'VIP Site' because it would give the TV a clear view of the lift-off. Before returning to the LM, he placed on

In taking his Station 10 panorama, Jim Irwin caught Dave Scott heading his way with the telephoto Hasselblad. The feature on the horizon is Hill 305. While Scott rephotographed the rille, which is to the left, Irwin inspected the boulders near the Rover.

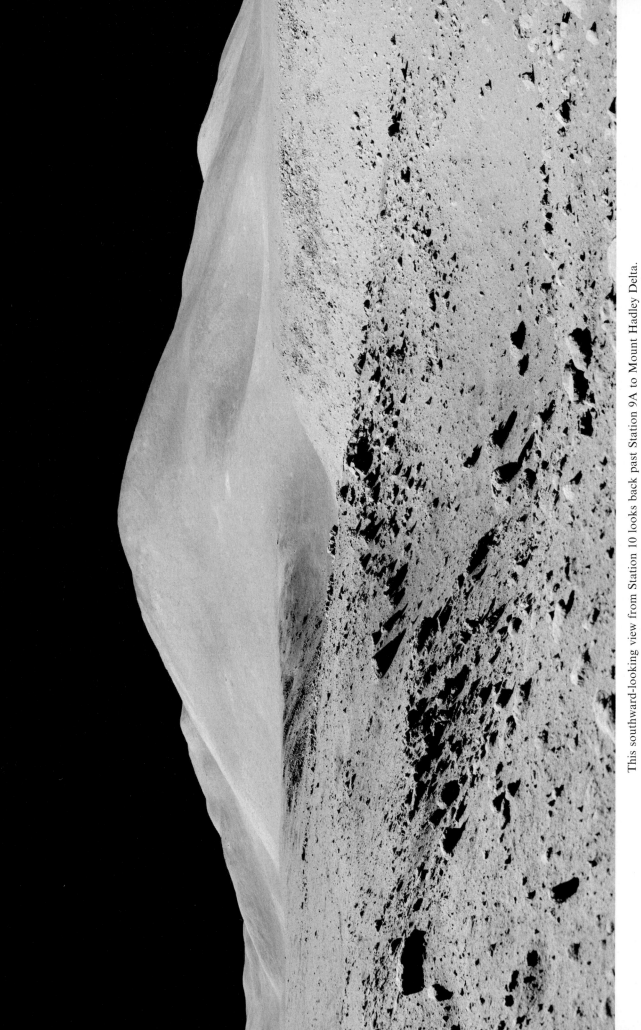

This southward-looking view from Station 10 looks back past Station 9A to Mount Hadley Delta.

the ground beside the Rover a statue and a plaque commemorating 'fallen astronauts'.

"I've noticed a smile on the face of the professor," Allen called, as Scott prepared to follow Irwin up the ladder. Lee Silver was their mentor for field geology. "I think you may have passed your final exam."

"Okay!" Scott laughed. His exploration of Hadley-Apennine at an end, he took a final look around, then left the surface. A few hours later, Falcon severed its links to its descent stage, lifted off and, astonishingly rapidly, disappeared into the black sky.

NO LONGER UNKNOWN

As the first of the 'J'-class missions, Apollo 15 marked a tremendous 'step function' in the capability of the program.

The Hadley-Apennine site could not have been visited earlier, because it required the rules of site selection to be relaxed. Up to this point, to be certified, both the site and the run in from the east had to have been surveyed at high resolution. One of the Lunar Orbiters had snapped Hadley-Apennine as a 'scientific' site, but the resolution was a barely acceptable 20 metres. Being so far north, it had not been overflown by any earlier Apollo. A greater obstacle, however, was the fact that the run in over the Apennines had only patchy coverage. In order to land in the valley confined by two major massifs and a rille, the LM had to make a plunging descent at an angle almost twice as steep as on previous missions as soon as it had cleared the backbone of the mountain range, with no room for a departure from the nominal track. Yet it was the topography that made this such an attractive site to the scientists. The availability of a rille on a mare embayment of a valley in the mountains peripheral to the Imbrium Basin made Apollo 15 the first 'multiple-objective' mission, and the Rover made it feasible.

Scott and Irwin spent 18.5 of their 67 hours on the Moon outside. In addition to setting up the ALSEP, they roamed up to 6 km from the LM and covered a total distance of 28 km. Their geological objectives were, in order of priority, to sample:

1) the Apennine Front;
2) the rim of the rille;
3) the craters of the South Cluster that had excavated the mare plain; and
4) the possibly volcanic nature of the North Complex.

A selection of frames from the Rover's TV coverage of Falcon lifting off, leaving the descent stage behind.

As the surface expedition drew to a close, the Rover was parked east of the landing site in order to provide TV coverage of Falcon's lift-off.

Three views of the Apollo 15 mothership Endeavour during the final phase of the rendezvous. Note the Scientific Instrument Module (right).

The Apennine Front

On Earth the process of orogeny (mountain building) is the result of plate tectonics, and it takes millions of years to produce a mountain range. The lunar surface, in contrast, was shaped by bombardment, and the ranges of mountains that ring the basins were pushed up instantaneously. The Apennines form the southeastern rim of the Imbrium Basin, and are among the highest lunar peaks. The steep basin-facing slope of this range is the Apennine Front. To appreciate the significance of this mountain range, it is necessary to consider its position between the Imbrium and Serenitatis basins.

As a small planet whose surface has been developed by geological processes, the Moon can be studied by stratigraphic analysis, which is based on the Principle of Superposition. This has revealed a complex depositional sequence. In particular, the Imbrium Basin was made shortly after the Serenitatis Basin, and the Imbrium ejecta sculpted features in the Serenitatis Basin. Much of this sculpture was masked by the lavas that later formed Mare Serenitatis. The stratigraphy showed that prior to the formation of the Imbrium Basin, the terrain now occupied by the Apennines was a layer of Serenitatis ejecta resting on the pre-Serenitatis crust. The impact that made the Imbrium Basin opened radial and concentric crustal fractures, and the blocks that were beneath the Serenitatis ejecta were displaced, tilted and thrust up as massifs. It is likely that much of the overburden remained on the summits, but some of it would have slid down-slope into the valleys between the massifs. In crater formation, the most deeply excavated rock is deposited on the rim. In basin formation, there is also a 'base surge' of semi-molten material that is excavated from far below the surface, and this would have blanketed the Apennines. Nevertheless, sampling a large crater on the flank of an Apennine massif might produce an excavated piece of the ancient crust, which was expected to be anorthosite. It was this prospect that made the Apennine Front Apollo 15's primary geological objective.

The first target was the rim of St George, the 2-km crater on the northern flank of Mount Hadley Delta, but its debris was already broken down and mixed into the regolith by the ongoing process of gardening. The one boulder that they found was clearly an interloper. It was the sparse coverage of the eastern flank of the mountain that prompted Scott to halt the second traverse short of its nominal turning point in order to focus on Spur Crater, and this proved to be a wise decision because it was at Spur that they found a piece of ancient anorthosite.

The fact that most of the samples from Mount Hadley Delta were breccias was in no way a disappointment, because structural composites have their own story to tell. Most had clasts consisting of a coarsely-grained feldspathic rock. Several fragments may be Serenitatis ejecta, but the association is weak. In the case of Apollo 14 most of the rocks were 'two-rock' breccias made of clasts that were themselves fragments of older breccias. The breccias from The Front were simpler, indicating that whereas the Fra Mauro Formation was shallow crustal material that had already been heavily processed by impacts, the homogeneous 'base surge' material which blanketed The Front originated from much deeper.

It was near Spur that the 'green boulder' was found. This was a breccia that had acquired a coating of greenish material along one side. This coating consisted of tiny spherules of glass. Although regolith that is shock-melted by an impact forms glass, this either pools and solidifies in the floor of the crater, or is splashed out and coats the ejecta. But in this case the microscopic green droplets came from a fire fountain. The faults in the floor of a newly formed basin would provide routes for deep lava reservoirs to reach the surface. A fire fountain could be expected as a precursor to an extrusion because, as the magma in the reservoir made its way up through the crust, it would liberate the volatiles in solution and these would blast out under pressure from fissures, spewing a mist of magma that would shock-cool as it fell back to coat the surface. On the airless, low-gravity Moon the plume from a fire fountain would likely have risen several hundred kilometres and coated a wide area. The pyroclastic glass was green as a result of the magnesium-rich silicates it contained. Because the material in the fire fountain would have ascended through the crust so rapidly that the lava would not have had time to evolve chemically, it was a welcome sample of the olivine and pyroxene reservoirs of the mantle – indeed, it was the most 'pristine' igneous sample of such material so far recovered. Although the pyroclastic was 3.3 billion years old, this corresponded to the end of the Imbrian Era. The green boulder at Spur may well have been excavated 1.2 billion years ago by the formation of Autolycus or Aristillus in the Imbrium Basin to the north of Hadley-Apennine.

The rille and the plain

In studying the sinuous rilles, the photo-geologists

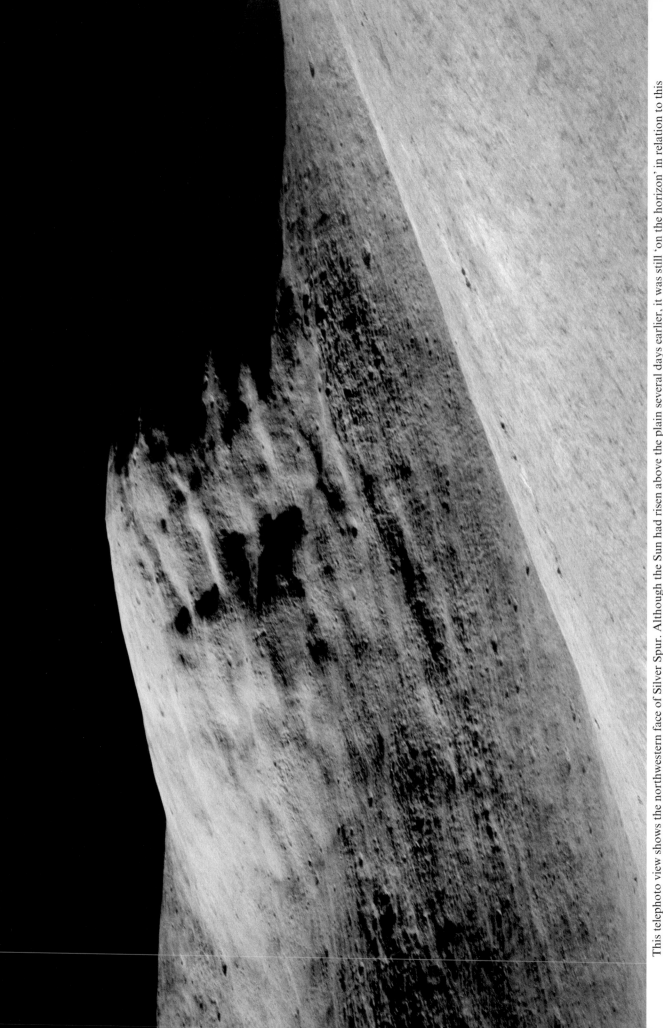

This telephoto view shows the northwestern face of Silver Spur. Although the Sun had risen above the plain several days earlier, it was still 'on the horizon' in relation to this section of Silver Spur, and while the slope is heavily ribbed it is not as steep as it looks. The feature in the foreground (right) is the flank of Mount Hadley Delta.

had used Lunar Orbiter imagery to map both their topographic and stratigraphic context. The fact that they cut through terrain rather than 'dropping' it, and at the turns their inner and outer rim arcs did not have the same radius, meant that they were not the result of extensional stresses. But what were they? Although Apollo 11 had found the lunar rock to be anhydrous, John Gilvarry insisted that the rilles were cut by running water. Jack Green thought they were cut by pyroclastic flows. Most geologists considered them to be lava channels. However, they were considerably larger than their terrestrial counterparts A low-rate terrestrial lava flow, if laminar, yields its energy to the surrounding rock and rapidly congeals, but a vastly more effusive lunar vent would be a *turbulent* flow that would not only not cool, but would melt the wall of its channel, scour it clean, and excavate a rille. Hadley Rille had produced a 100-km-long canyon that was over a kilometre wide and one-third of a kilometre deep.

The average thickness of the regolith on the Palus Putredinus plain was estimated to be 5 metres, but in a narrow strip paralleling the rille there was a shallow slope as the regolith diminished to essentially zero at the lip, at which point

the slope sharply steepened. The thinning of the regolith had enabled progressively smaller craters to excavate rock, but because there was insufficient energy in such small impacts to do more than shatter the bedrock, the blocks on the rims of these small craters were almost certainly *in situ*. On the lip, there were outcrops of true bedrock. This meant that variations in the chemical composition of samples taken there could be unambiguously related to the stratigraphic record.

The bedding in the rille wall was essentially horizontal, indicating that there had been no significant subsequent tectonic forces. The outcrops were heavily fractured and large blocks had broken off and travelled down-slope, in most cases becoming lodged in the finer scree that had accumulated on the lower slope. Some large blocks had reached the floor of the rille, but otherwise this was remarkably clear. The fact that there was no outcropping on the wall below St George indicated that this face of the rille was scoured massif rather than mare in-fill, which was the case immediately opposite. Indeed, there was evidence in the wall below Elbow of the contact where the mare in-fill had covered the material that had slumped off of the mountain into what had then

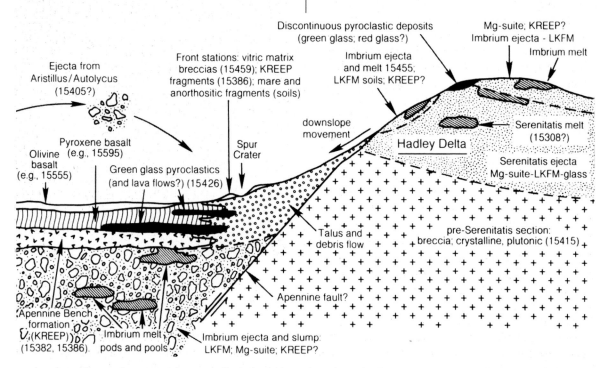

A schematic northwest–southeast geological cross-section through the Apollo 15 landing site showing the complex transition between mare and highlands (modified from Spudis and Ryder, 1985; Swann *et al.*, 1972; Swann *et al.*, 1986). In the mare (left), post-Imbrian basalt lavas overlie a thick deposit of Imbrium ejecta. In the highland area (Mount Hadley Delta, right), older (pre-Imbrian) ejecta from the Serenitatis Basin overlies ancient crust. Numbers refer to specific collected samples that are representative of the various units inferred to be present. Courtesy the Lunar and Planetary Institute and Cambridge University Press.

been a valley. The shallow ridge set just back from the eastern wall of the rille was elevated near Elbow. This suggested that as the lava turned the corner, it had splashed over the lip to build up a 'shoulder'.

The 'high lava' mark on Mount Hadley implied that surges of lava flowing down the rille had flooded the plain and then drained away again, leaving only a thin scum on the surrounding mountains. Remarkably, these flood marks were 75 metres above the current level of the plain. However, this probably includes the fact that the plain has settled isostatically since it formed. This withdrawal is suggested by the shallow trench that runs along the base of Mount Hadley Delta and around the base of the Swann Range – although it is conspicuously absent in the vicinity of Elbow, where the rim has been built up. If the trench is the result of the newly solidified plain contracting, the fact that it still exists indicates that there has been little down-slope movement off the massifs in the intervening aeons. Nevertheless, the fact that Scott and Irwin were able to drive across the trench without difficulty meant that there had been enough in-fill to smooth out its profile.

One issue to be determined was whether the mare plain was (a) a monolithic slab of basalt that represented a single massive eruption, or (b) the cumulative result of a succession of extrusions. The fine detail of Mare Imbrium implied a succession of flows. The visit to Hadley Rille would be the Apollo program's only opportunity to observe the vertical structure of the mare. The debate was settled when Scott reported distinct outcrops running along the far wall of the rille.

Stratigraphic analysis gives the relative ages of features, not their absolute ages. What was needed was ground truth – some rock that was clearly associated with the rille. This was provided by Scott and Irwin's sampling of the rim. The basalt ranged from 3.3 to 3.4 billion years old. Most of it was pyroxene-enriched, but the darker material at the very lip was olivine-enriched. This meant that immediately before the vent dried up it issued a chemically distinct magma, implying chemical evolution of the reservoir. To determine which lavas had flowed across the plain, they sampled Dune, several kilometres away. This impact had penetrated to a depth of 100 metres (the crater was not that deep of course, but it would have excavated from that depth) and its rim was littered by large blocks that included both extrusive and hypabyssal basalts, the latter being magma that crystallised at shallow depth. The pyroxene-rich basalt was found at each site, indicating that this

Two views of sample 15415, the 'Genesis Rock', and Dave Scott examining it in the Lunar Receiving Laboratory.

had covered a wide area, but the olivine-rich basalt was found only at the rille, and so was from one of the later flows that barely spilled over the lip.

To summarise the conclusions from the missions thus far: the Imbrium Basin was made about 3.85 billion years ago, the Tranquillitatis site was engulfed by lava over the next 200 million years, and Procellarum 500 million years after that. The history of the basalts on the rim of Hadley Rille indicate that the eastern part of Imbrium was still active at this late stage. This was a great advance

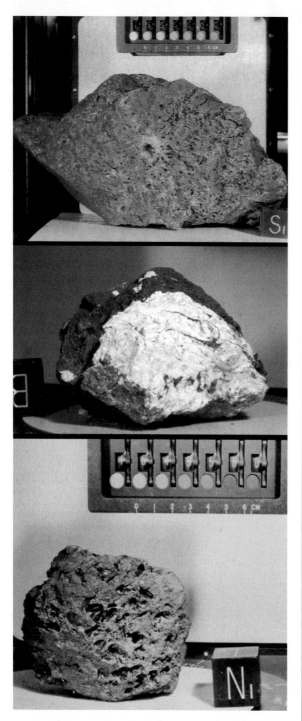

Apollo 15 samples: 15555, the 9.61-kg 'Great Scott' vesicular basalt collected at Station 9A (top); 15455, the 0.94-kg 'black and white' breccia with noritic clasts from Spur (centre); and 15495, a 0.91-kg coarsely grained gabbro with vugs lined with pyroxene that was collected near the Rover at Station 4 (bottom).

in the state of our knowledge of the Moon's early history. Of course, much of this had been predicted. The problem was that so too had much else, and the task was to sift the wheat from the chaff. It was astonishing how little direct evidence was needed to settle some issues. Although Scott knew as soon as he saw the far wall of the rille that Palus Putredinus had been built up by a succession of lava flows, there remained the issue of its thickness, and estimates varied from a 'thin' veneer to a stack as much as 25 km in thickness.

The Apennine Bench

A patch of light material within the Imbrium Basin is interposed, stratigraphically speaking, between the arc of the Apennines and the crater Archimedes. This material has embayed the Apennines, forming a stark contrast of terrain in which the massifs rise straight off the level plain. This 'light plain' is the Apennine Bench Formation, and the conventional wisdom held that it was impact-melt from the creation of the basin. A later impact produced Archimedes and its secondary craters. Later still, a succession of mare lava flows submerged most of this terrain. The Apennine Bench Formation was not a specific sampling objective for Apollo 15, but it was hoped that fragments of it would be recovered, and indeed many fragments of KREEPy non-mare basalt were found. However, because these did not show the kind of inclusions that would be expected of a shock-melt, Paul Spudis concluded that if they really did represent the light plain, then this had to be an aluminous basalt whose extrusion predated that of the darker Mare Imbrium. The only *large* KREEPy sample was the boulder at St George, and it held a contact between a mare and a non-mare material in which the plagioclase and mafic silicates were in equal proportions. The KREEPy fragments[47,48] were dated at 3.83 ($\pm$0.04) billion years, but the more mafic pyroxene basalt was 550 million years younger. It is therefore believed that the Apennine Bench Formation resulted from an extrusion of aluminous lava immediately after the Imbrium Basin was formed, and is light-toned because it came from a shallow magma source. That is, with so much of the crust having been cleared away as ejecta, isostatic forces would have caused the basin floor to rise, cracking the remaining crust sufficiently to allow shallow lava to ooze out and veneer the floor of the basin. Evidently, there was little activity in the eastern basin for the next 500 million years, at which time a series of extrusions almost completely engulfed the light plain, leaving only the part that we see today. This mafic basalt came from deeper reservoirs, so was darker and denser than the material on which it settled. It therefore *depressed* the floor, opening faults around the rim of the basin. When a vent in the Apennines issued lava, this exploited the faults and produced Hadley Rille. The embayment is shallower than in the basin, so it is possible that the 350-metre-deep rille cut right through the dark mare into the underlying aluminous basalt, and that its scoured floor is part of the Apennine Bench Formation. Scott joked about sampling the rille floor, but as Irwin countered, if they had driven down into it they would never have managed to get back out again.

The North Complex

In addition to the outcropping that Scott and Irwin saw in the uppermost 60 metres or so of the rille's western wall, a close study of overhead imagery hinted that there was a buried hillside poking out of the wall adjacent to the North Complex, which is a cluster of crater-scarred hills on the plain several kilometres east of the rille.

Although the North Complex was given the lowest geological priority, this was in no way due to a lack of interest in it. The primary objectives were to gain the ground truth required to assess rival theories, and sampling The Front, the rille and the plain would yield insights that would be more readily extrapolated to the broader context. But the North Complex was where the *discoveries* were likely to be made. Scott and Irwin were convinced of its value as soon as it was offered as a possible sampling target for this multiple-objective mission. To appreciate its significance, a recap is in order. When an impact excavated the Imbrium Basin, the 'base surge' of material from the deepest point would have blanketed the rim, making low rolling hills in the valleys between the massifs of the Apennines. Many of these hills were submerged by the succession of lava flows that subsequently embayed the Apennines, but a few still have summits projecting above the mare, the North Complex being an example. Noting that some of the depressions in the complex appeared to have been partially flooded, Gerry Schaber proposed that it was temporarily submerged with lava. The fact that the North Complex was not only darker than the hills further to the north but also slightly darker than the surrounding mare prompted Mike Carr and Keith Howard to suggest that it was thinly mantled by a dark pyroclastic material. The hills had been excavated by ejecta from Autolycus and/or Aristillus. The largest crater, Pluton, was littered by blocks as

much as 10 metres in size that showed a range of albedos, and it is possible that the light-toned blocks were intra-massif material.

Unfortunately, owing to the time devoted to the Heat-Flow Experiment, Scott and Irwin were unable to visit the North Complex. In fact, the final EVA started several hours later than planned, and, in retrospect, it is a pity that the LM's lift-off was not postponed by two hours (one lunar orbit by the CSM) in order to allow the excursion to run to its originally planned duration. If Allen had provided him with the slightest of openings Scott would certainly have seized it, but Houston, reluctant to 'stretch' this first 'J'-class mission too far, opted to stick to the scheduled time of departure.

The seismic network

The Apollo 15 seismometer, coordinating with those of Apollos 12 and 14, enabled the epicentres of moonquakes to be triangulated with greater precision. Most of the seismic events were of only magnitude 1 or 2 on the Richter scale, so weak that on Earth they would be ignored as part of the ongoing rumble in the dynamic crust. But the Moon is so inert that these were 'major' events. The low seismicity – barely one-billionth that of the Earth – implied that the outer mantle of the Moon is too cold to be plastic, and without mantle convection there can be no tectonic forces to deform the crust. The seismic data from Apollos 12 and 14 had suggested that the maria were 25 km thick and rested on about 40 km of anorthositic gabbro. It was initially presumed that this outer layer was all basalt, but further investigation established that there is actually a thin veneer of basalt resting on a thick deposit of an intensely brecciated material. Although the increased resolution from adding the Apollo 15 seismometer indicated that there was a distinction between the crust and the mantle, it was not possible to tell whether this was a sharp discontinuity at a depth of 55 km or a transition between 25 and 70 km. The pattern of seismicity suggested that there was some fluid deep in the mantle. However, this was probably confined to isolated pockets. It was not immediately apparent whether there was a liquid core at the centre.

The lunar 'heat engine'

The objective of the Heat-Flow Experiment was to measure the temperature and the thermal properties of the regolith in order to determine the rate at which the interior of the Moon leaked heat. Although the sensors could only be emplaced at a depth of 1 to 1.5 as opposed to the planned 2.5 to 3 metres, the data provided an astonishing insight into the state of the lunar 'heat engine'. The heat loss from a planetary body is related both to the rate of internal heat production and to the internal temperature profile. The heat can come from long-lived radio-isotopes and from a molten core, if one exists. It was also hoped that the data would refine the Moon's bulk chemical composition.

The data showed that in sunlight the temperature at the surface of the regolith at the mid-northern latitude of Hadley-Apennine rose to 380K. During the lunar 'day', therefore, there was a net heat-flow *into* the surface from solar irradiation. At sunset, the temperature fell to 100K, and during the 'night' heat was radiated to space. The thermal conductivity was strongly dependent on the temperature. Actually, because the efficiency of the radiative transfer process between the fine powdery particles of the surficial regolith was proportional to the cube of the absolute temperature, heat was able to flow more readily into the immediate subsurface during the 'day' than it was lost during the 'night'. Overall, however, the regolith strongly inhibited the flow of heat. In fact, it so efficiently damped the thermal variation experienced at the surface that the temperature at a depth of 0.5 metres was a constant 220K. Furthermore, the temperature at a depth of 1 metre was almost 40K higher as a result of the leakage of internal heat. However, this increase in temperature with depth was not expected to run very deep. The powdery regolith was such an efficient insulator that not only did it prevent the 'diurnal' variation penetrating more than half a metre, it inhibited the heat from below from leaking out, causing it to accumulate just beneath the surface. In effect, this 'warm zone' was storing up the heat that was efficiently conducted by the bedrock. Obviously, this would not occur in an area with a very thin layer of regolith.

On Earth, depending on the location, the temperature rises 1 to 3 degrees per 100 metres of depth; the best data having been obtained from 4-km-deep mines in South Africa. The heat-flow is 6.3×10^{-6} watt/cm^2. Although the flow of heat through the crust is trivial in comparison to the effects of solar heating, it provides the energy to drive the processes of plate tectonics. Prior to Apollo 15, remote-sensing studies of the lunar surface using microwaves had given heat-flow estimates in the range 1 to 5×10^{-6} watt/cm^2. The 'cold' Moon theory of Harold Urey required a value considerably less than the lower end of this range. A value near the top of the range would imply a Moon so 'hot' that it would be difficult to explain why a body with a mass 1.2 per cent that of Earth should yield so much heat. The early data from Hadley-Apennine showed a heat-flow of 3.3×10^{-6} watt/cm^2 (± 15 per cent). Although the value was adjusted several times as various 'corrections' were applied to remove sources of bias, it settled at 3.0×10^{-6} watt/cm^2 (± 20 per cent), which was within the uncertainty of the original result. If this value was typical of the lunar crust, it meant there had to be considerably more radiogenic heat than most people had expected. The problem with a 'hot' Moon was that it would make the crust so pliable that it would not be able to resist isostatic forces. As the Moon's interior is thermally differentiated, the mafic magma that upwelled had a higher density than the crust that it settled on. The fact that the mare-filled basins are correlated with the 'mascons' indicates the crust is sufficiently rigid to support these slabs. In an effort to resolve this conflict, a 'warm' Moon theory was proposed in which the heat-flow derived not from a molten core but from radioactives concentrated in a semi-molten layer englobing the solid core. This was consistent with the identification of extremely deep seismic epicentres. But this was a tremendous extrapolation of data from a single site. To determine whether Hadley-Apennine was typical, the experiment was assigned to the next Apollo mission, which was to land in a totally different type of terrain.

6

Surprise at Descartes-Cayley

SEEING RED

In 1964 the significance of the Moon's basin structures had only just been realised, and almost all 'sculpture' was attributed to the formation of the Imbrium Basin. On stratigraphically mapping the peripheral hummocky terrain, Dick Eggleton judged it to be Imbrium ejecta and named it the Fra Mauro Formation. On seeing rolling light-toned plains just beyond, he mapped them as part of the Fra Mauro. But in 1965 Don Wilhelms drew a distinction, and mapped the latter as the Cayley Formation (a name derived from the fact that the first patch that he studied was near the 10-km Cayley Crater). Nevertheless, it was still thought to be Imbrium ejecta. As the 'cold' Moon theory lost favour, the Cayley was reassessed as volcanic. When the dark maria were found to be lava enriched by mafic silicates from the mantle, it became evident that the mare flows would have been of such low viscosity that they readily spread out to form vast smooth sheets. The Cayley lava was thought to be enriched with silica and so more viscous. It was supposed that it had extruded from fissures in the highlands and filled in the low-lying areas. The Cayley was therefore widely referred to as 'upland-fill'.

When Eggleton had met light-toned domical hills south of the crater Descartes in the Central Highlands, he mapped them as an atypical patch of the Fra Mauro. But Dan Milton argued that their light tone meant that the previous terrain had been masked by volcanism. In the absence of evidence to the contrary, it was presumed that these hills were extrusions of very viscous silica-rich rhyolite. Initially referred to as the Material of the Descartes Mountains, such hills became the Descartes Formation. By analogy with terrestrial geology, therefore, when it came time to select a landing site for Apollo 16, the Descartes Formation was so widely believed to be rhyolitic that on their geological map Wilhelms and Jack McCauley

drew it in red, the standard colour to represent volcanism.

The volcanic interpretation was so well established that inconvenient observations were explained away. For example, the absence of wrinkle ridges and flow fronts on the Cayley was interpreted as evidence that these plains were deposits of pyroclastic ash rather than lava flows. Verne Oberbeck said the extremely subdued rims of the craters on the Cayley implied that it was ancient. He also said that the crater morphology implied that the regolith was much deeper than on the maria. But he interpreted it as a plain of some indeterminate material whose thick regolith was thinly veneered by ashflow tuff, which is welded pyroclastic. Despite the fact that the Cayley embayed the Descartes, Don Elston and Eugene Boudette said the fact that the Descartes was less cratered meant that it must be younger than the Cayley. Believing a mantra which said brighter meant younger in the highlands, Jim Head and Alex Goetz took the fact that the Descartes was brighter than the Cayley to imply that it might belong to the Copernican Era, which in turn held out the prospect that the lunar 'heat engine' was still active.

THE PERFECT PLACE

As a result of the first four Apollo landings, geologists had inferred that the Imbrium Basin was formed some 3.85 billion years ago, and were confident they understood the major events that occurred during the half-a-billion years that followed; this being the interval in which most of the basins were filled by up-welling lava. As a result, they were keen to send Apollo 16 to sample a site in the highlands. The 'dream site' was Tycho Crater in the Southern Highlands, but the flight dynamics team ruled this out as being too expensive in terms of propellant. This limited the

The Apollo 16 landing site is marked by a cross. The Kant Plateau is on the right, and to the left of the scarp at its western edge is the hilly Descartes Formation, and then (top and left) the 'plains material' of the Cayley Formation. The objective was an east–west embayment of the Descartes by the Cayley. The furrowed domical hill to the south of the target is Stone Mountain, and Smoky Mountain lies to the north.

site selectors to the Central Highlands. An early suggestion was the Kant Plateau. This was favoured by the petrologists because it appeared to be a block of primitive crust which, although it had been cratered, had not been masked by volcanism. The geologists rejected it for just this reason – they were *seeking* volcanism, and suggested landing in one of the valleys of the Descartes Formation that had been embayed by the Cayley Formation, with a view to sampling both types of terrain.

By specifying only the contact between the two formations, this criterion left the choice of landing site wide open. The only way to investigate the vertical structure of the Cayley would be to undertake a radial sample on a 'drill hole' crater. This not only narrowed the options, but led the selectors straight to a pair of ray craters: one each at the base of two mountains set 10 km apart on opposite sides of an east–west embayment. It was the perfect place for a 'J'-class mission because the first traverse could sample craters near the landing site on the plain, the second could explore the southern mountain and its ray crater, and the third could go north. By virtue of its geological context, this site was called Descartes-Cayley.

On Apollo 14, Stu Roosa was to have taken high-resolution photographs of the general area using the Hyflex Lunar Terrain Camera, but this failed just as the target came into view. He hastily snapped a set of frames using a Hasselblad fitted with a 500-mm lens, and then took additional pictures on later passes to facilitate stereoscopic analysis. Elston and Boudette based their geologi-

John Young, Apollo 16 commander.

The Apollo 16 mission patch.

Charlie Duke, Apollo 16 lunar module pilot.

Ken Mattingly, Apollo 16 command module pilot.

cal mapping on these pictures. By enlarging the pictures to the limiting resolution and enhancing the contrast in order to emphasise the surface relief they charted the craters, hills, lineaments and albedo variations. In the spirit of the prevailing volcanic interpretation, they saw scarps that seemed to indicate flow fronts, hills that might be cinder cones and irregular pits that could well be maars. Accordingly, astronauts John Young and Charlie Duke were taken to inspect a variety of terrestrial equivalents so that they would know what to look for, and Farouk el-Baz tutored Ken Mattingly to search for signs of volcanism while he waited in orbit.

BETTER LATE THAN NEVER

As Mattingly prepared to circularise his orbit after releasing the LM in the descent orbit, he encountered a problem with Casper's SPS engine, and this led Houston to postpone Orion's descent. But the issue was resolved after several hours of detailed trouble-shooting.

"Orion, you're 'Go' for PDI," Jim Irwin announced. Following standard practice, having flown as LMP on the previous mission, he was serving as CapCom for the next landing.

As a result of the delay, the LM was running a little south of track and was 16,000 feet too high, but its PGNS would be able to correct this during the early phase of the powered descent.

Although this was to be the first landing in the Central Highlands and the line of approach passed over the Kant Plateau, there was no need for an extra-steep descent profile.

"21,000 feet," Duke reported.

"You're 'Go' at 8 minutes," Irwin confirmed.

"I can see the landing site," Young announced. Standing on his toes against the pull of his restraint harness, he was able to peer out of the bottom of his triangular window and see the dazzlingly white ejecta of South Ray Crater, just to the west of Stone Mountain. In contrast to Dave Scott before him, Young had no impression of coming in 'long', and he was relaxed. Every mission had its share of troubles, and with a little luck his were behind him. He moved the focus of his attention northward to await his first sight of Flag and Spook, a pair of 300-metre-sized craters situated in the centre of the valley about 1 km beyond where he wanted to land.

"Right on profile," Duke confirmed. Everything was perfect.

At an altitude of 7,000 feet the PGNS pitched Orion upright for the final phase of the descent. Young studied the plain, checking for the landmarks that he had seen on the model used in the simulator. Because they were late the illumination was slightly different but, amazingly, everything was right where it was meant to be. His vision was restricted to ahead and to the south, where there was a cluster of craters which, from his perspective, resembled an inverted 'V' with one leg anchored on the flank of Stone Mountain and the other leg pointing right at his target, which was a pair of tiny craters that they had named Double Spot.

Young monitored the LPD as the PGNS flew them down. On reaching 4,000 feet he decided that the computer was heading for a point 600 metres north of and 400 metres west of Double Spot, so he applied a series of corrections to realign the track and shorten the trajectory.

Although Duke's function was to serve as Young's eyes in the cabin, reciting the instrument readings so that his commander could stay 'outside', Duke added a brief glance out of his window to his cycle, studying the approach to Smoky Mountain in order to assess the trafficability.

Although the 5-metre resolution of Roosa's pictures was sufficient to select a landing point and plan traverse routes, it had not enabled the coverage of blocks around the ray craters to be measured. Terrestrial radar studies suggested that the ejecta would be sufficient to prevent the Rover from driving to the rim of North Ray Crater. Young and Duke had inspected the site from orbit, but had not been able to gain any insight. But now Duke had an excellent view with the Sun casting long shadows showing the nature of the terrain. "It looks like we're going to be able to make it, John," he observed delightedly. "There's not too many blocks up there."

"I'm going to take over, Charlie," Young announced as they passed through 300 feet.

"Here comes the shadow," Duke called 10 seconds later.

Young rotated Orion slightly to the right to view the shadow, to enable him to use it to provide a sense of scale in the final few seconds of the vertical descent.

At 200 feet Duke warned that they were dropping a little too rapidly. Young cut the rate of descent to 5 fps. Duke extended his cycle once again to include a look out of his window. "There's a perfect place over here, John."

At 80 feet, the DPS plume started to blow dust, but there were enough small rocks poking up through it to enable Young to retain his sense of situational awareness. Or almost: there had been a 15-metre-diameter crater directly ahead; where had it gone?

"You levelled off!?" Duke observed, as Young hovered at 40 feet, seeking the best spot on which to land. "Let her on down," Duke urged.

As Young resumed the descent, he added a little forward velocity, just in case he was over the crater.

When Duke called "Contact", Young followed Scott's advice and waited for a second before he shut off the DPS. Nevertheless, Orion slammed down onto the surface with a jolt.

"We don't have to walk far to pick up rocks, Houston," Young said immediately. "We're among them!"

"Old Orion is finally here, Houston," Duke explained belatedly.

"It's not flatland, though," Young added. In fact, it was a *lot* more undulatory than he had been led to expect. As they correlated the ridge that they could see through their respective windows, it became evident that this arced around them – they were in the centre of a subdued crater about 100 metres wide. What they did not discover until they ventured outside, was that the rear footpad was a mere 3 metres beyond the rim of the 15-metre crater that Young had lost sight of. When he had hovered to select a spot on which to land, he was directly over the crater, and had narrowly missed landing on its rim.

As he waited for Houston to clear them to stay, Young assessed the Cayley from this new perspective. "We've got a lot of rocks." At 4 per cent, the coverage was double that on the mare plain at Hadley-Apennine. "Some are very white," he noted. This might mean ancient anorthositic crust. "I see one with some black; it could be a breccia – that's if you believe such a thing."

After powering down Orion, Duke resumed this through-the-window Site Survey. "About 30 metres out, there is a secondary crater with a metre-sized block in it. The top 3–5 per cent of the block is black-and-white; below that it's solid white." He was struck by the proliferation of two-toned rocks. "These black-and-white blocks are all over the place." There was nothing that was clearly the type of volcanic rock he had been led to expect.

In the Backroom, the geologists were simply happy that Orion had finally made it to the surface. There were likely to be a few breccias at every site, so they were not concerned. If they had been able to see what Young and Duke could see, however, they would have been alarmed.

A PLAN IN TATTERS

The 6-hour delay in landing had disrupted the plan, which was for Young and Duke to conduct the first EVA immediately. To add a full 7-hour excursion to what had already been a long day would be impracticable. The plan had been to go out on each of the following two days, then lift off early on the third and final morning. Instead, the crew retired and left it to Houston to rework the schedule.

There were several constraints. Orion had been consuming power, coolant and oxygen throughout the additional time spent in orbit, and this defined its endurance. There was also the matter of the SPS. It successfully made the circularisation burn, but the longer that Orion spent on the surface, the further the Moon's rotation would carry it from the plane of Casper's orbit and the greater would be the plane change needed preparatory to the rendezvous. The flight dynamics and spacecraft systems teams argued for Orion to depart on time, but the scientists were appalled at the prospect of

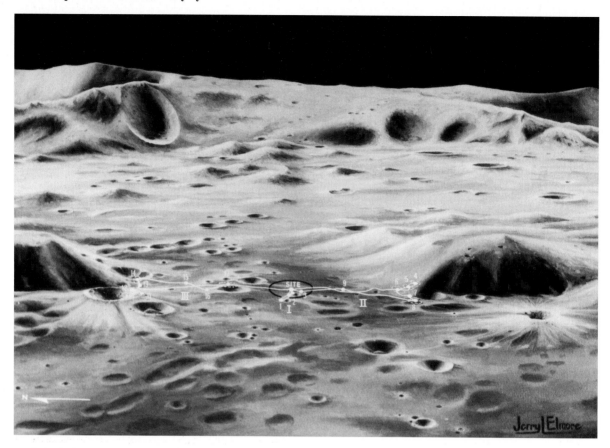

A depiction of the Descartes-Cayley landing site, showing the planned traverses: the first to craters on the plain a short distance to the west; the second south to the flank of Stone Mountain; and the third north to Smoky Mountain and North Ray crater.

losing the final excursion, which was to visit North Ray and Smoky Mountain. In these discussions, it was simply taken for granted that Young and Duke would wish to complete their full surface program.

BEAUTIFUL ROCKS

Tony England, the mission scientist, was to be the CapCom for the surface activities. After allowing Young and Duke a full night's sleep, he read up the revised schedule. The lift-off had been slipped 6 hours in order to permit the third EVA. This was the most that Orion's resources could be stretched. Young and Duke were delighted to hear the decision. Although the final EVA would have to be cut to 5 hours, this was better than losing it entirely.

Unlike Apollo 15, on which the first EVA included a long drive and the walkback constraint required the first traverse to be conducted prior to deploying the ALSEP, Young and Duke's first drive was to be to a cluster of craters only a

kilometre away and so it had been decided to start with the ALSEP.

"John, hurry up!" Duke called when Young paused on the porch. He was not as deferential towards his commander as Irwin had been to Scott.

"I'm hurrying," Young laughed. A fault in Orion's high-gain antenna had obliged the cancellation of the MESA TV, so Young became the first astronaut to step onto the surface undocumented. "There you are," he mused in doing so, "mysterious and unknown Descartes-Cayley Plain. Apollo 16 is going to change your image." With that, he stepped off the footpad and took a look around. "Oh, look at those beautiful rocks!"

As soon as Young was safely on the surface, Duke began to ease out through the hatch to join him. Free of Orion's confinement, he was staggered by the vista. "The view of Stone Mountain, Tony, is *superb*." At 530 metres elevation, this peak was a mere hill compared to the massifs of the Apennines. It was about 5 km to the south, and Smoky Mountain was the same distance away to the north, on the other side of the valley.

As Young worked at the MESA, Duke went to take a look behind the LM – and saw the crater behind the footpad. "Look at the hole we almost landed in!" If they had landed 10 metres short, they would have had to abort as Orion tilted back at 30 degrees while it slid down the wall of the crater. As he inspected the area, it dawned on Duke that Young had managed to set down on the only flat patch in sight. "This old Cayley Plain is really something," he enthused in his South Carolina drawl.

Young inspected the Rover on the side of Orion. The walking hinge had popped, so he reset it. With the LM standing within 3 degrees of vertical, they were unlikely to suffer the difficulties that had faced Scott and Irwin – and so it proved. Leaving Young to configure the vehicle, Duke took a panorama of the site from the opposite side of the crater in which they had almost landed.

As had Scott, Young found that one of the Rover's batteries gave a 'zero volts' reading but, as then, this proved to be only an instrumentation problem; the battery was fine. He took the vehicle on a test drive around Orion and parked by the MESA, where he remarked upon a rock that had "all kinds of dark clasts in it", which made it a breccia.

While Young set up an ultraviolet camera in Orion's shadow to take astronomical images, Duke installed the Rover's TV, the quality of which was an improvement on the previous mission.

As Gordon Swann had returned to academia, Bill Muehlberger was running the Backroom, but this was only a change of management because most of the team was the same. While Young and Duke worked, Ed Fendell panned and zoomed the TV to enable the Backroom to examine the site.

Young and Duke planted *Old Glory* and took the traditional 'tourist shots'. Young exploited the low gravity to score a 'first' by jumping to give his salute a metre or so *above* the lunar surface.

DEPLOYING THE ALSEP

Duke affixed the two ALSEP pallets to the carrying handle, hoisted it, let it slide back into the crook of his arms, and set off walking in a westward direction. He had barely gone 20 metres when one of the pallets slipped off the bar and tumbled into a small crater. As Duke veered the other way off-balance, he managed to grasp the bar before the other package could hit the ground. Undaunted, he retrieved the fallen pallet, which

included the RTG, made a rudimentary effort to dust it off and reattached it to the bar. So much for the instruction to avoid getting any dust on the instruments!

On climbing the ridge that marked the rim of the eroded crater in which Orion had landed, Duke paused to rest. From this vantage point it was evident that the Cayley was undulatory on the several-hundred-metre scale, and that the local horizon was a succession of ridge crests. The map indicated that South Ray Crater had blanketed the area with bright ejecta, and he had no difficulty believing this because there were angular white and black-and-white rocks everywhere. Continuing, he detoured past a 2-metre-sized block that was a dark-matrix breccia with light-toned clasts containing "some very fine, submillimetre-sized crystals". Where were the volcanics!? After 100 metres, he turned south to a "fairly flat" spot atop a ridge, where it looked as if it would be possible to arrange the instruments in the requisite pattern between the rocks and small craters. Looking back, there was no doubt that Orion had landed in a crater, as most of its descent stage was out of sight. Joining him, Young pointed out "even the craters have craters", and expressed the opinion that the Cayley must be a very old surface. As it turned out, he was right.

Duke immediately started to drill the first hole for the Heat-Flow Experiment. The drill stem had been improved to prevent its flutes from clogging so, in contrast to Scott's struggle, Duke encountered no difficulty. He let the drill penetrate at its own speed. The first stem rapidly penetrated the surficial regolith, slowed dramatically, cut through the obstacle, and ran to its full depth in under a minute. He fitted the second stem, and watched it enter the ground in just 30 seconds. Even though detaching the drill was no problem, it took longer to attach the new stems than it did to drive them in; it was remarkable.

"Mark has his first one on the Cayley Plain," Duke reported delightedly, once he had emplaced the sensor string in the first hole. He was referring to the experiment's Principal Investigator, the Columbia University geophysicist Marcus Langseth.

As Duke started on the second hole, Young, having set up the Central Station and the Passive Seismic Experiment, set off to deploy the mortar of the Active Seismic Experiment. Unfortunately, as he walked past the Central Station his left boot caught a cable that was not laying quite flat on the ground. Although Fendell was following Young's progress, the damage was done before England

Orion on the Moon. Beyond, John Young is preparing to take the Rover for a test drive.

Colour Section

The commemorative plaque on the main strut of Eagle's forward landing leg.

This picture of **Buzz Aldrin**'s boot print in the regolith shows that although it was almost a powder the material was sufficiently cohesive to retain the imprint of the boot's tread.

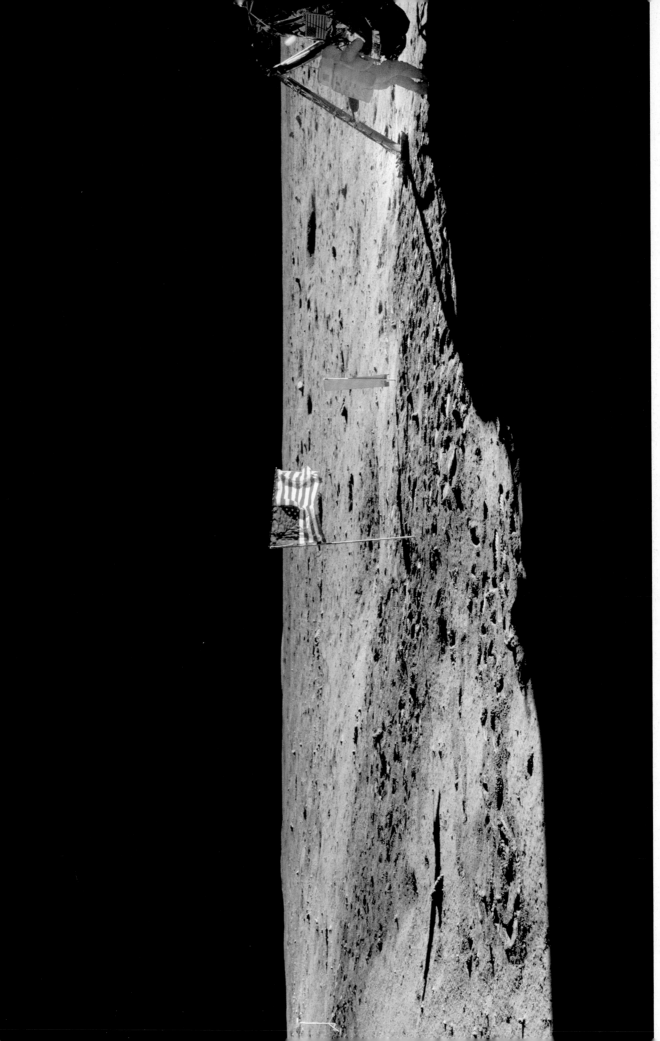

A northward-looking view showing the television camera, the US flag, the Solar Wind Collector and Neil Armstrong working at the Modular Equipment Stowage Assembly.

This spur-of-the-moment picture by Neil Armstrong of Buzz Aldrin became the iconic image of the Apollo 11 mission.

Buzz Aldrin retrieves the Early Apollo Surface Experiment Package from Eagle's Scientific Equipment bay.

Buzz Aldrin stands by the deployed EASEP – the Passive Seismic Experiment in the foreground and the Laser Ranging Retro-Reflector beyond.

A view out of Buzz Aldrin's window after the moonwalk, showing the flag, the TV camera and lots of boot prints.

Pete Conrad just after exiting Intrepid's hatch.

As soon as he was on the lunar surface, Al Shepard stepped around the north side of Antares and raised a hand to block the Sun as he inspected the ridge on the crest of which Cone Crater was situated.

An eastward-looking view of Antares with Cone Ridge in the background.

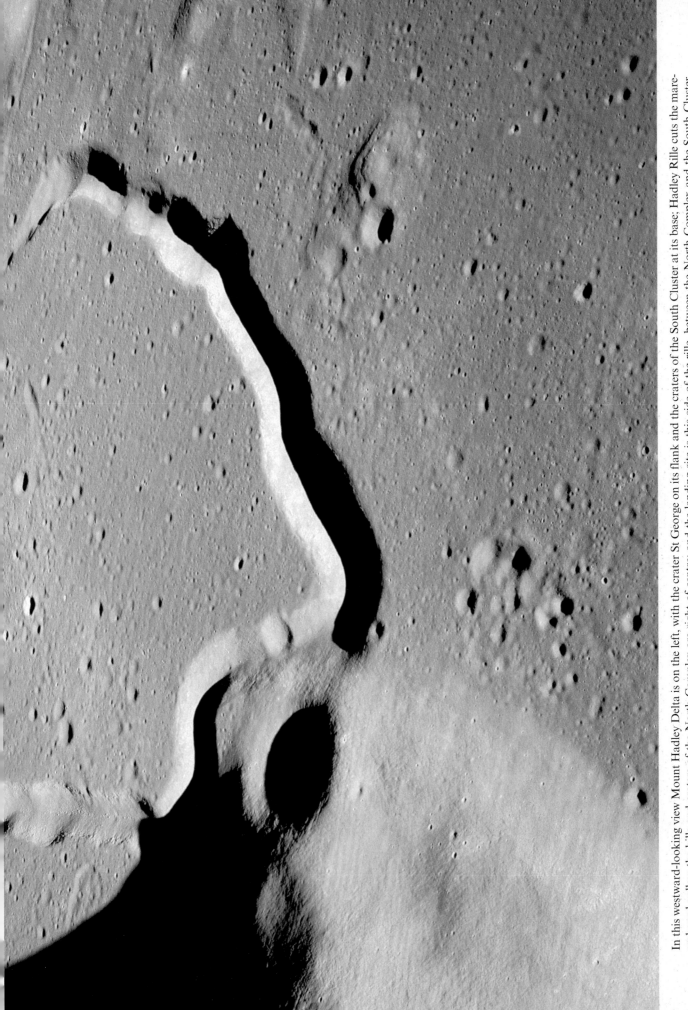

In this westward-looking view Mount Hadley Delta is on the left, with the crater St George on its flank and the craters of the South Cluster at its base; Hadley Rille cuts the mare-embayed valley; the hills and craters of the North Complex are right of centre; and the landing site is this side of the rille, between the North Complex and the South Cluster.

This northeast-looking view of Jim Irwin at the Rover after the first traverse shows the 4,000-metre-high Mount Hadley in the background, its western face in shadow.

Jim Irwin salutes the flag on the Plain at Hadley, with Falcon and Mount Hadley Delta as a backdrop. The Rover's TV is watching Dave Scott take the photograph.

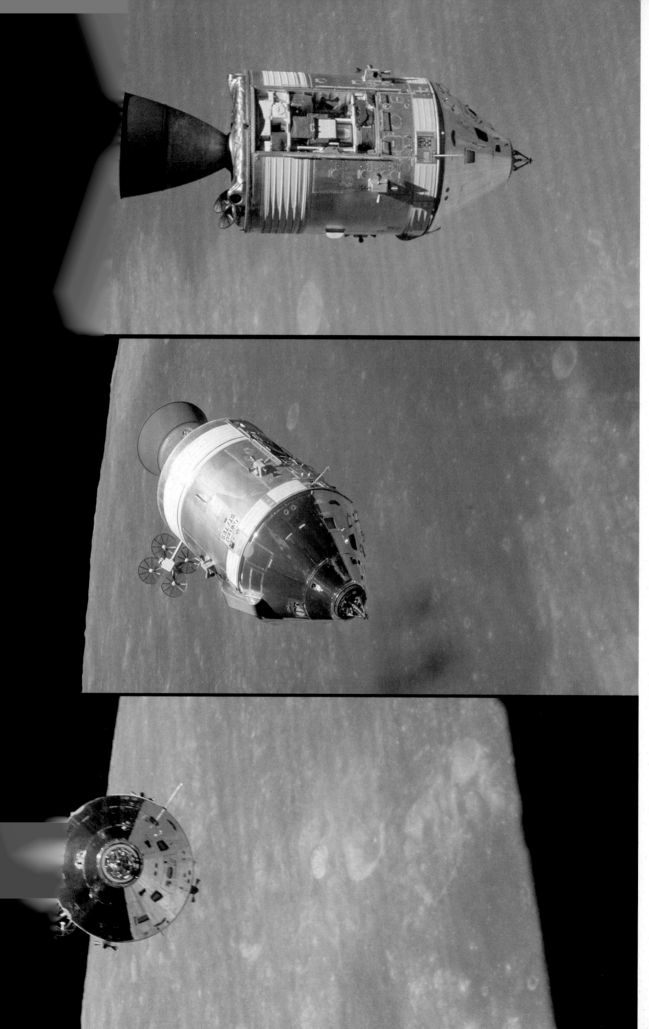

CSM Endeavour manoeuvres in the final phase of the rendezvous.

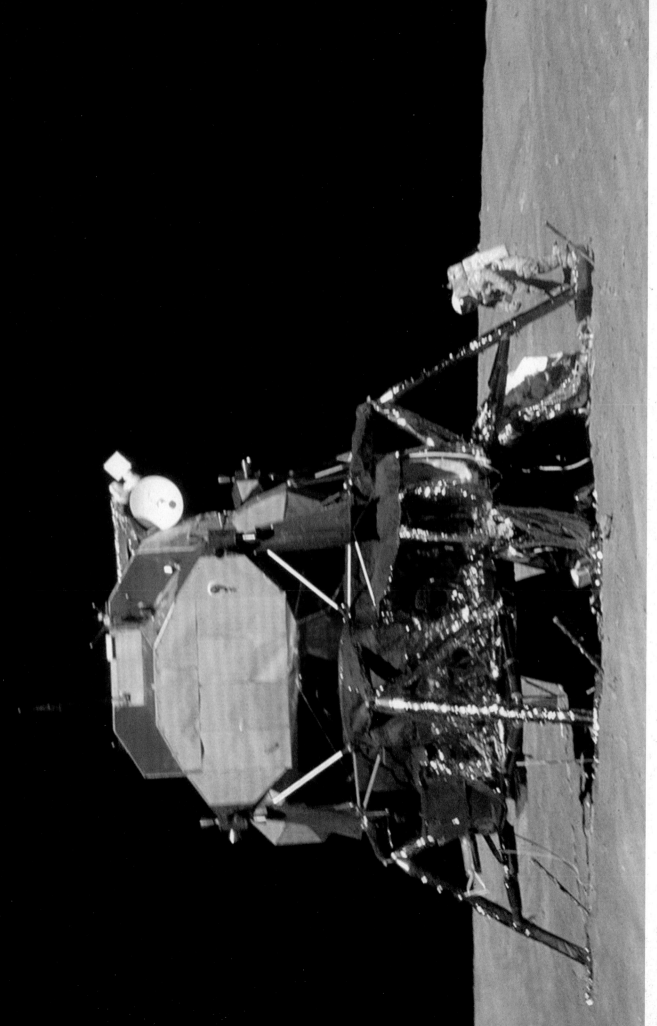

Orion on the Cayley plain.

The Ultraviolet Camera was set up in Orion's shadow, and left to work through a programmed set of astronomical targets.

In saluting the flag, John Young exploited the weak lunar gravity to jump about a metre off the ground.

John Young by the Rover.

John Young after the final traverse. By this point in the program, moonwalkers were admirably equipped for field geology: note the cuff-checklist, the shin-pocket for the hammer (absent), the chest-mounted Hasselblad and bag dispenser, and the sample carrier on the side of the backpack; together with the tools on the Rover. In the foreground is the gnomon, minus its pendulum.

The beautiful valley of Taurus-Littrow. Contrast this confined landing site with the open plain selected for Apollo 11. The Apollo 17 mothership, America, can be seen in the centre of the frame, set against the base of the South Massif.

In taking 'tourist' shots of the flag raising ceremony, Gene Cernan held the camera low and aimed high over Jack Schmitt's shoulder in order to illustrate that even on the lunar surface one is never far from the 'Good Earth'.

A westward-looking view of Jack Schmitt using the rake at Station 1.

The 'orange soil' on the rim of Shorty crater at Station 4.

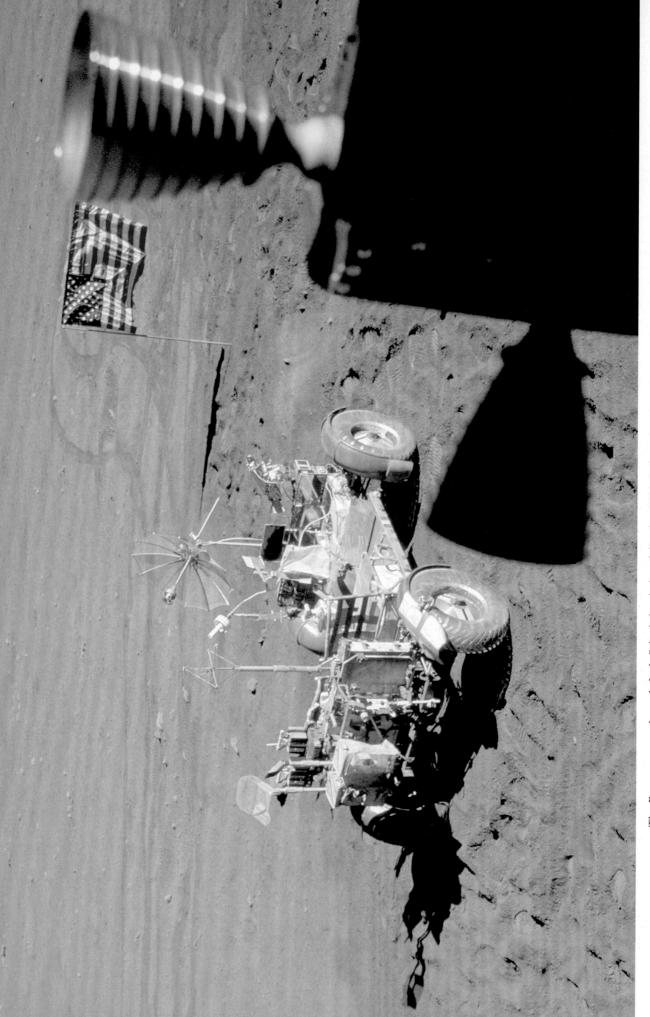

The Rover as seen through Jack Schmitt's window following EVA-2, featuring the *ad hoc* replacement right-rear fender.

Gene Cernan prepares the Rover for the third traverse.

Jack Schmitt tends the Rover at Station 7.

Jack Schmitt at conclusion of the third traverse.

Gene Cernan standing by the flag after the third traverse.

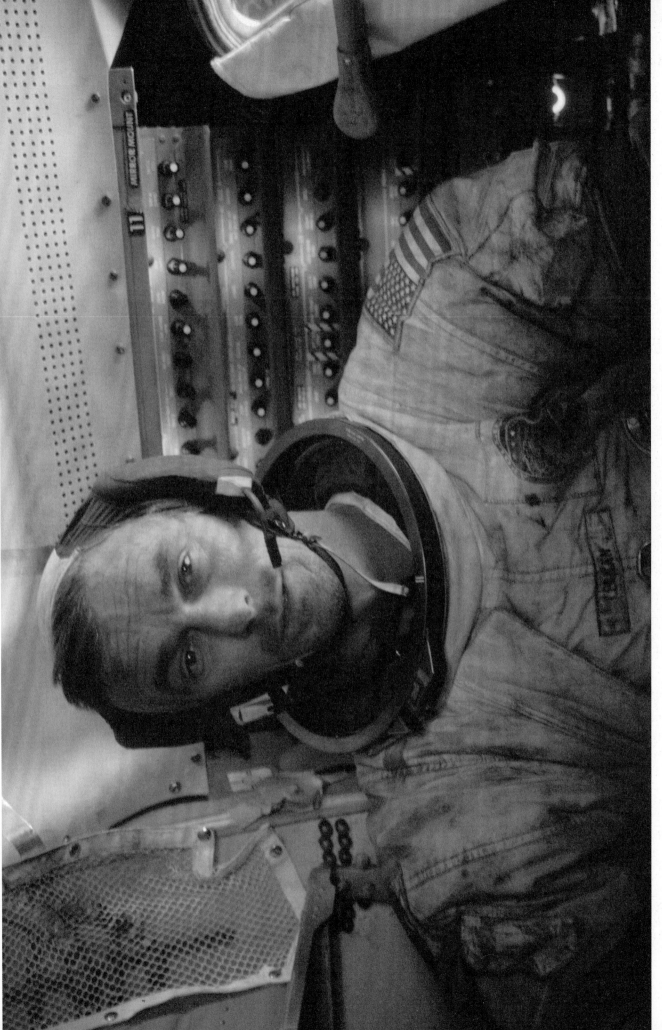

Gene Cernan in Challenger's cabin after the final EVA.

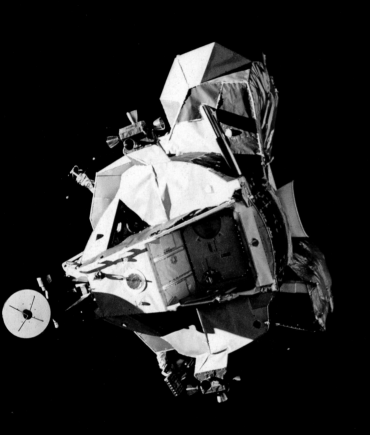

The ascent stage of Challenger after its rendezvous with CSM America.

In saluting the flag, John Young exploited the weak lunar gravity to jump about a metre off the ground.

could issue a warning, and the Backroom watched in horror as the cable was yanked from the Central Station.

"Charlie," Young called plaintively. "Something's happened here."

"What's happened?"

"I don't know." All Young had felt was the resistance as he advanced his foot. On turning around, he realised what had happened. "Here's a line that's pulled loose."

"Oh-oh," Duke sighed.

"What line is it?" Young asked.

"That's the heat-flow," replied Duke, seeing that the line from his experiment was laying loose. "You've pulled it off."

"God almighty!!!"

"I'm wasting my time," Duke noted. To continue with the second hole would be pointless. "I'll go do the deep core." This was to have indicated the material that the heat-flow sensors were measuring; now it would be merely an interesting sample of the regolith on the Cayley. "Rats!" he cursed in frustration.

Remaining silent, Young went to set up the mortar.

To Duke's astonishment, the first half-metre stem of the core sank into the ground in just 5 seconds. Because such rapid penetration risked destroying any fine layering in the sample, he decided to restrain the drill in future. As he tried to detach the drill, the stem, which was barely projecting from the ground and was free to rotate, did so. This threw Duke off balance, and he stumbled and almost fell to his knees. Using an improved wrench, he easily detached the drill to fit a succession of segments. After the difficulty that Scott had encountered in extracting his deep core, the treadle had been fitted with a jack. The tube rose only a few centimetres with each throw of the lever, but it emerged from the ground smoothly. As a result, barely 15 minutes after he had begun, Duke had secured a 2.3-metre core. With the fault in the vise fixed, he had no difficulty breaking it down. As an impromptu soil-mechanics experiment, he dropped the now-superfluous rammer into the hole and was surprised that it went all the way in. The Backroom took note of the fact that the hole had not collapsed, and recommended that on the next mission a 'freebie' experiment should be inserted into the hole.

Meanwhile, Young had set up the Lunar Surface Magnetometer and was about to conduct the 'thumper' portion of the Active Seismic Experiment. Having finished the core, Duke joined him. The three sensors were to be emplaced 45 metres apart in a line running west from the Central Station. As Young unreeled the line, Duke put his boot on the conical sensors to drive them into the ground. There were indicators at 3-metre intervals, and as he walked back along the line Young placed the thumper on each indicator in turn and fired a charge. This experiment had been performed by Ed Mitchell, and had successfully probed the intensely brecciated Imbrium ejecta of the Fra Mauro Formation. It was a powerful tool with which to probe the structure of the Cayley, but it was time-consuming to conduct.

After taking a pan of the ALSEP site, Duke, his chores done, undertook a little opportunistic sampling. He started with a small spherule[1][*] of black glass a few centimetres across. His second target was a typical white rock[2] that proved to be brecciated anorthosite. As his tongs were not to hand, he put his boot beside the rock, dropped to one knee and grabbed the rock before his suit forced him back upright. To sample a patch of light-toned regolith surrounding a brilliantly white rock[3] he had to go get his scoop. Scott and Irwin had considered scooping soil and pouring it into a bag to be a team effort, but Young and Duke had rehearsed solo-sampling, and it was paying off. The light-toned patch of soil was dust eroded from the extremely friable rock, which was another brecciated anorthosite.

FLAG AND SPOOK

By design, Orion had landed east of a cluster of 'drill hole' craters at which radial sampling was likely to yield clues to the vertical structure of the Cayley. At just less than a kilometre away and 350 metres wide, Spook was the nearest and largest, but because the plan also called for a visit to Flag, slightly smaller and half a kilometre beyond, the walkback constraint required that they visit the westernmost crater first. The plan was therefore to drive just north of Spook, between it and a smaller crater named Buster, continue on to another small crater dubbed Halfway, and then swing southwest to establish Station 1 at Plum, a small crater near the southeastern rim of Flag. Depending on what they saw on the way out, Station 2 on the return trip would be at either Spook or Buster.

Amazingly, by the time that Young and Duke

---

[*] Information pertaining to each sample is given on page 369.

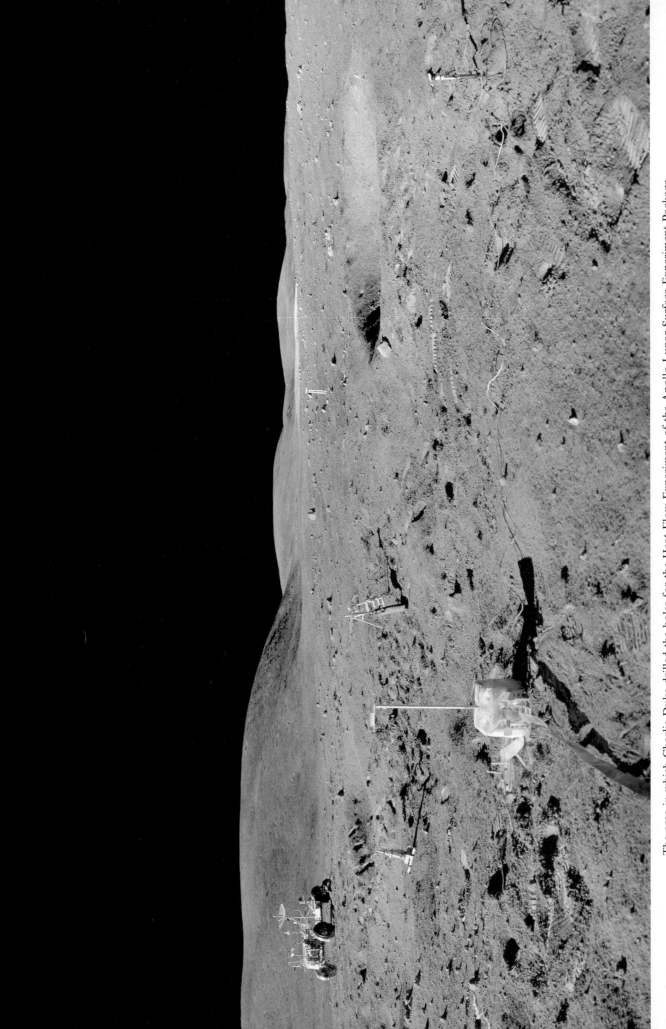

The area in which Charlie Duke drilled the holes for the Heat-Flow Experiment of the Apollo Lunar Surface Experiment Package.

In addition to the Central Station and the Passive Seismic Experiment of the Apollo Lunar Surface Experiment Package, this northeasterly panorama illustrates the fact that Orion was sitting on the floor of a 100-metre-wide eroded crater – the descent stage is partially hidden. The low dome of Smoky Mountain lies in the background.

John Young in among the instruments of the Apollo Lunar Surface Experiment Package.

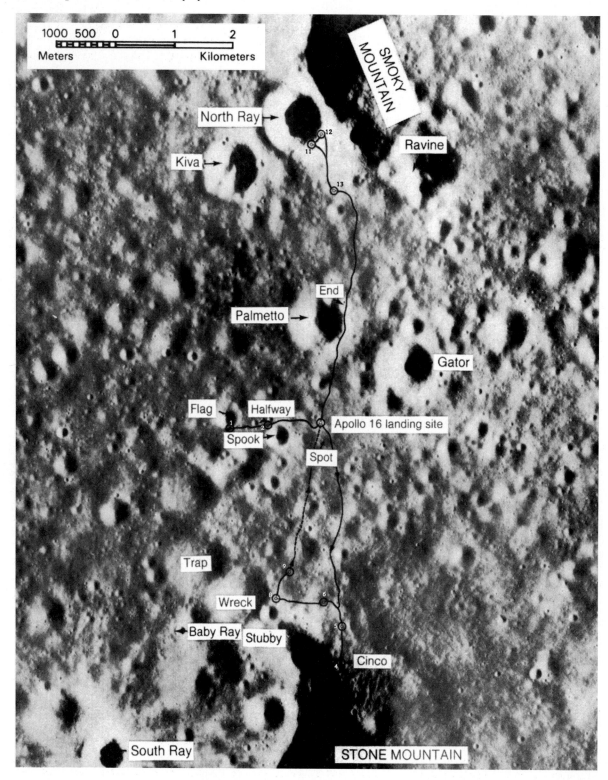

The actual Rover traverses and sampling stops made by John Young and Charlie Duke.

were ready to start (exactly 4 hours into the EVA) they were 3 minutes *ahead* on the timeline. It was fortunate that they had not already slipped behind, because their drive over the undulatory surface was slower than expected owing to the glare of backscattered sunlight in the zero-phase. Young paused at the crest of each of the many ridges in case it was one of the scarps that the map showed crossing their route. As it turned out, there were no scarps; they were simply ridges that had appeared treacherous when the contrast of the overhead imagery had been boosted to highlight the topography.

Duke tried to correlate the map with what he could see, but even though he knew his starting point it was difficult to navigate. He provided a running commentary for England's benefit. The rock litter was predominantly angular fragments of dark-matrix white-clast breccias up to a metre in size. After 300 metres, the coverage of cobbles actually increased, suggesting that the landing site was on the eastern fringe of a ray. In addition to being undulatory on the several-hundred-metre scale, the Cayley was heavily pocked with both old eroded and small relatively fresh craters, the freshest of which had brilliant white material in their walls. There was no time to stop to collect samples, but Duke used his chest-mounted Hasselblad to snap pictures at 50-metre intervals in order to document the route for post-flight analysis.

In a lull in Duke's commentary, England relayed a question from the Backroom relating to the impromptu sampling that Duke had done at the ALSEP site. "Those rocks that you collected, were they *all* breccias?" Duke explained that the rocks had been dusty, but confirmed that they had indeed all seemed to be brecciated. England tried a different tack. "Have you seen any rocks that you are certain *aren't* breccias?"

"Negative!" Duke replied emphatically.

As they passed through a field of boulders, the 'nav' put them within 150 metres of Spook's northeastern rim, but they could not see it. But on cresting the next rise it was directly ahead. The crater was extremely eroded and surprisingly clear of debris. Duke saw no evidence of a bench in its wall to suggest that it had penetrated bedrock. Buster was 100 metres to the right and, in contrast, its rim was littered with blocks. It was evident that if they were to stand any chance of sampling material excavated from deep within the Cayley, then they would have to put Station 2 on Buster's rim. As they drove on, Duke reflected that the mottled appearance of the rocks ejected by Buster implied that they were breccias.

On finding their checkpoint, Young veered left and, no longer in the zero-phase, picked up a little speed. Although the plain sloped away to the southwest, this did not help to locate their destination because the local horizon was always too close. A few minutes later, a 40-metre-wide pit opened up before them. It was not the kind of crater to be surprised by when driving at full speed! It was about the correct size for Plum, and

its walls were as steep as Duke had expected, but, if this was Plum, where was Flag!?

"Let's start working," urged Young. He was only too happy to trade driving for sampling.

As part of the procedure for powering down the Rover, Duke read off the 'nav' data. The fact that this put them 1.1 km from the LM confirmed that they were in the wrong place. After debating the merits of sampling this crater against spending time searching for Plum, they decided to push on. Realising that he had strayed too far south, Young made a correction and several minutes later another crater opened up. With Flag just beyond its far rim, there could be no doubt that this crater was Plum, and the 'nav' confirmed it. But it had taken 24 rather than the planned 11 minutes to reach Station 1, which meant that they would have to hustle.

SAMPLING PLUM

As they disembarked on Plum's eastern rim, England asked whether they were still in the ray. Duke was positive that they were not, because he had noticed a transition about 50 metres back where the coverage of cobbles diminished. This was excellent news, because it suggested that most of the material on Plum's rim would be local excavation.

Flag's subdued rim was astonishingly clean, but there were blocks poking through its steep wall, and it was peppered by craters up to 10 metres wide, several of which had exposed white material. Although Plum was only one-tenth the diameter of Flag, its wall was just as steep and it had a prominent bench about 3 metres down that made it impracticable to see the bottom of the pit. A bench usually indicated the level of the bedrock, but in this case it more likely marked a change in the nature of the regolith.

"This is spectacular," Young assured England as he aligned the TV antenna. This done, he retrieved the rake and went to the planned spot, a crater's diameter north of Plum, to start a radial. The majority of the fragments in the swath were less than the spacing of the tines and slipped out, and all he got were friable clods of consolidated regolith. By the time Duke joined him with the scoop, Young was documenting his first rock sample.

"That old thing?" Duke said dismissively.

"It's as good as any," Young replied.

"It looks like it's gonna come apart!"

"It might," Young agreed. "There's three or four samples right there we can get."

"They're all covered with dust," Duke pointed out.

England asked whether the rocks all looked the same – it would be better to go for variety.

"They're angular," Young added. Out of the ray he had expected to see rounded rocks, so this fact alone made them worth sampling.

"Here's one with a white streak," Duke noted.

They took them all, placing them into a single sample bag, then took a soil sample for context. One fragment[4] proved to be a partially melted breccia with a dark matrix and light clasts, and another[5] was crystalline feldspathic material. Next they took an angular rock[6] with white inclusions that were visible through its thick cover of dust; on studying it more closely, Duke reported seeing greenish phenocrysts sparkling in the grey matrix.

While Young and Duke collected these samples, Fendell panned and zoomed in response to prompts from the Backroom, and it was not long before they spotted a rock on Plum's eastern rim that had a 'white patch' on top. This could be either a dark breccia with a clast or, tantalisingly, a crystalline rock with a cleavage face that was reflecting the Sun. They told England to suggest the astronauts take a look at it "if they happened to venture over that way". As Young and Duke worked in towards Plum on their radial, they collected a piece of glass[7] and "a good-sized" light-matrix dark-clast breccia[8] that displayed lineations.

"You guys look like you're having a ball," England said, when Fendell resumed his watch on their activities.

"We are!" Duke assured.

"It really is fun," Young added, more reflectively. They had trained to recognise all manner of volcanic rock, but there didn't seem to be any of it around.

Back at the Rover, Duke noticed that Young's boots had exposed an "absolutely white" substrate beneath the dark surficial regolith.

"Do an exploratory trench," Young prompted.

"Look at that!" Duke exclaimed, after he had used his scoop to dig down into the substrate. To follow up, he scraped a proper trench and took samples from different depths.

With time running out, Duke set off running around Plum's eastern rim to inspect a flat boulder on the southern rim. As Young followed at a slower pace, he shuffled right by the rock that had so fascinated the geologists.

At a metre and a half in size, the boulder that Duke had spotted was the largest block in the vicinity. Being large and almost buried in the rim,

it was a sure bet that this rock came from the deepest part of the pit, and as such was precisely what they were looking for.

"This thing has greenish-black clasts in it," Young reported as he examined it.

"Let's see if we can get a piece," Duke urged. The rock was intensely fractured, so it might readily yield a chip to the hammer.

Previously, the hammer had been stowed on the PLSS of the astronaut who would *not* use it, where it would be accessible to the one who *would* use it. This arrangement was fine so long as the crew remained together. However, Young had decided to carry his hammer in his shin pocket to facilitate solo-sampling. The geologist in him was only too delighted to strike the rock. A single solid blow to one of the fractures was sufficient to detach a conveniently sized fragment.[9] Duke described it as a grey-matrix breccia containing light and dark clasts a millimetre or so across, together making up about 5 per cent of the rock. Turning it over, he saw a crystalline clast about 5 mm in size. Like the other rocks in the neighbourhood, this one was pretty shocked.

Watching the sampling at full-zoom from the opposite side of the crater, England questioned their identification of the rock as a breccia. If it had been excavated from deep in the Cayley, it ought to be volcanic!? Duke handed the rock to Young. "It's a breccia," Young agreed upon seeing the inclusions. "Or a welded..." he began, as he studied the matrix. But the words died on his lips – he had been about to say ashflow tuff, which is a congealed pyroclastic deposit, but changed his mind. "It's a breccia," he confirmed. But since he could see "two or three different types of clast" it was an interesting one.

England pointed out that they had only 9 minutes remaining on the timeline for Station 1, so while Young took a pan from Plum's southern rim Duke went further south to look at some angular rocks. Having left the gnomon on the boulder, he dug his scoop in the ground to provide a reference for the sampling documentation, and then, because the rock was far too big for his scoop, he used the scoop as a prop and dropped to one knee in order to snatch the rock,[10] which proved to be a light-matrix breccia with dark clasts and was coated with glass. It was too big for a sample bag, so he put it straight into Young's SCB as a loose sample.

Seeing the astronauts returning around the crater, England seized his opportunity. "There's a rock that has white on the top, and we'd like you to pick it up as a grab sample." He meant that

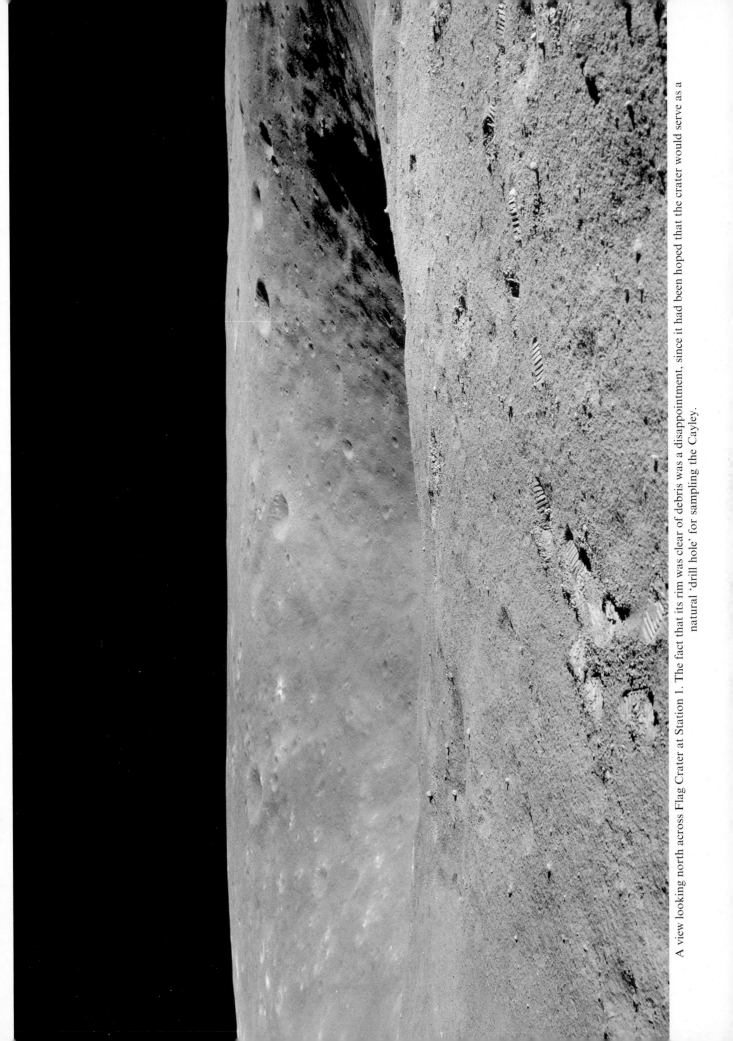

A view looking north across Flag Crater at Station 1. The fact that its rim was clear of debris was a disappointment, since it had been hoped that the crater would serve as a natural 'drill hole' for sampling the Cayley.

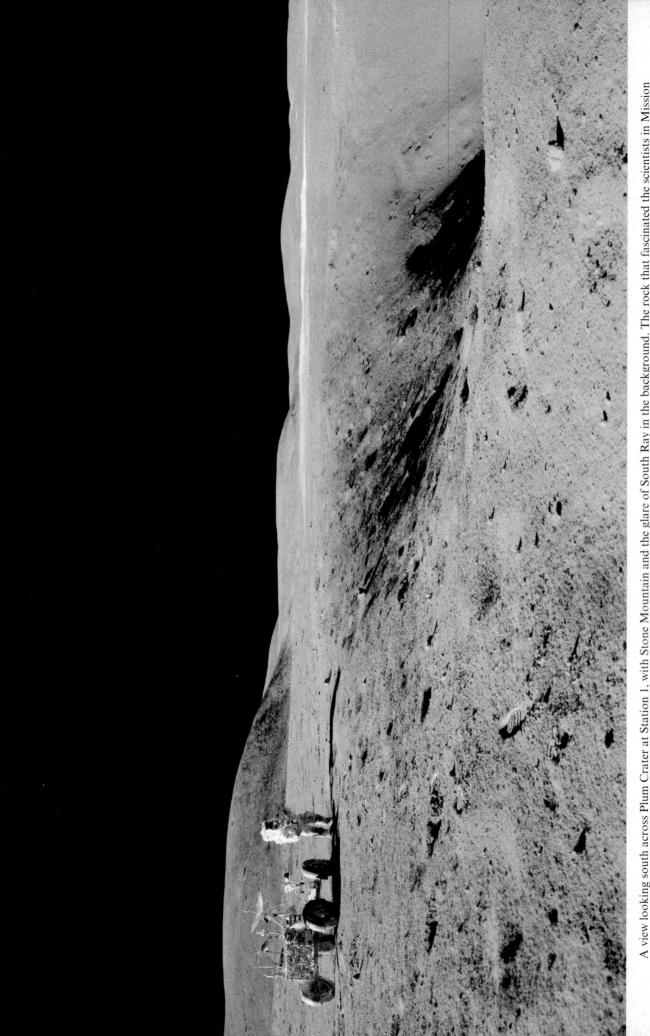

A view looking south across Plum Crater at Station 1, with Stone Mountain and the glare of South Ray in the background. The rock that fascinated the scientists in Mission Control is located just above the end of Young's shadow.

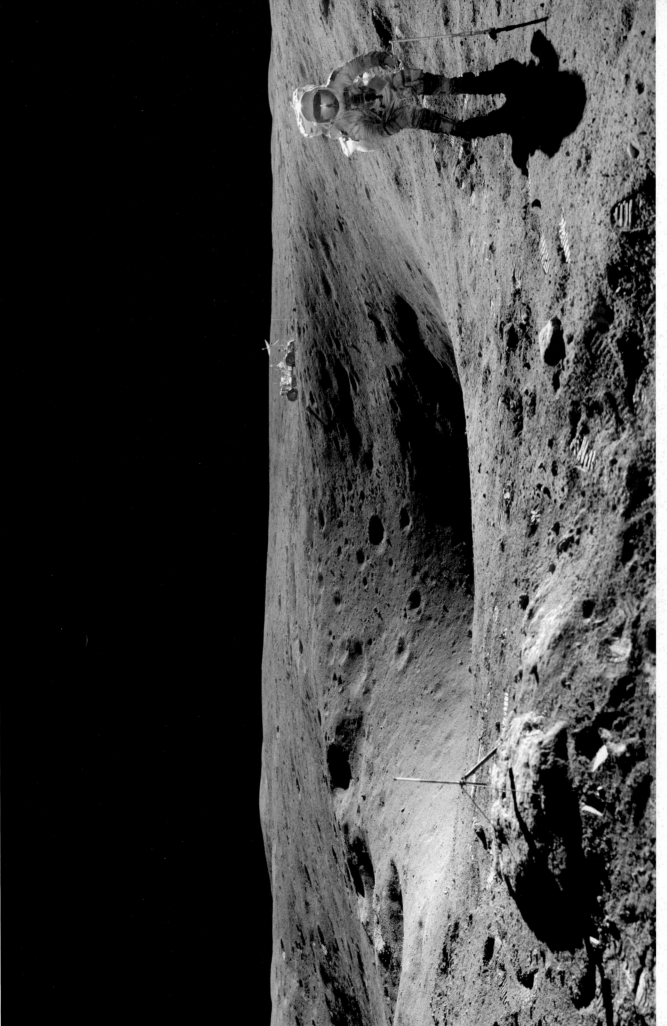

The septum between the craters Flag and Plum.

John Young takes his hammer to the boulder on the southern rim of Plum Crater.

Apollo 16 mission scientist Tony England in the Mission Operations Control Room as John Young and Charlie Duke work on the lunar surface.

because the rock's context had been established by the TV, they need not photograph it.

"This one?" Duke asked as he approached the rock. He paused alongside it, and pointed with his scoop to be sure that he had the correct one. England confirmed the identification.

"Are you sure you want a rock *that* big?" Young enquired, pointing out that it was probably a "20 pound" rock; if Houston really wanted a sample, maybe it would be better to hammer a fragment off it.

In fact, the Backroom had not realised how big the rock was until the astronauts reached it and gave a sense of scale. There was a limit on the mass of lunar material the LM could lift from the Moon. While the Backroom argued the wisdom of taking it, England, aware only of the certainty of the initial request, replied positively: "Go ahead and get it."

The rock was dusty, but as he stood alongside it Duke noted that he could see "some beautiful crystals" glinting in the Sun. Even so, it was clearly a breccia. As before, he leaned on his scoop and eased down to his knees in order to dig the rock from the ground by hand. After rolling it against his knee, he nearly stumbled backward as he stood up. Being on the inner slope of the crater, he was in a precarious position. "If I fall into Plum Crater getting this rock, Muehlberger has *had* it!" Once he was safely away from the crater, Duke scraped the dust to inspect the 'white' that had attracted Houston's attention. He initially thought that it had glass in it, but then saw that the clast was shocked crystalline rock. "It's a sort of a

feldspar-looking crystal." There might not be any volcanics laying around, but the Cayley breccias included clasts of igneous rock. Dubbed the 'Big Muley', this rock[11] is the largest sample returned by any Apollo mission and, as Duke diagnosed, is a grey-matrix breccia with large light clasts.

The plan called for a reading at Plum using the Lunar Portable Magnetometer, but as they were now running 24 minutes late this time-consuming procedure was deferred.

"I can't wait to get back to Buster," Duke said as they left Plum and headed back east.

Because they had arrived at Plum late and then spent longer there than scheduled, Station 2 had to be cut to 20 minutes, which was less than half of the time allotted. It would have been possible to extend the EVA to regain this slippage, but for the fact that Duke's PLSS was consuming its water coolant at a higher-than-nominal rate. There would be little point in him switching to a lower setting, as he would overheat and then be obliged to use an even higher setting to recover. They would just have to make the best use of the time that they had left.

Driving east was easier than driving west, not least because Young now knew that there were no scarps in his way, and he ran at full speed, bouncing across the smaller craters and dodging the larger rocks. It was a "real sporty" ride. "You can't believe how up-and-down this is, Tony," Duke noted.

Seeing an opportunity to check off another question from the Backroom, England enquired whether the Rover left different tracks when in and out of the ray material. Young just laughed. In suits with no peripheral vision, and strapped tight into their seats, they couldn't turn around to look at the tracks behind them. He could have looped back in order to take a look, but was not going to waste the time. Persisting, England suggested that maybe they could see the tracks made by the front wheels. "It's a cloud of dust, Tony," Duke chided. He couldn't look to check, but it was a sure bet that at this speed the Rover was leaving a veritable rooster tail in its wake.

BUSTER!

As they approached their destination, England suggested that it would be best to park on Spook's rim, so that the Backroom could check it out while they worked at Buster, but Duke said it was Buster that would "turn them on". In the event, because Buster had a raised rim, Young parked by Spook, which, as with Flag, had a rather subdued rim suggestive of great age.

England recommended that they take the deferred magnetic field reading, take the planned pictures using the 500-mm lens and get "a couple of samples of Buster".

Leaving Young to set up the magnetometer by the Rover, Duke first took the pan from between the craters, and then used the "big camera" to shoot Stone Mountain and South Ray, which were visible beyond Spook. South Ray was a gleaming white ellipse to the west of the mountain. As on the Apennine massifs, the hilly Descartes displayed striking lineaments. As these tended to become less distinct when viewed from close up, the mountains were best photographed from afar. However, most of Duke's long-range shots were smeared by camera shake.

"Wowee!" Duke laughed, as he scrambled onto Buster's rim. "This is some sight, looking down into this beauty." Because the TV could not see it, he provided a brief report: "The bottom is *covered* with blocks up to about 5 metres across. The blocks seem to be in a preferred orientation, northeast to southwest. They go all the way up the wall on those two sides, and on the other sides you can only barely see them outcropping." He noted that the distribution of the ejecta was similarly asymmetric, there being only small fragments where he was, on the southeastern rim. It was not just the large scale that displayed this preferential alignment. As he studied the rocks on the floor of the crater, Duke saw that many of the blocks seemed to be mutually aligned.

An asymmetric distribution of debris would normally indicate that a crater was a secondary, with the shattered projectile littering the far wall and extending down-range. In this case, however, the coherent arrangement of the blocks suggested that Buster was a primary, and that the projectile had shattered a large slabby block buried within the Cayley. How had such a slab come to be buried? There was no time to speculate. The geologists could study the block distribution in his pictures at their leisure.

As Young would require most of the available time to operate the magnetometer, Duke had brought a spare SCB with him to assist solo-sampling. The soil on the rim *looked* soft, but it was pretty well indurated, so he set the bag down right on the rim and set to work.

Meanwhile, Young was experiencing trouble with the magnetometer. Shepard and Mitchell had taken such an instrument up Cone Ridge, and it

had been tricky to use. The sensor package was set on a tripod and linked to its control unit, which was on the Rover, by a 20-metre cable. Unfortunately, the cable would not unreel properly, and then it continued to twist, disturbing the package, which had to be in a specific orientation with respect to the Sun. The instrument had a trio of orthogonal sensors, so three datasets were needed for each 'measurement'. Frustratingly, the sensors had to be left to settle, which obliged Young to stand idle.

Duke's objective was to sample both typical and exotic rocks. However, Buster's rim was so dusty that he did not find out what he had selected until he had finished the prior-to-sampling documentation and lifted a rock. In the event, his first sample[12] appeared to be igneous, and his second,[13] taken from alongside, was a light-matrix breccia with anorthositic clasts. Moving just inside the rim, he took a sample[14] that proved to be a larger fragment of this type of breccia.

At this point, England warned that they would have to leave as soon as Young had finished taking the LPM data, so Duke retrieved the SCB and moved 20 metres back towards the Rover to sample a slabby fist-sized rock. After depositing the SCB, he used his scoop to flip the rock into the air and tried to catch it, but he succeeded only in batting it back and forth several times and it fell back to the ground. Flipping the rock again, this time he ran forward and managed to trap it against his shoulder. As Fendell followed this juggling, laughter could be heard from the controllers. Duke scraped the dust and exposed a gleaming white rock.[15] "It's very friable," he pointed out. "It's the most shocked rock I've ever seen." Because the "pure white" matrix did not display any sign of inclusions, he decided that it was "not a breccia". In fact, it was a breccia, but an 'anorthosite-in-anorthosite' breccia in which fragments of anorthosite were bound together by a matrix of pulverised anorthosite.

"The geochemists are always telling is how little rock they need," England joked as pieces of the rock flaked off in Duke's hands.

Although, Young had finished stowing the LPM, Duke lingered to take another sample for his rudimentary radial; it proved to be another seemingly igneous rock.[16]

Although Duke's suit was using coolant water faster than intended, the walkback constraint became less strict as they approached the LM and it had been decided to relax the margin, and so the EVA was shortened by only 8 minutes. Although this time had to be 'used' at the landing

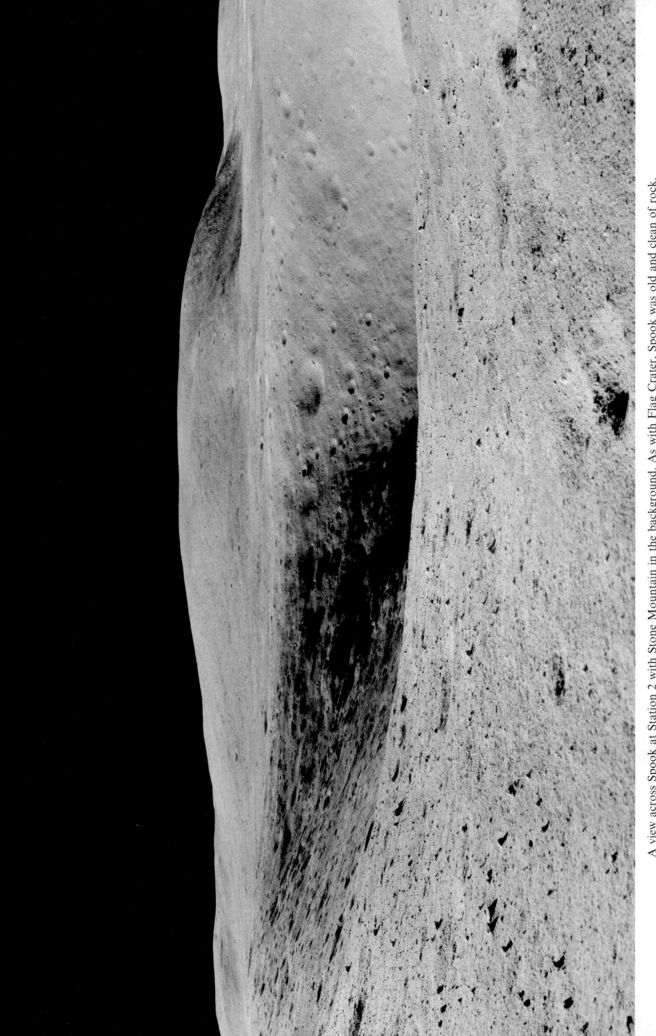

A view across Spook at Station 2 with Stone Mountain in the background. As with Flag Crater, Spook was old and clean of rock.

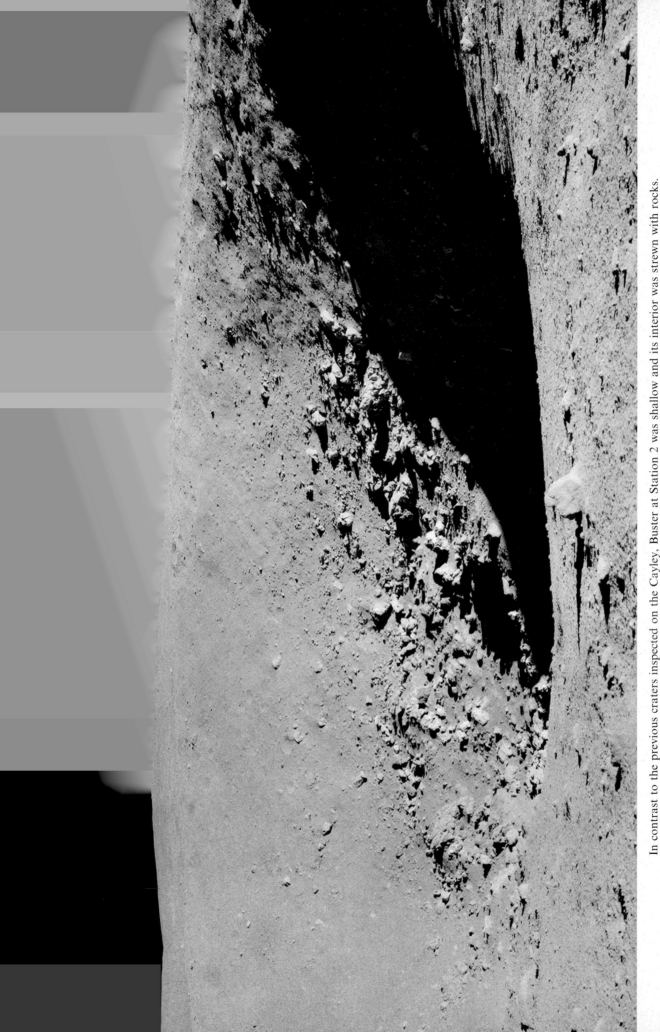

In contrast to the previous craters inspected on the Cayley, Buster at Station 2 was shallow and its interior was strewn with rocks.

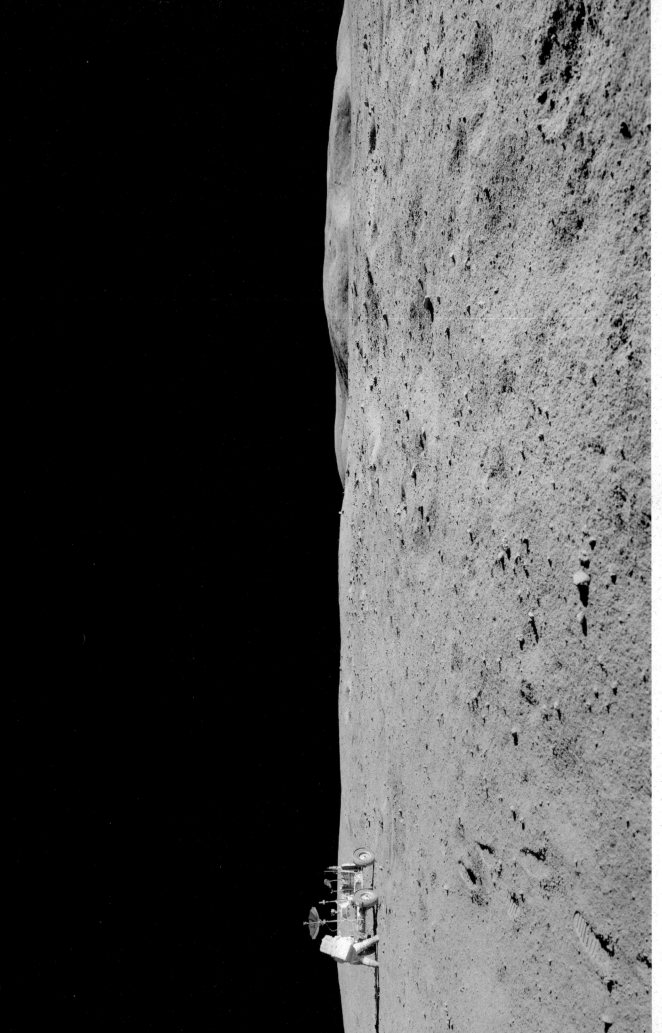

A view north from Station 2, with Smoky Mountain in the background.

site, Station 2 was allowed to run well beyond the "20 minute" mark.

As a result of his antics, Duke was so covered in dust that Young teased him that there was likely to be enough dirt within the folds of his suit to provide a regolith sample.

On the return trip, their contented mood saw them laughing and joking. Given the things that could have gone wrong, they were doing well. Apart from the loss of the Heat-Flow Experiment, the deployment of the ALSEP had been nominal, and they had begun the traverse right on time. They had lost a little time driving down-Sun in unknown and possibly treacherous terrain, but had managed to visit both of the sites planned. If the ALSEP had put them significantly behind, then they might not have made it to Plum at all. In fact, if the Rover had not deployed properly, or had proven unserviceable, they would have been limited to a walking traverse with Buster as its sole objective.

"We need to stop for the Grand Prix," Duke reminded, as they approached the ALSEP site. They were to repeat the engineering assessment of the Rover that was done by Apollo 15 but not recorded due to a problem with the 16-mm movie camera. Young let Duke disembark and retrieve the camera, and then he went to conduct the test.

As Duke tracked the Rover's motion, he commentated in his usual excitable style. "He's got about two wheels on the ground. And as he turns, he skids. The back end breaks loose – just like on snow. When he hits the craters, and starts bouncing, is when he gets his rooster."

They were to record about 4 minutes' worth of film, but Young completed the test in less than half of that time. Duke suggested that he repeat it, and Young obliged, this time making a series of impromptu sharp turns. "Hey, that's great!" Duke called delightedly. "When those wheels really dig in, you get the rooster."

Concerned that Young might be being a little too rough on the vehicle, England called a halt. Perhaps it would have been wiser to schedule the potentially damaging test for the end of the final traverse, just to play safe.

"Man!" Duke exclaimed as Young ran straight across a small crater. "That was *all four* wheels off the ground."

"I have a lot of confidence in the stability of this contraption," Young reported as he finally drew to a halt alongside Orion. It pitched and rolled as it bounced in the low lunar gravity but, having such a low centre of mass, it was unlikely ever to flip over.

As Duke walked back, he noticed that the

A frame from the 16-mm movie taken by Charlie Duke of John Young putting the Rover through its paces for the 'Lunar Grand Prix'. When travelling at full speed, the plume of dust thrown up by the rear wheels was a veritable 'rooster tail'.

angular rocks of the ray crossing the landing site appeared to be "different" from those that he had seen out west: many of them were coated with glass. Ruminating upon it, he concluded that the rocks at the ALSEP were predominantly grey-matrix dark-clast breccias, whereas those at Buster were very shocked – so much so that the one he had juggled had actually crumbled in his hand. This reinforced the likelihood that, even though the ray ran across the Buster/Spook site, the rocks he had sampled at Buster were local.

Young's opinion of the Cayley was blunt. "There ain't nothing 'plains' about this place." It certainly wasn't flat. "I don't know whoever thought this was a plain," he laughed dismissively as he went to service the UV camera.

Duke took advantage of the lull to grab a rock that had attracted his attention prior to leaving the LM – it appeared to be crystalline, and was coated with glass.[17] On recalling that England had asked whether he had seen any rocks at the landing site which he was certain were not breccias, he announced that it was "not a breccia". He examined it carefully, and pronounced it to be "an igneous, plutonic rock". It was anorthosite. Almost football-sized, it was too large for a sample bag, so he took it to the MESA to be placed into a rock box.

A few hours later, as they ate their dinner, Young and Duke reflected upon their day, and worked through the Backroom's list of geology questions.

"I'd like to give you what I think are the three major areas that we saw today," Duke began. After confirming that the landing site was under South Ray's influence and that Plum was not, he turned to Buster. "I don't think it's a secondary, because the rocks were truly shocked, and I can't see a secondary doing that." He had no reason to suspect that the Cayley might itself be extremely shocked rock and that a secondary impact had merely excavated it.

When Mattingly asked Houston whether the first day had turned up any surprises, he was informed: "they found all breccias; they found only one rock that might be igneous."

"Is that right!?" Mattingly laughed. "Well, it's back to the drawing boards, or wherever geologists go."

This, indeed, was beginning to look like what the geologists would have to do. The Cayley was clearly not what it had been interpreted as. The first step in photo-geology was to map all the morphological units in terms of their surface characteristics. This was done *objectively*. The stratigraphic relationships gave the depositional sequence. The result was then *subjectively* interpreted in terms of the processes that might have been active. As the 'light plain', the Cayley *was* a distinct morphological unit, but its volcanic interpretation seemed to be wrong. If it was brecciated material, then the 'upland fill' was more likely to be ejecta from a basin-formation event. This was something for the Backroom to ruminate on overnight.

HEADING SOUTH

The primary objective of the second traverse was to sample Stone Mountain in order to determine whether it was a silicic lava dome. After sampling a 'drill hole' crater on the mountain, Young and Duke were to drive off the mountain along a 2-km line that crossed the contact between the Descartes and the Cayley, and then drive home through the eastern fringe of the South Ray ejecta blanket. Several locations on the northwestern flank of the 530-metre hill had been shortlisted as possible stations, but the plan was flexible because the resolution of the overhead imagery was too poor to show which crater had the blockiest rim. The highest option was Crown, about 100 metres in diameter located about two-thirds of the way up, but the most likely was one of a cluster of five smaller craters half way to Crown that had been dubbed the Cincos.

While Duke configured the Rover for the traverse, Young tilted the UV camera right over in order to photograph Stone Mountain. Its 20-degree field of view would not offer any greater resolution than the telephoto Hasselblad, but it was hoped that the ultraviolet-sensitive film would detect emission from any gas escaping from the supposedly volcanic pile. On his way to join Duke, Young paused to lift a fist-sized "very white rock with a black glass layer and a lot of zap-pits". To have suffered so many impacts by micrometeoroids, it must have been on the surface for a long time. It was too large for a bag, so he put it into one of the spare SCBs on the Rover. This first sample of the day was a light-matrix light-clast breccia.[18]

Although it took about 40 minutes to complete their chores, they were once again able to set off essentially on schedule. Today, by driving cross-Sun, they hoped to be able to make excellent time in the traverse.

Not only was the terrain undulatory on the local scale, the Cayley itself sloped down to the

The Ultraviolet Camera was set up in Orion's shadow, and left to work through a programmed set of astronomical targets.

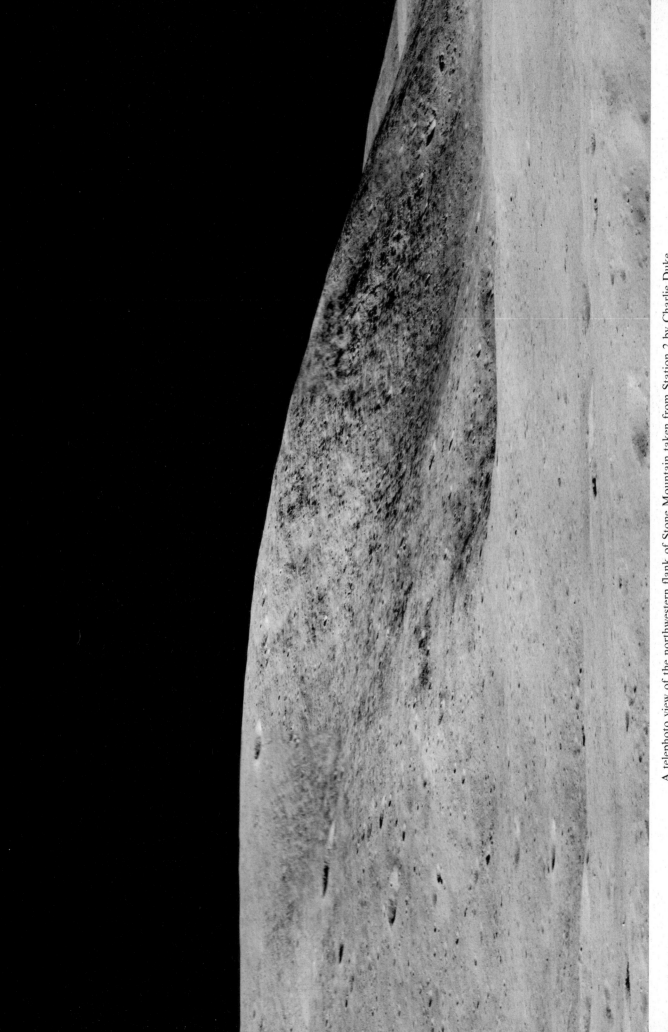

A telephoto view of the northwestern flank of Stone Mountain taken from Station 2 by Charlie Duke.

A view of Orion taken by Charlie Duke at the start of the second excursion.

southwest at an angle of about 5 degrees. The major ridges ran directly downslope, and so were also aligned on this heading.

Several hundred metres south of the landing site, they descended into a shallow valley. "There sure are a lot of rocks," Young observed. Duke pointed out that there were clearly two populations: those that were angular and fresh-looking, and those that were rounded and much dustier.

Looking ahead, Duke studied the next ridge, about a kilometre away. This was Survey Ridge, and they were to drive up onto its crest and follow it southwest for half a kilometre or so before descending into the next valley, which would take them to the base of the mountain. The approach to the ridge was saturated with secondary craters several metres in size, prompting Young to slow down to 6 kph. "Its getting rougher," Duke reported. "A lot more hummocky." Since setting off, he had been assessing the coverage of the cobble-sized rocks, which seemed to be an indicator of being within the ray. "We haven't got out of this cobble field yet," he told England. In fact, the coverage markedly increased, and they encountered a number of metre-sized rocks that Duke reported to be breccias. They were penetrating the dense *core* of a ray. "The top of Survey," Duke told England as they crested Survey Ridge and Young swung right, "has got a *lot* of secondaries."

"Take plenty of pictures!" It was frustrating that the TV could not function while the Rover was on the move.

As previously, Duke was taking pictures to document the route. When these were analysed, they showed that he was overestimating the coverage of cobbles. This was because on the low-slung Rover his foreshortened view made a sparse litter appear denser than it was. The pictures had the same angle of view of course, but they were able to be measured more thoroughly.

"It's very blocky," Duke emphasised. Not only were there more rocks on the crest of the ridge, there were large boulders too.

"You won't *believe* the blocks," said Young, slowing to walking pace in order to weave his way through.

Although driving along the ridge was slow, it offered the advantage of improved visibility, and Duke, free to sight-see, could see the core of the ray as a line crossing the undulations. "I can track it right in – up across the ridge that blocks out Wreck and Stubby, and *into* South Ray," he reported, referring to two of a chain of craters that

arced around the eastern side of South Ray, although at the moment his view of Wreck and Stubby was masked by a ridge.

"We've got to get out of this, Charlie," Young sighed. "We could spend the rest of the *day* in this ray." It had taken 10 minutes to run the half-a-kilometer section along the ridge.

In the valley south of Survey Ridge the terrain was, if anything, more undulatory, and both the number of larger secondary craters and the size of the rocks increased.

"There's a great split boulder!" Duke called. This had struck sufficiently violently to split open, yet the two pieces had barely separated. As soon as he could, Young increased speed.

"There's the Cincos, right up there," Duke pointed out.

Young considered the merits of Crown versus the Cincos. Although Crown was sufficiently elevated for mass-wastage on the slope to have thinned the regolith and it was large enough to have excavated the mountain, it was clearly an old crater with a subdued rim which – if Flag and Spook were anything to go by – would probably be clean. He would decide when he got to the Cincos.

Distracted by the view, Young drove over a rock that scraped along the underside of the Rover. "I've never seen so many blocks in my life," he muttered. This was no idle remark; he had visited the craters made by underground nuclear tests, some of which had *really* broken up the overlying terrain. To make up time, he accelerated to 12 kph and ran across instead of around the smaller craters, briefly turning the Rover into a low-flying spacecraft.

On the overhead imagery the rays appeared to have quite distinct edges. The one that crossed the landing site certainly had a noticeable western boundary. Although the blanket of rocks thinned as they neared Stone Mountain, Duke noticed that it extended up onto the mountain's flank, which was bad news because this 'pollution' would make it more difficult to find samples that were representative of the mountain. As they neared the base of the slope, he monitored the regolith for any indication of the contact between the Cayley and the Descartes, but there was nothing. Nor were there any rocks that looked as if they might have originated on the mountain.

Just as the 'nav' put the Rover 3 km from Orion, they finally started up the slope. The broad shallow dome was etched by a furrow that ran southeast to northwest (perpendicular to the relief on the plain, in fact) and its northwestern flank, where Young and Duke were heading, was terraced with a series of arcuate benches. One of their tasks was to sample the benches, on the principle that these would match the profile of the underlying mountain. The base of the hill was a pronounced 10-degree slope, but this soon yielded to the first bench – which they were to sample on their way back down.

"The block population is up again," Duke reported. It was apparent from the tone of his voice that he had not seen any difference in the rock type.

The slope beyond the bench was steep, but the Rover took it without difficulty. When they were beyond the next bench, England urged them to be alert for any fresh crater with a blocky rim. In fact, there was a promising 20-metre crater. "That should be a good Station 5," Duke ventured. He took note of the 'nav' so that that they would be able to return to this crater on their way down without wasting time. "We're *really* going up a hill," he laughed as the Rover clawed its way up the 20-degree incline.

"We're at the Cincos, Tony," Duke reported as they crested the next rise several minutes later.

Following their progress on his map, England urged them to drive on for another few hundred metres to reach Cincos-A, the largest of the cluster. "You're well ahead on the timeline," he encouraged. Almost one-third of the 30 minutes they had been 'on the road' had been spent in the debris on Survey Ridge.

"Let's go up," Young said.

"Where?" Duke asked.

"Right up there!"

"That's Crown," Duke pointed out.

"You don't want to go up?" Young asked. He had not seen anything yet to tempt him to stop.

"Why don't we stop *here*?" Duke indicated a field of large rocks that was directly ahead.

"We concur," England chipped in, satisfied that by now they must be at the point that he had circled on his map.

"Okay," Young acceded, and stopped to look around.

"Can't we get up closer, so we won't have so far to walk up-slope," Duke urged.

Parking on the slope would be tricky. But they had just passed a 15-metre crater, and if they parked there then its wall would cancel the slope of the hill. "Charlie, I want to go back down and park in that crater," Young decided. They would simply have to trek up the slope to search for the crater that had issued the rocks.

At this point, England said that Mattingly, who

was overflying, had just noted the Sun glinting off something on the northwestern flank of Stone Mountain.

"That's us!" Duke laughed.

As they drove 30 metres down to their chosen parking spot, Young and Duke got their first view back across the plain. "Tony," Duke called excitedly, "you just won't believe this!" At an elevation of almost 175 metres above the plain at the base of the hill, they were 50 metres higher than Scott and Irwin had reached on the flank of Mount Hadley Delta.

HIGH ON STONE MOUNTAIN

Duke's first task was to take telephoto pictures of where they intended to go on the final traverse. The combination of the facts that the plain sloped up to the north and North Ray was on a ridge meant that he could not see the crater, but he could see its rim, and right at the position that they planned to sample was "one huge boulder". It could only have been excavated from great depth. There were boulders running from North Ray right up Smoky Mountain's western flank in "a great ray pattern". As he had during the landing, he checked the route that they were to drive, and verified that the northern stations appeared to be accessible.

Next Duke turned his attention to the west. He could look right down *into* South Ray, about 4 km away and 750 metres in diameter. A kilometre to its north was a 150-metre companion dubbed Baby Ray. These craters were *so* bright that it was painful to look at them. The craters Trap, Wreck and Stubby formed an arc barring access to the ray craters. Now that he could see into it, Stubby's eroded rim did not appear to be much of an obstacle. But there were so many blocks close to South Ray that the Rover would have had great difficulty reaching that crater's rim. He followed a ray out across Stubby and then over Survey Ridge to verify that the rocks that he had observed during the drive south had indeed come from South Ray. Studying the rays closely, he realised that rather than there being alternating streaks of debris and no debris, there was actually a fairly even coverage, and the streaky pattern was due to alternating rays of predominantly light or dark material. There were actually narrow lines of dark blocks running in over South Ray's rim and down into the crater. There were clearly two populations of rock, presumably excavated from different depths. In fact, the overhead imagery from

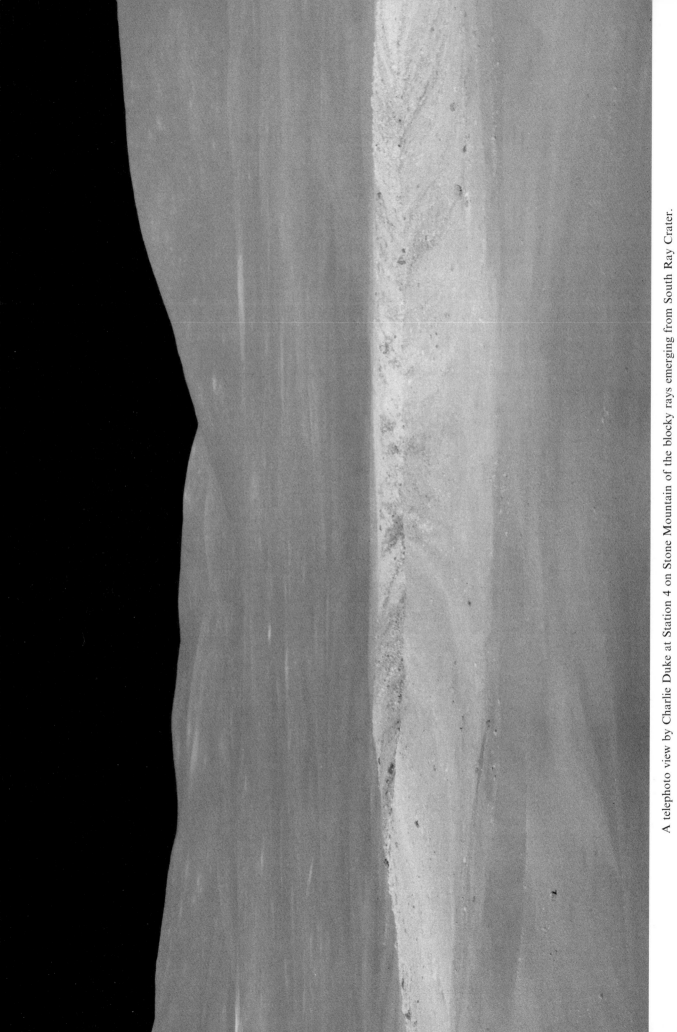

A telephoto view by Charlie Duke at Station 4 on Stone Mountain of the blocky rays emerging from South Ray Crater.

cameras in the CSM would reveal that South Ray had dug through 10–15 metres of regolith and a layer of dark rock 60 metres thick before bottoming out in lighter rock. Although they could not reach the rim, the plan was to sample the rays on the way home, so with luck they would gain some insight into the deep substructure of the Cayley.

On returning to the Rover, Duke saw that the TV camera was on him. "Look up-slope, Tony," he prompted. "You'll see the rock field that we're in." He stowed the telephoto, took the rake and went 10 metres up the hill to trawl for pebbles that would hopefully characterise the underlying mountain.

Having taken his first sample[19] – "a hard rock" that he thought was glassy but was so dusty it was difficult to be sure – Young joined Duke to assist with the rake. "Most of these have white clasts," he pointed out upon examining the dozen or so pebbles in the first swath.

"You think they're breccias, then?" England posed.

Knowing how much rested on his assessment, Young was reluctant to commit himself. He reiterated that the dust made it difficult to see. Perhaps 'clasts' had been too specific? He meant that he could see light-toned patches showing through the dust. "They could be shocked rock," he allowed. He cogitated, and then announced: "I think these rocks were laid here when South Ray came in."

"I don't get the impression that they're breccias," Duke offered. As he dug deep to extract a soil sample, he exposed a layer of white material. "Just like the Cayley," he reflected.

The next item on Duke's list was a soil-mechanics test using the penetrometer. As this would take at least 10 minutes, Young resumed solo-sampling, and collected a "splatter of glass"[20] that proved to be regolith splashed by impact-melt, and "a very angular" fragment[21] which was related to a rock nearby that had "a brecciated appearance".

Finished with the penetrometer, Duke went to take a double-core below the rim of the crater in which they had parked. "You know, John," he said. "I'm not sure we're getting Descartes."

"I'm not, either," Young agreed. "I wish I could say these rocks differed, but they don't." They were supposed to be sampling the mountain, but the rocks appeared to be the same as those on the plain, which made them South Ray ejecta.

"Do you see a blocky crater within walking distance?" England asked hopefully. Although Fendell had made the usual site pan, it was not clear whether there was a crater in the rock field up the hill, or whether it was simply the core of a prominent ray. According to his map, they should be close to Cincos-A, but unfortunately this crater was masked by a small ridge 80 metres to the east.

"How much time have we got?" Young asked.

As this site, Station 4, was to be the primary sampling stop of the traverse, it had been allocated 50 minutes. "About 22 minutes," England replied.

Using a determined stride to overcome the gradient, Young headed up the hill to find out whether there was a crater in the rock field. After 75 metres he came across a fairly fresh 25-metre-diameter crater that had a jumble of rock running out over its eastern rim. The distribution of the debris indicated it to be a secondary from South Ray. The variation in the tone of the blocks on the far rim suggested that they were breccias, but they were predominantly of the dark-matrix type. As fragments of the impactor, the rocks were useless for sampling the mountain. As the western side of the crater was totally clear, Young walked around it seeking a bench where he might be able to dig up an *in situ* sample, but the crater had not penetrated to the bedrock. In the hope that the northwestern section (opposite the spill of rock) would be less polluted by South Ray, he sat the solo-sampling SCB near a small crater on the rim and took a soil sample. As he did so, Duke started up the hill with the rake, having reasoned that their best chance of gaining Descartes material would be by raking the clear rim for fragments of rock that were working their way up through the regolith as a result of the on-going gardening process.

"I'm pooped!" Duke complained on reaching the crater. Young took the rake and got to work, gaining more than a dozen "really dust-covered" fragments in the first swath.

"I hate to tell you," Duke warned, "but it's indurated regolith." He meant that the fragments were not rocks, but regolith breccias.

"Clods of dirt," Young acknowledged in disappointment.

As Duke made his way back down to the Rover, the soil slumped beneath his feet and his boots left a train of deep imprints. After a moment's reflection about how the primary objective of the traverse was eluding them, Young followed suit.

"Cinco Crater, we bid you a fond farewell," Duke said sardonically as they pulled out of their 'parking lot' at the scheduled time. Perhaps they would have better luck lower down. "Tony," he called, "I think we ought to stop *away* from any boulders."

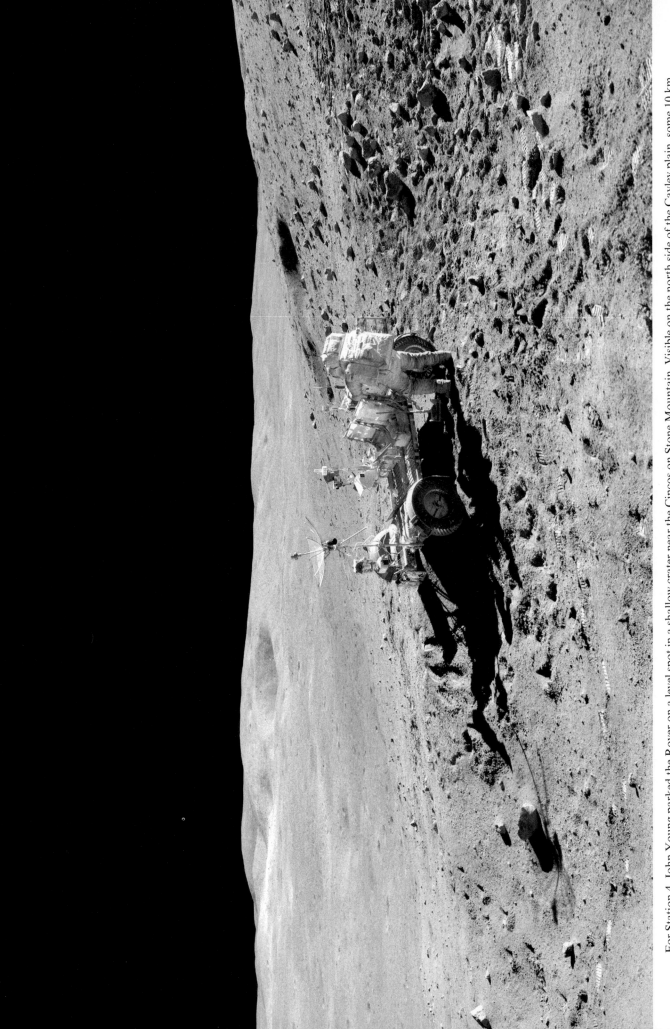

For Station 4, John Young parked the Rover on a level spot in a shallow crater near the Cincos on Stone Mountain. Visible on the north side of the Cayley plain, some 10 km away, is Smoky Mountain.

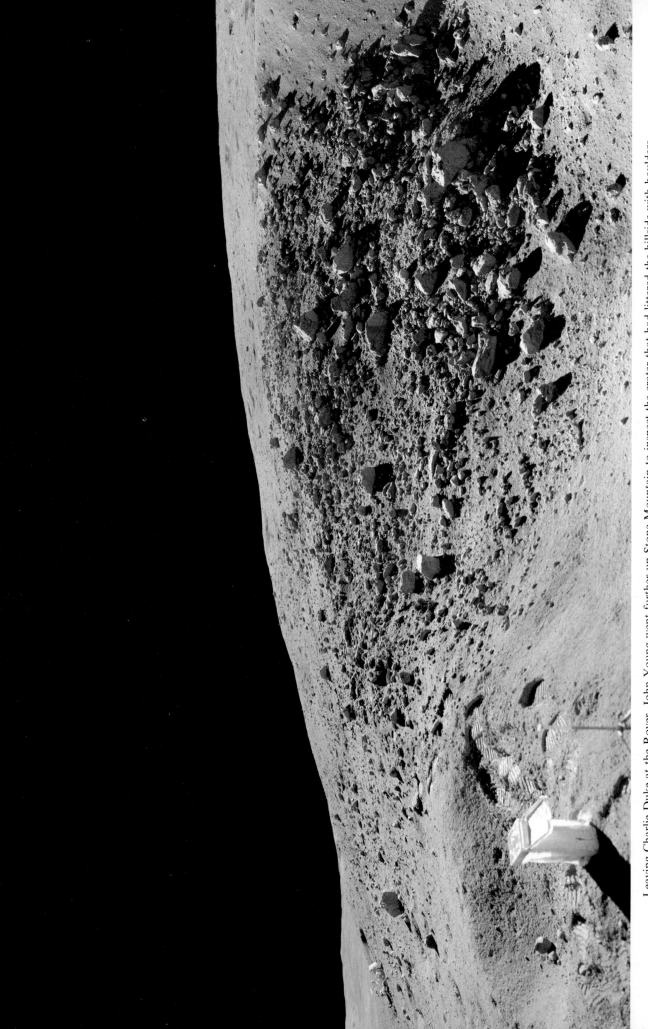

Leaving Charlie Duke at the Rover, John Young went further up Stone Mountain to inspect the crater that had littered the hillside with boulders.

"We agree," England replied. Astronauts were irresistibly drawn to boulders but, if these were pollution, to do this would confound the effort to sample local material. "What we're looking for," he emphasised, "is a primary." Since a high-speed impact would have vaporised the impactor, its debris would not be present to complicate the sampling.

"Suppose we give you a primary impact with no blocks?" Duke posed.

"It'll have to be blocky," England insisted. There was no point sampling a crater that had just rearranged the dirt. As yet, the Backroom did not appreciate the extent to which South Ray had mantled the flank of the mountain.

THE LOWER SLOPE

As Young 'free-wheeled' the Rover down the hill, Duke studied the crater that they had spotted on the way up. Although it was littered with cobbles, the paucity of large blocks suggested that it was unrelated to South Ray. "It might be a primary," he ventured. Sampling its 'protected' inner slope might pay off.

Young parked outside the crater's eastern rim in order not to disturb the sampling area but, after disembarking he decided that he did not want to park the Rover on the slope. Rather than climb aboard and drive to a flat spot on the rim of the crater, he and Duke simply hoisted the vehicle between them and carried it!

"The angular fragments are probably from South Ray," Duke mused as he took the site pan, "but maybe the rounded ones are working their way out of the regolith."

"A rake sample in the wall would probably be our best bet," Young said, having decided to ignore all the surface rocks. The rim of the crater would be material dug from the deepest point. Although in this case, this was clearly no greater than a few metres, it might be deep enough to be from below South Ray's mantle. He raked a half-metre glassy crater that was just inside the western rim, using this as a 'drill hole' into the rim of the larger one.

"This isn't rock," Duke observed of the first sample. The bag held friable clods from the creation of the small crater. They broke apart as he kneaded the bag. "It's gonna be a bag full of *dirt* when we get it home!"

Young raked a position inside the main crater, gaining several real fragments of whitish rock, most of which were angular and several coated

with glass. "They may have come from South Ray," he conceded. Instead of making the second swath alongside the first, he decided to make the first swath deeper.

"These are rounded!" Duke said delightedly on examining the second batch. After scraping off the thick coat of dust, he reported them to be white. Surveying the rim, he spotted "an old rounded rock" that was "badly beat-up".

"That's what we're looking for!" England prompted immediately.

They dug the rock out of the regolith and, although it was a little tricky, managed to lift it up using the rake. The rock[22] was dirty, but Duke noted straight away that one corner was thickly coated with glass. He scraped away the dust and saw the rock sparkle. "It has a crystal in it, a couple of millimetres in size!" Was this a clast? Or a phenocryst? He scrubbed off more of the dust. "It doesn't look like a breccia, Tony."

"Outstanding!!" England replied.

Young took the rock and studied the granularity of its light-grey matrix. "It seems to be a fine-grained crystalline rock."

Finally, they had found something that might be volcanic. The light-matrix would be consistent with it being from a rhyolitic magma. However, when the rock was cleaned and studied on Earth it proved to be a breccia, and the crystals were in dark clasts that were too small to have been distinguished by the astronauts through the dusty coating.

Enthusiastically resuming their survey, Young and Duke found a fist-sized rock[23] on the south wall that was "at least a quarter" glass-coated. However, it was so dirty that Young refused to speculate: "It defies description."

On the principle that where there was one good rock there might be others, Duke raked the site and, after gaining a few fragments, snagged a much larger rock that he dug out. Young described it simply as "a friable white rock".[24] It proved to be a light-matrix breccia with dark clasts, much like the earlier sample.

Playing a hunch, Young decided to try a vertical trench in the wall of the crater. In fact, this was the worst possible position for such an experiment because the slope of the hill augmented that of the wall, and as his boots sank deep into the regolith he was obliged to 'tread water' in order to overcome the slumping while he used the clamshell of the rake as a scoop. But he persisted because the going was good.

"It all looks the same," Duke said, as the trench approached a half-metre in depth. In fact, there

At Station 5, John Young rakes in a small crater on the flank of Stone Mountain.

was the occasional "white splotch" in the regolith and many clods of regolith breccia, but only a few real rocks, and these were subsequently found to be fragments of brecciated material.

Sensing that they were making no real progress, England noted that they had been on-site for half an hour, and proposed that Young should take a magnetometer reading even though this was not on the plan. Although Young scrambled back up onto the rim to make his way around the western side of the crater, Duke decided to be a little more adventurous and ran straight across the pit and used his momentum to carry him up the far wall. "That was fun!" he laughed, as he popped up beside the Rover.

As Young used the magnetometer, Duke sampled near the Rover, rapidly bagging a fist-sized rounded rock[25] that was so dirty that all he could say about it was that it had "white streaks" (it turned out to be another light-matrix breccia), a white fragment,[26] and a dark-matrix crystalline-clast breccia[27] that had been partially melted and turned into a glassy agglutinate.

Meanwhile, Young literally stumbled on a sparkling rock. Although the highly angular white rock was clearly from South Ray, it was not at all what he had come to expect of that crater's ejecta, which, thus far, had been exclusively brecciated rock. After finishing with the magnetometer, he lifted the rock and examined its matrix to see whether the sparklies were clasts in a breccia. On finding that it was the matrix that was glinting, he judged it to be plagioclase. A green hue indicated the presence of olivine. "It's the first one I've seen on Stone Mountain that I really *believe* is a crystalline rock, Houston," he announced optimistically.

"That's the best sample we've got," Duke agreed, after looking at the rock. "It's a crystalline rock if I've ever seen one."

"First one today," Young repeated, emphasising the paucity of rocks which might be igneous, to say nothing of material that might be volcanic.

"Call it the 'Great Young'," England teased, alluding to the 'Great Scott' that was snatched by Dave Scott just before leaving Hadley Rille.

"Oh, come on!" Young groaned. In fact, since the rock had a triangular shape that resembled the head of a rattlesnake it later became known as the 'Diamondback Rock'.[28]

By the time they had packed up, they had been at Station 5 for 50 minutes, which was about 10 minutes longer than intended.

As Young concentrated on driving down to the lowest bench, Duke searched for a suitable site for Station 6. At this point, the bench trended northeast to southwest. They could not go eastward because they were to head northwest for the next batch of stations, but there was a likely target in the other direction, and on reaching the bench they turned left. "We're driving west along the bench, trying to find a blocky crater," Duke explained to England, to enable him follow their progress on his map.

"What do you think?" Young asked, as they approached the crater.

At 20 metres in diameter it was by far the largest crater in the immediate vicinity. The fact that it was fairly shallow meant that it could not have dug up any bedrock, so the sparse litter of half-metre-sized angular rocks were evidently from South Ray. What attracted Duke, however, was an abundance of much smaller rocks displaying a wide variety of shapes.

"Park!" Duke recommended.

In order not to disturb the inner western wall, they parked on the eastern rim. As soon they dismounted, they noticed that the regolith was so consolidated that their boots barely etched its surface.

After aligning the TV antenna, Young turned around to discover Duke standing just behind him, hammering on one of the more rounded boulders. "What are you doing, Charlie!?"

"I'm trying to get a fresh surface, to find out what it's made of," Duke explained. On finally splitting the rock open, he was astonished to see a pure white matrix with large elongate black clasts laced with black phenocrysts and lath-like crystals.

"That's a two-rock breccia!" Young called, his interest piqued. "Let me get a piece!"

"It's too hard," Duke insisted, "but it looks like there's some more sample-sized fragments here." He returned to the Rover to fetch the gnomon, and when he withdrew it from its holder the inclinometer detached from the tripod section! They bagged two of the smaller rocks,[29,30] but these light-matrix breccias were clearly unrelated to the larger rock.

Young scraped an exploratory trench to see if there was any layering, and found nothing of interest. "What's that?" he mused, spotting a peculiar light-toned patch of soil on the western rim of the crater.

"It's a unique-looking white something-or-other," Duke laughed.

It seemed that an extremely frangible white rock had flaked completely away and coated the crusty regolith, so they scooped up a sample.

England prompted them to prepare to move out.

Station 6 was alongside a small crater on the lowest bench of Stone Mountain.

At Station 6, Charlie Duke "whacked" this boulder to get a chip of what he hoped would be crystalline rock.

"Maybe we can get another sample," Duke ventured.

"Just one more," England allowed.

On the southwest rim, Duke picked up a rounded rock[31] that looked particularly well weathered and had "white spots" showing through its cover of dust, and then, ignoring England, went on to a half-metre-sized "flat-topped" rock nearby. "Let me whack this thing," he said. "I can't pass it up." He delivered a savage hammer blow to a fracture on top of the rock and a chip flew off. "It's a *great* rock," he reported, to placate England. "The matrix is so fine-grained, I think it's crystalline."

Young looked at the fist-sized chip,[32] wiped it, and saw that the light matrix had white clasts containing needle-like black crystals. "It's gotta be a breccia, Charlie."

"We're going to have to move out," England repeated, frustrated that Young and Duke had extended the intended 20-minute stop by 10 minutes. At breakfast, he had told them that the schedule had been changed in order to enable the sampling at the landing site upon their return to be "beefed up" to collect sufficient rocks near the LM to enable the geologists to distinguish between the true Cayley and the South Ray debris. This had required deleting Station 7, the second of the brief stops designed to sample both sides of the contact at the base of Stone Mountain. In the event, by venturing west along the bench, it appeared that they had done Station 6 at a point midway between those two sites. However, the delays were mounting, and it was starting to look as if there would be the usual last-minute rush.

As they ran down off Stone Mountain, it was not clear whether they had achieved their objective.*

SAMPLING SOUTH RAY

Back on the plain, Young turned west to run north of Stubby's rim, heading for Wreck. The 16-mm camera usually faced forward, but because it would see nothing driving into the zero-phase, Duke was to have turned it left to record the view across Stubby. In the event, they found themselves in an uncharted depression.

"We're not going to be able to see Stubby, Tony," Duke warned. "Do you want us to go up and travel along the ridge?"

"Depending on trafficability," England, unable to assess the situation for himself, allowed hesitantly.

"I'd like to see into Stubby," Duke prompted.

"Why don't you press on up there," England encouraged.

"Want to John?" Duke asked.

The ridge was about 30 metres high, but the slope appeared to be manageable and there did not appear to be any obstacles on its crest. It was several hundred metres to the left, but the detour would take only a few minutes and, in any case, it ran straight towards their next target. "Might as well," Young agreed.

On climbing, they discovered that the ridge was steeper than it had looked and, to their astonishment, the Rover slowed to a crawl.

"Have you got full throttle?" Duke asked.

Young checked the instruments. "We've lost the rear-wheel drive!" He continued up until the Rover began to slip, and then turned to run back down. "I'd just as soon not go up this ridge, Charlie!"

The geologists' interest in Stubby was that it abutted Stone Mountain, and Duke was to have looked for whether the crater's eastern wall was embedded in the base of the mountain, or the base of the mountain encroached on the crater. From on the hill it had seemed to Young that the crater had cut into the mountain and that its eastern wall was unusually steep. The significance of this was that if Stubby both sat on the plain and had struck the mountain, then the mountain must predate the crater, whose appearance implied it to be ancient. If so, and the Descartes was silicic, then it could not be 'recent' volcanism. A photographic record of the eastern rim of Stubby from ground level was therefore sorely missed.

England assured Young and Duke that the Rover engineers were assessing failure modes, and that he would need Young to perform some diagnostics when they reached their next stop. In fact, the fault had made the flight controllers a little antsy. The worst-case scenario on a traverse was the loss of the Rover near its operating radius. They were currently about 3 km from Orion, and the terrain was quite rough. The main data on the consumption of oxygen and water for a sustained walking traverse was Shepard and Mitchell's ascent of Cone Ridge. A contingency team immediately began to plan how Young and Duke might walk home, in order that this procedure would be to hand if required. England casually

---

* In fact, it has *still* not been conclusively established that they managed to sample true Descartes.

asked how fast the Rover was going, now that it was back on what passed for 'level ground'. Knowing that there would be people arguing that it would be better to curtail the sampling and turn for home just in case the Rover were to lose its remaining drive system, Young assured England that they were doing a steady 7 kph.

Station 8 was to be in the core of one of the rays, as near as possible to South Ray. The chosen site was where Stubby abutted Wreck. Navigation was straightforward: Young simply ran west until he was obliged to swing north!

"How about stopping in the middle of these big boulders, John?" Duke suggested.

Young accepted his recommendation and parked a few metres north of a shallow 12-metre crater.

In addition to abundant cobbles, there were some boulders ranging up to 2 metres in size. "There's a boulder we can split," Young pointed out.

"I think we could turn that one over," Duke added. This would enable the cosmic-ray 'exposure age' of the underlying soil to be measured. It was hoped that doing so for a large rock in the core of the ray would enable the impact that made South Ray to be dated.

While Young raked, Duke went to do a double-core a diameter west of the nearby crater. At first the tube slid straight into the ground, but then it stopped dead. Duke hammered it repeatedly, but it did not penetrate any further. As his fingers succumbed to fatigue, the hammer flew out of his hand. "Rats!" he cursed in frustration. Although he attempted to lift the hammer by hand, it eluded him, and he returned to the Rover to fetch his tongs. After recovering the hammer, he drew the tube out of the ground, hit it several times to empty it, and then moved several metres over and drove it in a second time. However, even after 65 hammer strikes the tube was only half way in. "Ah, you dog!" he complained.

"I think we'll take it, Charlie," England said, to dissuade Duke from trying a third position. He watched as Duke, expecting the tube to be firmly embedded, took a good grip, tugged, and almost toppled over as it came straight out.

"There's not much material around here," Young reported after five swaths of the rake on the flank of the crater had yielded only a dozen pebbles. On the way back to the Rover he spotted a fragment of glass sparkling like a rainbow, so he bagged it.[33]

"I'm going to drive over there, Charlie," Young called. He wanted to reverse the Rover in order to find out whether it had lost its rear steering in addition to its drive. "Check this out."

"It's working," Duke confirmed.

"Still working, huh?" As Young methodically reconfigured the power distribution system to troubleshoot the fault, he realised that one of the switches was incorrectly positioned. It had evidently been nudged during their embarking and disembarking. The Rover was still fully operational.* "Sorry about that," he apologised to Houston as he joined Duke, who was inspecting a black block just outside the crater's southern rim.

"This is *some* rock," Duke enthused. The white clasts appeared to include small fragments of a fine-grained crystalline material. "It's a two-rock breccia."

"Is this the one you want me to turn over, Charlie?" Young asked in amazement. It was a metre-and-a-half long and even in the weak lunar gravity it must weigh half a tonne.

Duke gave the rock a trial push, and was delighted when it rocked back and forth.

"Charlie wants to turn that one over, Houston," Young pointed out sceptically.

Duke leant against the rock and shoved, but it refused to roll over. "Let's get a chunk of it."

When England suggested that they sample any fillet first, they bagged a scoop of this and then hammered a fracture in the rock. "The whole rock is coming apart!" Duke chuckled. Retrieving the biggest chip,[34] they saw that the fracture was penetrated by a vein of glass.

There was a white boulder several metres further south, so they went to try their luck with it. "That's crystalline," Duke said confidently, "it's no breccia." It was so covered by zap-pits that it had evidently been sitting there a long time.

"Absolutely great!" England said. "See if you can turn it over."

"We can't turn that over," Young insisted. It was much smaller than the first rock, but it had a broad base and a such low profile that they would not be able to gain any leverage.

"We might," Duke countered. "But I'd like to get a sample first."

"Where do you want to sample?" Young asked. Duke indicated one corner of the rock and Young hit the spot, exposing a gleaming white surface. "If that ain't pure plage!" On considering the friable

---

* It was not realised until later, but Young's reconfiguring of the power system had the side-effect of nulling the 'nav'.

Although it does not look very dense (because most of the rocks are further away than they seem, and correspondingly larger) Station 8 was located in the core of a ray of ejecta from South Ray on the Cayley plain.

The first boulder examined at Station 8 was a two-rock breccia veined with glass.

The second boulder examined at Station 8 appeared to be crystalline.

The third boulder examined at Station 8 was a breccia that had evidently once been in a semi-molten state.

state of the chip,[35] he added a caveat: "It could have been reworked."

"We'd still like you to try turning it over," England prompted. Young shoved the rock with his boot but it did not budge.

Looking around, Duke spotted another rock off to the east that might be more amenable, so they went to give it a try. This rock turned out to be further away than they thought, and hence correspondingly larger. It was highly angular and resting on a base a metre-and-a-half wide, making it impossible to roll. As it was a dark-matrix dark-clast breccia, they decided to sample it. Young delivered a few hefty blows, but to no effect. "A hard, hard rock," he mused. Duke examined the rock, suggested a likely cleavage plane, and Young dislodged a 21 x 15 x 15-cm chunk. While Duke retrieved the sample,[36] Young studied the scar on the parent. "It's a vesicular type of breccia," he announced. It had once been hot.

"Fifteen minutes," England warned.

When Young asked whether the Backroom wanted them to try another boulder, England suggested they collect as wide a variety of cobbles as they could. However, upon looking for a place to store the breccia chip, which was essentially a football-sized rock in its own right, they discovered that both of their SCBs were full, and so they returned to the Rover to reconfigure. However, when they were once again ready to sample, England refused: "It's time to go."

The final station of the traverse, Station 9, was to assess the difference between a ray and an 'interray'. "We're looking for an area with as little evidence of South Ray as possible," England reminded.

Young scoffed at the prospect of their finding *anywhere* that hadn't been polluted by South Ray. In fact, as they drove north the coverage of large blocks was actually increasing. Clearly, they had not sampled the densest part of the ray. "I think we're just too close to South Ray," he mused. It was a continuous field of debris of varying density.

"Pick the best you can," England encouraged.

Young drove through the debris until they found a less densely littered area, and then Duke looked for an isolated boulder for a novel experiment. They were to take a pristine sample of the surficial layer of the regolith at a site that was sheltered both from South Ray and from the gases of Orion's plume. The initial idea had been to find a sheltered spot at the base of one of the scarps expected to run across the plain, but when these had proven to be illusory it had been decided to seek a boulder in the interray – which was, at

least in principle, a contradiction in terms. As the coverage of blocks diminished, there was an accompanying switch from predominantly light-toned to dark-toned rocks, which suggested that instead of the interray being clear, it was simply populated by darker rock. It was impossible to say whether the regolith was darker because, when viewed at ground level, the appearance of the lunar surface was dominated by Sun-angle.

"See that big rock, John," Duke prompted. "There's no way to see the LM from there."

The rock was on the south-facing slope of a low ridge that would have protected it from their landing. Young parked about 10 metres away, in order as not to disturb it.

The original assumption had been that by virtue of being in the interray, the rock would predate the formation of South Ray, but since it was a dark-matrix breccia in a field of such rocks, there was every chance that it was South Ray ejecta. However, if by some chance it predated South Ray its northeastern side should have been sheltered.

The procedure called for Young to "sneak up" on the 'dirty' side of the rock and reach across to put an adhesive pad on a handle flat on the surface on the other side in order to collect the uppermost millimetre or so of material. In the weak gravity of the airless lunar surface, the dust scuffed up by a boot flew a ballistic trajectory far ahead, which was why he could not just walk up and sample the sheltered side of the rock. In training, he had treated this task lightheartedly and approached on tip-toes.

"Are you sneaking?" Duke asked.

Fendell was trying to locate the action. He first ran one way, and then gave up and reversed, but could not locate the astronauts. "We're missing the Great Rock Hunt," England moaned.

"Gotcha!" Young called, upon reaching the rock.

"Which way do we go?" England pleaded.

"Shh!" Duke chuckled. "We're sneaking."

Young reached across the rock and pressed the sampler to the ground.

"Got it!" Duke laughed, a split-second before Fendell found them. "It didn't even know John was coming!"

To give England a sense of what he had missed, Duke promptly re-enacted "the sneak", reviving the fun of their training, and then took a core sample of what was hopefully a relatively clean regolith derived from the underlying Cayley.

Young delivered a series of heavy hammer blows to the top of the rock. "It's a *hard* breccia!"

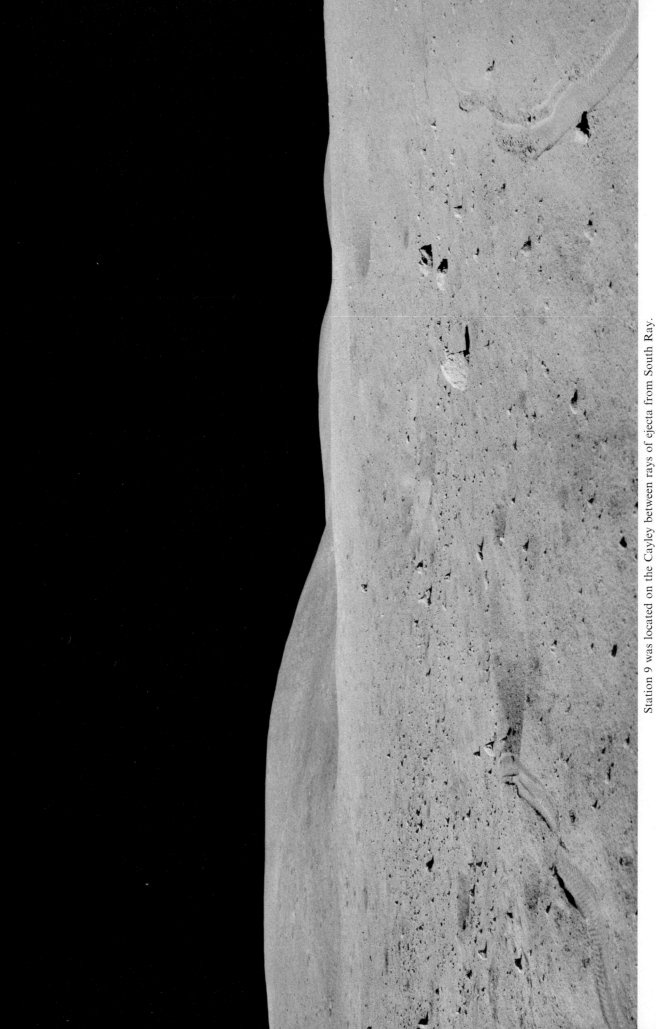

Station 9 was located on the Cayley between rays of ejecta from South Ray.

The 'locator' for the boulder at Station 9 that John Young decided to tip over.

After taking care not to kick up any dust, John Young took a pristine sample of the regolith alongside the Station 9 boulder.

After tipping over the Station 9 boulder, John Young sampled the material underneath, which the rock had shielded from the solar wind.

he observed, but eventually broke off a chip[37] of the matrix. Having noticed that the rock had moved when he leaned on it to take the special sample, he placed his knee against it and tipped it over. By rolling it into its shadow, he ensured that the newly exposed material would be illuminated.

"Charlie, I got it!" Young announced triumphantly.

"So you can not only sneak up on them, you can flip 'em over, huh!?" England chuckled. Unfortunately, he had missed this too – Fendell had been watching Duke coring, in order to compare his progress at this site with that at the previous station.

There was a sharp crest in the fillet, and the "caked-looking" soil was imprinted by the base of the rock, which was a coarsely grained crystalline material coated with glass. It took 13 hammer blows to break off even a tiny fragment[38] of the base.

"Real friable, isn't it," Duke observed sarcastically. Having finished his core, he took a soil sample from where the rock had been.

"We'd like you to pack up," England cut in.

Station 9 had been allotted 25 minutes, but to roll the rock it had been allowed to stretch to almost 40 minutes. They were about half an hour 'down', but because their suits were consuming coolant at a fairly moderate rate the EVA had been extended by 20 minutes; but this time would have to be spent at the landing site, not out here.

The direct route took them through some particularly hummocky terrain. It didn't seem to be so on the overhead imagery, but the entire area was overlapping, highly eroded old craters.

"I just can't get over how hilly this place is," Young reflected. He meant the plain itself, not just their immediate vicinity. "Anybody that ever called this place Cayley Plain really didn't know what he was talking about."

The irony was that if the Rover had not deployed properly, the objective of the second EVA would have been to walk to Station 8. With all the intersecting ridges, it would have been even more disorientating than the hummocky Fra Mauro was for Shepard and Mitchell; they would never have made it!

It was while driving through this area that they discovered the 'nav' was no longer working – it had lost its reference when Young had reconfigured the power supply. However, it was possible to navigate using the occasional glimpses that they got of Smoky Mountain, beyond the landing site.

This was not their only problem. While exchanging their SCBs at Station 8, the hammer

in Young's shin pocket had snagged and ripped away the rear fender of the Rover on Duke's side. It was only when they cleared the hummocky terrain and ran up to full speed that it became evident that the fenders were not optional extras but essential, because the rooster tail thrown on that side of the vehicle was now carried forward, with the result that by the time they got home they were covered in dust.

Back at the LM, Duke conducted soil-mechanics tests, using the penetrometer to measure the bearing strength and compressibility of the regolith with depth. Young tried a double-core. After a promising start the tube halted, and thereafter penetrated only a few millimetres with each hefty hammer blow. On finding the tube to be firmly stuck in the ground, he had to straddle it and really tug, and when it broke free his pent-up muscles lifted him clear off the ground. As Duke was still working with the penetrometer, Young did a little solo-sampling. England suggested that he look for "vesicular basalt", meaning some indication of a volcanic extrusion. "I just don't see any," Young countered. A few moments later he spied a small angular rock that was encouragingly dark, but proved to be a dark-matrix breccia.[39]

"We're going to have to pack up," England prompted finally.

As Young and Duke sorted their samples and cleaned up, it became apparent that they were dawdling to claim the record for the longest time spent outside; this duration having been set at 7.2 hours by their immediate predecessors; England obliged them, and then chivvied them inside.

"If this place had air," Young observed just prior to going up the ladder, "it would sure be beautiful."

Over supper, England relayed the Backroom's assessment. "I know that we didn't see exactly what we expected to see," by which he meant Descartes volcanics, "but we think you got everything that we went for."

DRIVING NORTH

As soon as Young and Duke were awakened for their final day on the Moon, they pursued a 'get ahead, and stay ahead' strategy in their preparations, in order not to jeopardise the northern traverse. Because their lift-off time was fixed, they were up against a 'hard stop'; there could be no extension. Even though the mission planners had managed to salvage 5 hours for this EVA, the route had been severely curtailed and much of the time had been allocated to packing up at the end.

It had been planned to survey the eastern rim of the large crater Palmetto *en route* to North Ray. Station 11 and Station 12 were to be set several hundred metres apart on this large crater's south-eastern rim, and then its outer ejecta blanket was to be sampled at Station 13 while driving east. Station 14 was to be at Cat, a small crater on the western rim of Ravine, the large crater in the southern flank of Smoky Mountain. After a brief stop on the plain, two small craters called Dot and, fittingly, End, were to be used to sample the northeastern rim of Palmetto on the way home.

As the primary objective was North Ray, the revised plan, which England read up over breakfast, called for the Palmetto drive-by on the way to North Ray, which was to be sampled as planned, after which they were to retrace their tracks, pausing only to take a few samples at the southern fringe of North Ray's ejecta blanket. The visit to this crater was important not just because it offered an opportunity to sample deep within the Cayley, but also because Young and Duke were likely to be the only Apollo astronauts to reach such a vast pit – which, at 1,000 metres across, was fully three times the diameter of Cone, which Shepard and Mitchell had narrowly missed viewing at Fra Mauro.

"We're going to concentrate on big-boulder samples," England pointed out. The bigger the rock, the nearer to its point of origin it was likely to be, and when Duke shot telephoto pictures of the North Ray area from Station 4 on Stone Mountain he had seen an enormous black boulder situated right where they were to do Station 12; it could have been excavated from a depth as great as 200 metres.

"We're about 10 minutes ahead," Duke announced soon after emerging from the LM for the last time, "and we're ready to move out."

"You fellows are really getting smooth," England acknowledged, complimenting their preparation of the Rover.

Unfortunately, Duke had been too optimistic, and by the time that they finally set off they were precisely on schedule.

When they crested the ridge about 100 metres north of the landing site, they were able to view their route. It was strewn with blocks up to half a metre in size, but the litter was nothing in comparison to Survey Ridge. "I think the boulder population is starting to thin," Young observed, 100 metres or so further on.

"They're getting smaller," Duke agreed. "It looks like we could just have left this ray." So, at last, they were free of South Ray's influence.

Duke inspected their destination. "There's a line of boulders that comes out of the northeastern rim of North Ray, and goes 'up' Smoky Mountain." This remarkably narrow ray extended all the way to the summit, and presumably continued down the far side.

As he ran off the 300-metre-wide ridge into a shallow valley, Young aimed for the eastern rim of Palmetto, which was now visible about 750 metres ahead. If they had been restricted to a walking traverse, their primary objective would have been to sample the small craters on Palmetto's rim.

Half-way to Palmetto they found a 400-metre rimless crater with such shallow walls that it was only 50 metres deep. Although it was clear of debris, Young veered around its eastern rim.

Duke continued to study their destination. He could see "two tremendous blocks" at the place on North Ray's rim that they were to sample, whose mottled appearance suggested they were breccias.

On approaching Palmetto they came upon another old eroded crater, but this one was asymmetric, with its western wall considerably steeper than its eastern wall.

As they pushed on, Duke pointed out that the terrain was becoming more rolling and – apart from the usual cover of small craters – smoother. Exploiting this, Young drove at full speed.

"It's a tremendous crater!" Duke announced, when they crested Palmetto's raised rim and he gained his first look into the cavity.

Momentarily distracted, Young drove over a rock that scraped the underside of the Rover. "Sorry, Charlie!" he apologised lamely. "I've gotta keep my eye on the driving." He swung east to avoid the litter of half-metre-sized blocks.

Trafficability in the valley appeared to be excellent, but the ridge on which North Ray sat began to look increasingly imposing.

In addition to the ray that ascended Smoky Mountain, Duke now spotted a line of boulders that ran across the ridge towards Ravine. In fact, this second ray petered out just short of Cat, so Station 14, if they had been able to undertake it, would have provided a 100-metre-long radial from the fringe of the ray to the material excavated from the mountain as a means of distinguishing the Cayley from the Descartes.

Upon drawing level with End Crater, Duke reported that it was not embedded in Palmetto's flank, but was on the slope down into the broad northeast-to-southwest orientated valley that ran to the base of the ridge on which North Ray was

John Young loading the Rover for the third traverse.

located. This slope was every bit as steep as the flank of Stone Mountain, where Young had free-wheeled the Rover, but here he exploited the exceptionally smooth surface and the rapidly thinning coverage of rocks to *accelerate* to 15 kph, and in so doing set a lunar land speed record. They were barely half-way to North Ray, but it was starting to look as if they would be able to 'get ahead' a little on the traverse timeline. A few hundred metres further on, however, the cobble coverage increased again when they encountered their first North Ray ejecta. Duke half-seriously suggested that they divert to the line of boulders that ran across to Ravine in order to do Station 14, but Young insisted on sticking to the plan: "Let's go up onto the rim, and see what we've got."

On closing to within one crater's diameter of the rim of North Ray, they hit a field of whitish rocks that ranged up to 3 metres in size and had a fractured appearance.

When a large depression appeared ahead, Young veered around to the east, where they encountered another cluster of blocks which Duke said had a "frothy" surface. These rocks were so distinctive that he suggested they should be the target of Station 13 on the way home. A hundred metres further on, they started up the 50-metre-tall ridge which, although "pretty steep", the Rover took in its stride.

Young slowed as they approached the crest. "I think we're about at the rim," he announced in anticipation.

"We've got you 500 metres from it," England countered; he was following their route on his map.

"We're climbing again," Duke noted a moment later. The flank of the ridge was terraced with benches.

"Well I'll be darned!" Young chuckled. This was turning into a repeat of Shepard and Mitchell's frustrating experience on Cone Ridge.

Now that they had a clear view of the base of Smoky Mountain and the ray across to Ravine, it was apparent that most of the boulders in the ray were grey, but some were white and a few were *gleaming*-white. Duke swung the 16-mm movie camera around to document the strikingly concentrated ray for the geologists.

Finally, the massive black boulder poked over the horizon and, with this point of reference, Young steered directly for where the crater should be, and a moment later the pit appeared ahead. Having made excellent time, they were 11 minutes ahead of schedule.

ON NORTH RAY'S RIM

As they parked on the southeastern rim of North Ray, Duke reported, "There's some beautiful white rocks right on the rim."

"If we could get both the white and the black rocks in one stop," England came back, "that would be really fine."

"That white one's a breccia!" Duke decided, having seen splotches in the nearest one. He had expected it to be crystalline. He peered over at the "big black boulder" and confirmed that it had a "speckled appearance" suggestive of white clasts. If *that* was a breccia then the volcanic theory was doomed.

Young went to look into the crater. "Does this thing have steep walls!" he exclaimed, awed that the rim fell away from him at 15 degrees and then plunged "all of a sudden" at 35 degrees. "I'm as close to the edge as I'm going to get, and that's the truth." Even though the crater was a kilometre wide, he could not see its floor. It was not clear whether the level of the transition in the slope marked a bench.

England read up the revised plan: they were to spend at most an hour surveying a 100-metre arc of the rim, this distance being designed to facilitate a stereoscopic pan of the crater using the 500-mm lens. And since they would no longer be able to visit Smoky Mountain they were to take similar long-shots of the mountain and the route they would have taken to Ravine.

"Man, is that a hole in the ground!" exclaimed England when Fendell panned the TV across North Ray.

As Duke began his first pan, he looked for stratigraphy exposed in the wall of the crater that might indicate a series of lava flows. "I see no bedrock," he reported. "All I see is boulders around the crater. There's nothing that reminds me of bedding, it is just loose boulders. Although it might very well be that it's so shocked that there is really bedrock there."

Young joined in: "The boulder layers are horizontally oriented and, of course, the walls are all covered with talus. Over on the north wall in particular, about one-third of the way from the top, there is a line of boulders that would lead one to think there is bedding there." His voice reflected his uncertainty though.

When England asked about the colour of these boulders, Young told him that they were dark. A moment later, England asked whether the white rocks resembled those that Shepard and Mitchell had seen on the rim of Cone, the crater that had

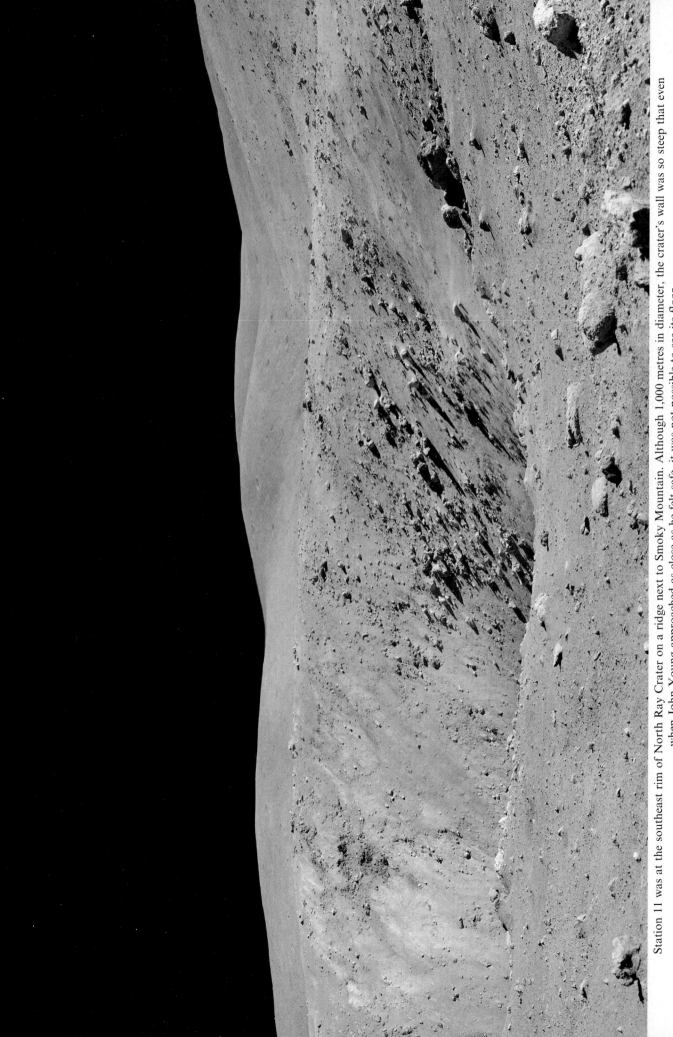

Station 11 was at the southeast rim of North Ray Crater on a ridge next to Smoky Mountain. Although 1,000 metres in diameter, the crater's wall was so steep that even when John Young approached as close as he felt safe, it was not possible to see its floor.

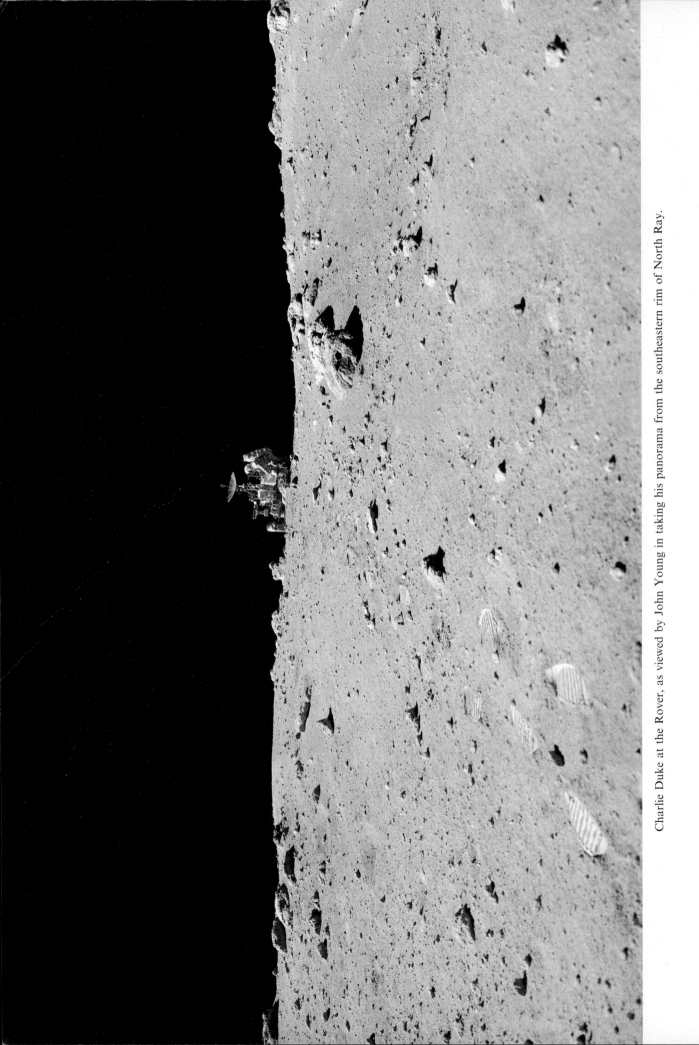

Charlie Duke at the Rover, as viewed by John Young in taking his panorama from the southeastern rim of North Ray.

excavated the brecciated Fra Mauro. Duke opined that the albedo range of was less pronounced than the two-toned rocks at Cone. The Backroom was starting to reassess the Cayley as a thick deposit of brecciated material, similar to the Fra Mauro Formation but with a smoother texture.

Young went to inspect a nearby half-metre-sized white block with dark mottling. "This is *definitely* a breccia." Indeed, some of the clasts had darker specks, indicating that they were composites in their own right. He lifted a loose fragment[40] as a grab sample. "It's very friable. It looks shocked." It had an extremely rough texture. "The clasts are sticking out of it." He studied the ground-mass. "The white matrix doesn't have any crystalline structure that I can recognise."

Finished his photography at the southern site, Duke set off north, stopping about 65 metres around the rim from the Rover to take his stereo shots.

Young took a spare SCB and went away from the Rover to start sampling. Since he had become quite adept at kneeling to pick up rocks by hand, he did not bother to take his tongs. Because there was so much white rock laying around, he selected "a black type" as an exotic.[41] Having no tools, he set the SCB alongside the rock to give a sense of scale in the documentary pictures. "Well, I was wrong," he sighed, as he scraped at the thick coat of dust. It was a white rock but, having 50 per cent black clasts, was much more clasty than his first sample. It was extremely shocked and very friable. "It's taken a heck of a beating." He popped it into the bag and went to get a larger rock[42] which, although clearly a white breccia, had dark clasts in it that were several centimetres in size. He wasn't sure what the clasts were. "They're either a very fine-grained black breccia material, or a dense basalt-like rock."

Having finished his photography, Duke spotted a "really shocked" rock that was so friable that it was disintegrating *in situ*. It was so "pure white" that it looked like chalk. He went to get a sample bag from Young, scooped up the white-matrix white-clast breccia,[43] put it in Young's SCB, and returned to the Rover. Remaining, Young sampled[44] "really black glass" lining the floor of an unusually deep small primary pit and found a rock[45] with "a right angle" on it which suggested that it had broken off a parent at the intersection of two orthogonal cleavage planes.

"We'd like big boulder samples too," England prompted.

Duke replied that he was going to look at a cluster of large white blocks situated about 50 metres west of the Rover.

As Young started back to the Rover, England asked how far away the big black boulder was. Fendell had zoomed in on Young while he was sampling, and the rock had appeared to be just beyond him. Indeed, at times, it had seemed as if Young had been alongside the boulder and, to the astonishment of the Backroom, had pointedly ignored it. When Young replied that it was "about 150 metres" behind him, the scale of the boulder finally dawned on the watching scientists.

On the way to his target, Duke bagged a dark-matrix light-clast breccia[46] with a "hackley looking" surface.

As Duke reached the white boulders, which were light-matrix dark-clast two-rock breccias, Young joined him with the rake, which he promptly used to scoop some of the prominent fillet that had evidently been shed by the highly eroded boulder. Closing in to take a sample of the rock, Duke found his boots sinking 15 cm into the fillet. The rock's friable matrix[47] was so finely-grained that he felt obliged to call it "aphanitic", meaning that the uniform ground-mass was so fine that the individual crystals were indistinguishable. To complete the sampling process, he chipped one of the clasts.[48]

Meanwhile, as Young raked pebbles in the regolith nearby he found a rock[49] that appeared to be a piece of the matrix shed by the boulder, so he bagged it as a bonus.

"How's our time going?" Duke asked England.

"You've got 25 minutes." In fact, they had been granted a 10-minute extension.

"Let's head out for the big rock," Young prompted.

"Okay!!"

"We think that sounds like a great plan," England added.

The Backroom heartily approved; the black boulder was by far the biggest rock that any astronauts had yet encountered.

"It may be further away than we think," Young warned.

"No!" Duke countered, lest someone decide that it was *too* far. He immediately set off north. After a few seconds, Young ran after him.

They stopped to take a rake sample in the vicinity of where Duke had taken the northern pan, then set off again. It was only when the astronauts began to be masked by the crest of the ridge that the Backroom realised that their target was half-hidden.

"Look at the size of that rock!" Duke chuckled. Although he had not reached the rock yet, it was already filling his visor. "The closer I get to it, the bigger it is!!!"

In documenting the sampling of a large white breccia at Station 11, Charlie Duke included the "big black boulder" some 200 metres away around the eastern rim of North Ray that they had decided to make their next objective.

John Young rakes the shallow regolith on the eastern rim of North Ray. The Rover is parked in front of a large white breccia, and from this vantage point the horizon is only 100 metres away.

By the time the astronauts finally reached the rock, only their heads were visible above the ridge in the TV view. "Well, Tony," Duke said, "there's your 'house rock' right there."

The rectangular block was 20 metres long and 12 metres high. It rose straight off the ground without a fillet, and its essentially vertical face was heavily textured. At its southern end, and separated from it by a narrow gap, was smaller fragment of the same material about 5 metres in size. It was not clear whether this was a piece that had landed alongside, or a piece which snapped off when the boulder landed. The large boulder was named the 'House Rock' and its companion the 'Outhouse Rock'. Even though these rocks were strikingly different to anything previously encountered, the schedule allowed only about 15 minutes to examine them.

"I'd have to call this a black-matrix breccia," reported Young, confirming Duke's earlier suspicion. The fact that such a large block must have come from the deepest point of North Ray's excavation banished any lingering hope that the dark material in the crater might be volcanic.

On examining the base of the rock, Duke initially took it to be coated with glass, but then he noticed that the glass penetrated deep into fractures, which indicated that the rock had once been sufficiently hot to have undergone partial melting.

As Young walked around the rock, assessing its texture, he spotted a shatter cone. "I'll be darned." He hammered off a fragment[50] that appeared to be crystalline and heavily reprocessed.

Moving on, they found a white clast which – at 3 metres across – was larger than most of the rocks they had seen elsewhere; in all likelihood, this was a rock that had become bound up in the matrix. They broke off a grapefruit-sized chunk[51] of it that proved to be a breccia with dark clasts.

"Did you see a permanent shadow?" England enquired. One of the items on the agenda for this final excursion was to find a rock whose situation was likely to cast a permanent shadow that could 'trap' and accumulate any volatiles there might be in the lunar 'atmosphere'.

"We've got an east–west split," Duke offered, referring to the narrow gap between the House and Outhouse rocks. It was impossible to be sure that this was permanently shaded, but it was the best option in their immediate vicinity.

"Go ahead!" England urged. Duke poked the rake in between the rocks to scoop a sample of regolith and then, prompted by England, moved away from the rock to take a reference sample, but the regolith proved to be very thin and he was barely able to scrape any up.

"You're going to have to leave after this," England announced.

"There's a real frothy rock, John," Duke said to Young, pointing to a small rock at their feet.

Young picked up the glass-coated breccia[52] as a grab sample. "Okay, let's go back."

"I'll be right with you," Duke promised; he snapped a few more pictures and then set off after his commander.

It was evident that if there had been substantial volcanism in the area, it was masked by the breccias. And there were clearly two types of breccia. The light-matrix type was well-rounded and poorly jointed. The darker matrix type was more angular and had prominent jointing. Some were almost totally coated with black glass. The light-matrix breccias were so shocked as to be friable, which was why they were so well eroded; they were predominant and evenly distributed. The dark-matrix breccias were very hard, and confined to fairly well-defined clusters. There was white anorthositic rock, but there was absolutely no sign of either the dark basalt or light-toned rhyolite that had been expected.

"I guess we're a little late," Duke apologised as they reached the Rover. They had exceeded their extension by 8 minutes; this overrun would have to be taken off the next station since there was no margin of flexibility.

SHADOW ROCK

"Station 13 will be down your tracks," England reminded, "in the midst of the big boulders that you described on the way up."

On starting down the ridge, Duke expressed his surprise at how much steeper the slope appeared going down than coming up. In fact, one section of the ridge plunged 35 metres in only 150 metres. It was so smooth, however, that Young let the Rover accelerate to 17 kph, and thereby set another speed record. He drove straight to the largest of the rocks selected for Station 13 and parked 12 metres from it.

"We have a rock about 3 metres across," Duke informed England. "It looks like that great big one we sampled on the rim." He meant that, like the House Rock, it was a dark-matrix breccia.

"We'd like John to take a magnetometer reading," England said. "And Charlie, you go sample."

In the original plan, this was to have been a

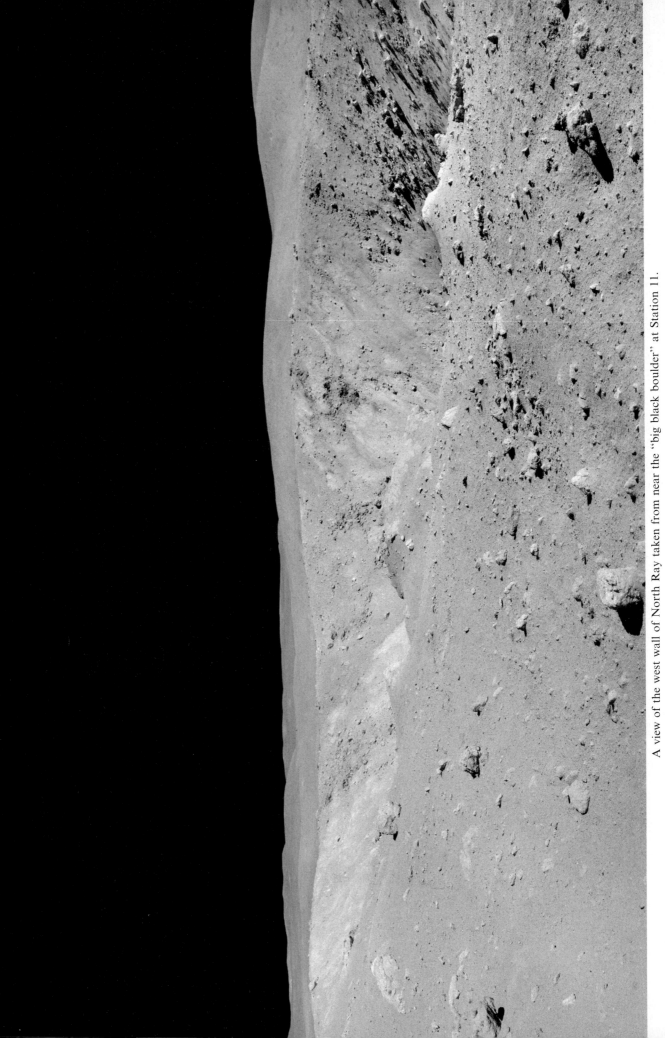

A view of the west wall of North Ray taken from near the "big black boulder" at Station 11.

Hammer in hand, Charlie Duke is sampling the "big black boulder" at Station 11. He has leant the rake against the rock, and because his sample bag dispenser had become detached from its mount he had slipped his finger through its attachment ring in order to continue to carry it.

brief no-TV stop in order to take a pan and literally grab a few small rocks to sample the outer fringe of North Ray's ejecta blanket, but on the cancellation of the visit to Smoky Mountain it had been upgraded to include a rake sample.

While Young erected the LPM, Duke took a site pan and then raked; the regolith proved to be thicker than on North Ray's rim, and was full of pebbles. "While John's doing the LPM," England said when Duke had finished, "we'd like you to hammer that rock, Charlie."

In making a walk-around inspection of the intriguingly shaped boulder, Duke saw what might be a permanently shadowed spot below an over-hang – this led to it being named the 'Shadow Rock'. "That's a perfect shadowed soil sample," he announced. He took up position alongside, preparatory to poking the rake into the shadow to scoop a sample of the regolith. On realising that there was a hollow under this end of the rock, he reached his free hand up to the overhang to provide support as he reached the rake all the way into the hole. It was a sure bet that the material in this pit had been frozen for millions of years. Ideally, such a sample would have been stored in a SESC in order to guarantee that it would arrive at the LRL in pristine condition, but he did not have one handy and had to settle for a standard sample bag.

"We'd like you to spend the rest of your time hammering on that rock," England directed.

Duke, happy to oblige, chipped a few tiny fragments[53,54] from high up and then bent to get another from lower down, whereupon he lost his balance and slipped against the rock. "Oh, rats!!" Then he discovered that he could not get back up. "John, I'm trapped."

Having taken the magnetometer reading, Young was in the process of reeling-in the instrument's balky cable. Trapped!? "What do you mean?"

"I'm against this rock."

"You can't get up!?" Leaving the cable strewn across the ground, Young ran over to help Duke to his feet.

Duke had been successful in his sampling, however; having gained a chunk[55] of rock that was too large to go into a sample bag. In any case, he was running out of bags. This realisation served to reinforce the fact that this final day was fast drawing to a conclusion.

"It's the same as that black one, up on the rim," Duke reiterated, after examining the rock. "Except that the one up there didn't have any of these holes in it." When they had driven by on the way up, it

was the presence of the holes that had prompted him to describe the rock's texture as "frothy".

"They look vuggy," Young mused. "Although they're rounded." Vugs tend to be asymmetric. What was puzzling was that they extended deep into the rock.

"They look like drill holes," Duke mused.

Young had an inspiration. "They look like the holes that you get in rock where you have a *venting* of gas that comes up through a ..." he faltered, trying to recall the proper term for a hole made by gas being vented through semi-plastic rock under great pressure.

"Vesicle pipes," Duke finished the thought. Like the House Rock, this breccia had once been extremely hot.

"That's it!" Young confirmed.

When England invited them to try to roll the block, Duke dutifully gave it a push but it was solidly situated.

As he strapped into the Rover, Duke stared at Ravine Crater. Its wall was very steep where it had excavated the flank of Smoky Mountain, but undulatory on the Cayley side. In contrast to the pollution of Stone Mountain's flank by South Ray, the ejecta from North Ray seemed to be confined to narrow rays. At Cat Crater, it might have been straightforward to unambiguously sample Descartes material. Smoky was their 'lost opportunity', just as the North Complex had been for Scott and Irwin. The problem with Casper's SPS engine had taken its toll, but by carefully husbanding Orion's consumables they had managed to squeeze in a trip to North Ray. As Young raced home, he had the satisfaction of knowing that, as he had forecasted, Apollo 16 had changed the image of the Descartes-Cayley.

BACK TO THE DRAWING BOARD

The 'instant science' started to flow even before Young and Duke had left the lunar surface. It was apparent that the Cayley was not a volcanic plain. The status of the Descartes was uncertain, because it was not established that they had managed to sample it. Nevertheless, it was clear that if Stone Mountain was volcanic, this was masked by South Ray. Although Smoky Mountain had not been sampled, there was nothing at North Ray to suggest that it was volcanic. It did not take long for another theory to emerge: the Cayley was 'fluidised fragmental ejecta' from a basin-forming event that had 'sloshed' across the highlands, transforming the valleys into light-toned rolling

On the way back from North Ray, it was decided to stop to inspect a large boulder that was out in the open, and this was designated Station 13.

Charlie Duke explores the shaded end of the 'Shadow Rock'.

John Young by the Rover after the final traverse.

John Young after the final traverse. By this point in the program, moonwalkers were admirably equipped for field geology: note the cuff-checklist, the shin-pocket for the hammer (absent), the chest-mounted Hasselblad and bag dispenser, and the sample carrier on the side of the backpack; together with the tools on the Rover. In the foreground is the gnomon, minus its pendulum.

plains. When the Cayley rocks were examined, they proved to represent three types of breccia: regolith breccias, fragmental breccias and impact-melt breccias. The chemical composition of the impact-melt breccias resembled similar rocks taken from Fra Mauro and Hadley-Apennine.

When Don Wilhelms distinguished the Cayley from the Fra Mauro in mapping in the mid-1960s, the significance of the basins had only just been recognised, the Imbrium Basin was the focus of attention, and all manner of sculpture was attributed to it. At first, the Cayley was presumed to be Imbrium ejecta, but as the 'cold' Moon theory lost favour the Cayley was reassessed as volcanic. A few days of field geology had revealed it to be ejecta after all! However, it was evidently not Imbrium ejecta. The Orientale Basin cannot be appreciated from Earth because most of it lies beyond the limb. It is the youngest of the Moon's basins, and would have sprayed ejecta across much of the Nearside. When the gleaming white boulder in the ray at Station 8 was dated at 3.76 billion years old, this ruled out Imbrium as its source. It may have come from Orientale, but the case must be regarded as tentative.

One lesson of the Cayley experience was that in studying the Moon the geologists had relied too much on drawing analogies with Earth. The Moon does not share the Earth's endogenic geological history. The Apennines adjoin Mare Imbrium, but their origin was not realised until their relationship to the Imbrium Basin became evident. This lunar mountain range is not at all like its namesake in Italy. Its peaks were not uplifted gradually as a result of slow but irresistible deformation of the crust, and they are not volcanic edifices, they are shattered crustal blocks that were displaced by the shock of a major impact. Although such events must have occurred on Earth early in its history, those structures were destroyed long ago. Similarly, there must once have been ejecta blankets on Earth analogous to the Fra Mauro. The value of the Moon's ancient surface was that it provided insight into the early history of our own planet. But to seek insight into the Moon in terms of terrestrial processes had proved to be a risky business. Another important lesson was that the lunar geologists had settled for a single 'working hypothesis' and therefore, as a community, had not been receptive to contradictory evidence.

New seismic insights

For the seismic team, the advantage of selecting an easterly site for Apollo 16 was that it meant the new passive seismometer would be far removed

Sample 61016, the 11.73-kg 'Big Muley' that Charlie Duke recovered from the rim of Plum Crater.

Sample 64435, a 1.08-kg breccia from Station 4 on Stone Mountain.

Sample 66055, a 1.31-kg light-matrix breccia from Station 6.

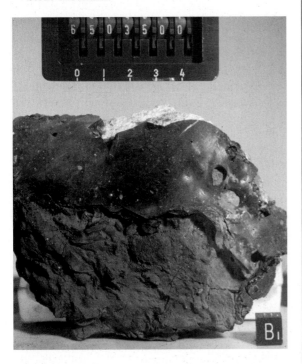

Sample 65035, a 0.45-kg glass-coated breccia from Station 5.

Sample 68115, a 1.19-kg fragment of the 'veined' breccia at Station 8.

from the remainder of the network, and this enabled the sources of the episodic 'tidal' moonquakes to be refined. Of the 22 epicentres identified, all were deep, and five were 800–1000 km below the surface. On 17 July 1972, a rock of about 5 tonnes in mass hit the Moon. It was the largest strike ever recorded by the network. The extremely strong signal showed the crust to be 60–65 km thick at Procellarum and Fra Mauro, and slightly thicker (75 km) in the highlands. It also established there to be a sharp discontinuity, rather than a gradual transition. As regards the thickness of the maria, it took a long time to gather the data, but it was established that far from being 25 km thick, the basalt flows are a thin veneer resting on top of a deep deposit of intensely brecciated rock. In fact, the maria account for a mere 1 per cent of the lunar crust, and the crust is just 12 per cent of the volume of the Moon. By good fortune the meteor struck the Farside, and analysis of the manner in which the seismic energy

Sample 68416, a 0.18-kg fragment of the 'white boulder' at Station 8 that was dated to 3.76 billion years.

Sample 68815, a 1.83-kg dark-matrix breccia chipped from the vesicular boulder at Station 8.

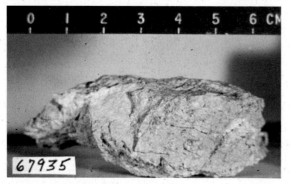

Sample 67935, a 0.23-kg chip from a shatter cone on the House Rock at Station 11.

propagated through the interior indicated that the Moon has a molten zone, but with such limited data it was difficult to gauge its size.

The Active Seismic Experiment had two parts, one of which was undertaken by the astronauts, and the other after they had departed from the Moon. The probing by the 'thumper' charges revealed the regolith to be 12 metres thick; somewhat thicker than at the maria sites. Unfortunately, Young had encountered difficulty stabilising the mortar, and as the third of its four charges was fired this upset the alignment. With its orientation unknown, there would be no value in firing the final charge. The experiment showed there to be a layer of brecciated rock beneath the regolith, and the absence of any reflection indicated this to be at least 75 metres thick. The dark breccias excavated by North Ray from the periphery of the Cayley plain suggested that it was several hundred metres thick.

The Cayley Formation

The Cayley was sampled at widely spaced sites on a north–south axis right across the valley. The only evidence of stratification was in the walls of the ray craters, which showed distinct populations of breccias derived from layers of light-matrix and dark-matrix material. The crucial step towards understanding the Cayley was the realisation that it is several hundred metres thick, which is so substantial that a new name, the 'subregolith', was coined to distinguish it from the highland basement. In general, the Cayley breccias were derived from plutonic anorthosites (which is to say, the crustal material from the ancient magma ocean) and feldspathic gabbros. The X-ray fluorescence instruments in Casper's SIM bay found the site's aluminium-to-silicon and magnesium-to-silicon ratios to be typical of the highlands. Silicon is used as the reference because it is the most common element in the crust. The significance of these particular ratios was that the plagioclase crust is rich in aluminium, but not in magnesium (the magnesium sank into the mantle, and formed the mafic silicates that characterise the dark lava that was later extruded as the maria). Many of the Cayley breccias had suffered shock-melting. Some crystalline impact-melts were so rich in alumina that a new category was coined: very-high alumina (VHA). The matrix-rich clast-poor VHA impact-melt resembled basalt, but it had not been extruded, and so was not evidence of volcanism. Basalt *was* recovered as fragments in the regolith, but it undoubtedly originated elsewhere and was thrown in. Although comparatively low in alu-

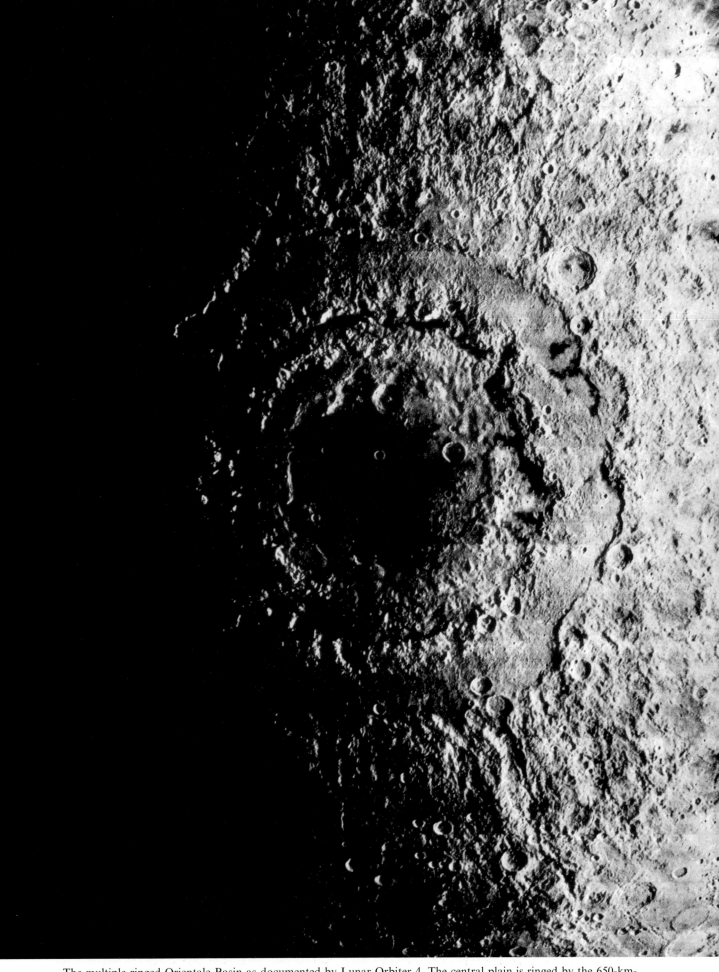

The multiple-ringed Orientale Basin as documented by Lunar Orbiter 4. The central plain is ringed by the 650-km-diameter Rook Range and by the 1,000-km-diameter Cordillera Range, with ejecta extending far beyond. The Imbrium Basin would once have looked very similar to this.

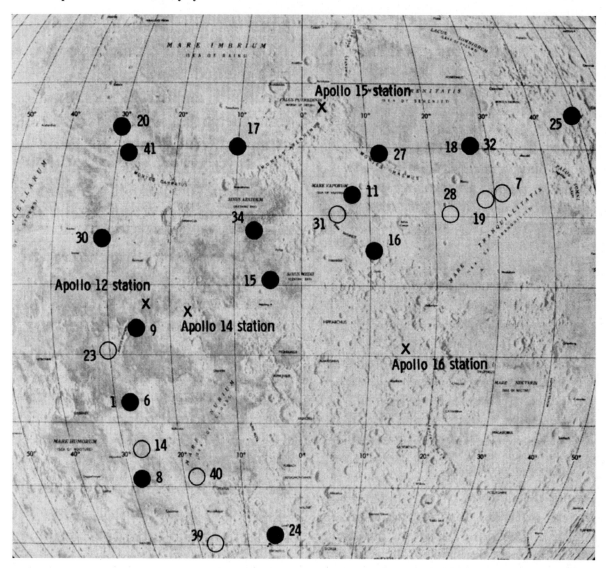

The most active epicentres of seismic energy inferred from the instruments of the Apollo Lunar Surface Experiment Packages. A solid circle indicates a source whose depth was reliably determined; most of the order of 800 km. The numbers are arbitrary identifiers.

mina, the thermally processed breccias were otherwise chemically similar to the LKFM-melts of the western sites, indicating that they likely originated there.

Although the light-toned plain sampled by Apollo 16 turned out to be different to expectation, the fact that it seemed to be typical of the Cayley formation suggested that the conclusions could be extrapolated. As for the ray craters used as 'drill holes' into the Cayley, North Ray is 30–50 million years old, and South Ray is a mere 2–3 million years ago.

The Descartes Formation

Apart from having fewer craters with diameters exceeding 100 metres, the cratering density on Stone Mountain was found to be comparable to that on the Cayley plain, implying that the two are of a similar age. The biggest crater in the area explored by Apollo 16 was Crown, but the traverse did not extend to it. Like the other primaries, it was very subdued, and thus old. Given that South Ray had mantled the flank of the mountain, it was unfortunate that no recent primaries had deeply excavated it. The large craters in the sampling area were all secondaries derived from South Ray. The nature of the Descartes was therefore difficult to resolve, not least because (despite the innovative procedure of sampling sites potentially 'protected' from South Ray) it was difficult to be sure that the

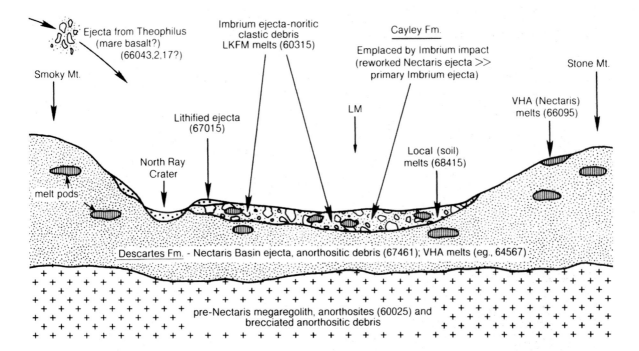

A schematic geological cross-section through the Apollo 16 landing site showing the complex interrelationships of different units of impact-produced ejecta excavated from large mare basins (modified after Spudis, 1984; Stoffler *et al.*, 1985). The Descartes Formation is mostly ejecta from the older Nectaris Basin. The overlying Cayley Formation, emplaced by the younger impact event, consists mostly of reworked material from the Descartes Formation with some Imbrian ejecta. Both the Descartes and Cayley Formations are inferred to rest on an older (pre-Nectaris) megaregolith composed of a thick layer of complex debris from many older impact events. Numbers refer to specific collected samples that are representative of the various units inferred to be present. 'LM' shows where the Lunar Module landed. Courtesy the Lunar and Planetary Institute and Cambridge University Press.

samples were local. Nevertheless, some conclusions were evident.

First, there was no evidence of volcanism. The breccias from the mountain were more friable and lighter in albedo than those from the Cayley, but there was no real chemical difference. Stone Mountain is probably a mound of brecciated rock and, as such, an ejecta product.

Although it was not possible to drive onto the western ridge to ascertain whether Stone Mountain had encroached on Stubby Crater's rim, when Young looked down at it from Station 4 he gained the impression that the crater cut into the base of the mountain, which meant that this predated the crater, the eroded state of which ruled out the mountain being of Copernican Era – as some people had speculated on the basis of the empiricism that in the highlands 'the lighter the surface, the younger it is'; so this rule-of-thumb had to be discarded.

Just as the 500-mm lens was a great advance in surface photography, it is amazing that Mattingly was the first CMP to take binoculars to assist his orbital survey. Even as his colleagues were discovering that the Cayley was not what they had been led to expect, he reported that it looked like "a pool of unconsolidated material which has been shaken until the surface is relatively flat". As he circled the Moon every two hours, he methodically noted the character of the surface, and concluded that there was nothing distinctive about the Descartes; it fit right into the Imbrium structure.

In 1979 George Ulrich proposed that the Descartes was the equivalent of the Fra Mauro from the Nectaris Basin, some 300 km to the east. The stratigraphic mappers had split the pre-Imbrian into pre-Nectarian and Nectarian. Paul Spudis and Odette James independently found some of the Apollo 16 samples to be 3.92 billion years old, and it is possible that this dates the formation of the Nectaris Basin. Although the link is weak, it is consistent with the stratigraphic evidence that the Descartes formed prior to the Cayley.

7

The beautiful valley of Taurus-Littrow

THE FINAL TARGET

The site selection for Apollo 17 took place before Apollo 16 flew, so was influenced by how the 'crucial events' in lunar history had already been resolved or were likely to be addressed by Apollo 16. The site for what would almost certainly be the last visit to the lunar surface for a generation, was therefore hotly contested.

Tycho Crater in the Southern Highlands and Tsiolkovski on the Farside were sites of 'special interest', but were impracticable operationally. Copernicus's central peak offered the prospect of sampling material from deep within the crust, but Apollo 12 had tentatively dated this crater by its rays, which made it 'partially known'. The Humorum Basin is only partly flooded, and the 90-km-diameter crater Gassendi sits between the rim and the northern shore of its mare. The floor of the crater had risen isostatically and been badly fractured. A landing in this crater would have been able to sample its central peak, date the crater's formation, and possibly shed light on the basin.

But the selection process was driven by the imperative to refine the timescale over which the lunar 'heat engine' had been active. Because the formation of the Imbrium Basin was now well understood, as was the lava up-welling that flooded most of the basins over the ensuing half-billion years, the objective was *late* volcanism. The Marius Hills and Davy Rille were no longer considered worthy of a 'J'-mission. The leading candidates were Gassendi and Alphonsus. It was into Alphonsus that Ranger 9 had made its spectacular televised dive. The floor of this 110-km crater offered a variety of tempting features, including small dark-halo craters that appeared to be volcanic cones. At first it seemed that Alphonsus would be selected, but suddenly the focus of attention turned to the southeastern rim of the Serenitatis Basin.

The sculpture etched on the southern rim of the

The crater Gassendi, as photographed by Lunar Orbiter 5. Some of the features on its floor appeared to be of volcanic origin.

The Apollo 17 mission patch.

Serenitatis Basin indicated that this predated the formation of the Imbrium Basin. The upwelling of lava into Serenitatis began prior to it doing so in Imbrium. The empiricism for the maria was 'the darker the surface, the younger it is', and there was a dark patch in the southeastern portion of Mare Serenitatis. It looked as if a visit to this area would sample both ancient and very young volcanism. But where to land?

Before it was redirected to Fra Mauro, Apollo 14 had been set to sample the dark mantle at a network of ridges and rilles 60 km west of the crater Littrow, which was located in the mountains just outside the eastern rim of the Serenitatis Basin. Such a site would have been well suited to a single-traverse 'H'-mission, but was unworthy of a 'J'-mission. The selectors were therefore obliged to study where the dark mantle had embayed the mountains that form the basin's rim. Al Worden surveyed the area on Apollo 15, and reported several small dark-halo craters in a valley in the Taurus Mountains, south of Littrow. Furthermore, he observed dark streaks on the flanks of the mountains that suggested they were coated by pyroclastic from fire fountains. In February 1972 Apollo 17 was assigned the task of landing in this

Jack Schmitt, Apollo 17 lunar module pilot.

Gene Cernan, Apollo 17 commander.

Ron Evans, Apollo 17 command module pilot.

A regional view of the Apollo 17 landing site, situated south of the crater Littrow, in a valley in the Taurus mountains on the southeastern rim of the Serenitatis Basin. The orbital view (below) shows the South Massif at lower left, with its fan of 'light mantle' radiating across the valley floor toward the North Massif. Notice the scarp crossing the valley and running around the western flank of the North Massif. The East Massif is lower centre. The Central Cluster is the group of craters on the floor of the valley. The right of the frame shows the low domes of the Sculptured Hills, over which Challenger would have to pass in making its approach.

valley, named Taurus-Littrow, to seek proof that the lunar 'heat engine' was still active. It promised to be a fitting finale to the program.

A MULTIPLICITY OF OBJECTIVES

Like the Apennines, the Taurus massifs are up-thrust crustal blocks. Although they appear to be only half as tall, this is in part due to the fact that they have been more thoroughly embayed.

The landing site was just east of a cluster of 600-metre craters on the flat floor of the 10-km-wide valley. A small peak to the west was named Family Mountain, but the other massifs were assigned obvious names: North, South and East Massifs. The lower domical hills running off to the east by the northern flank of the East Massif appeared so etched that they were named the Sculptured Hills. The relationship of these hills to the basin rim was unclear, but it was hoped that Gene Cernan and Jack Schmitt would find out.

The evidence from Apollo 15, in the form of orbital and surface photography, was that the summits of the Apennine massifs had shed most of their regolith cover by mass-wastage, with the result that the contouring near the summits of both Mount Hadley and Mount Hadley Delta represented the actual shape of the underlying blocks. Intriguingly, Worden had spotted boulder tracks on the massifs around this Taurus valley, and there was therefore the prospect of Cernan and Schmitt sampling rocks whose tracks could be traced to specific outcrops. It was therefore considered likely that Apollo 17 would reveal more about the pre-Serenitatis crust than Apollo 15 had about the pre-Imbrium crust. Although the quest for recent volcanism had been the motivation for selecting this site, the task of characterising the massifs was made the primary objective.

The fact that the embayment followed the topography so well suggested that it was a lava flow. The faint streaks on the massifs implied pyroclastic fountains, and several dark-halo craters strengthened this case by suggesting the sites of the vents. Prior to Apollo 16's discovery that the Cayley was not volcanic, the consensus was that the floor of the Taurus-Littrow valley was an ancient basin in-fill that was later mantled by dark pyroclastic. Having learned a lesson with the Cayley, the geologists now openly embraced a range of working hypotheses. In fact, to cover themselves, they included the possibility that the floor of the valley might be fluidised brecciated ejecta; but no one really believed it. Accordingly, the geological maps described the valley in-fill morphologically as the 'subfloor unit'. Determining its actual nature by inspecting the stratigraphy in the walls of the large craters and by sampling the rocks that were excavated was the second objective. As for *when* the putative pyroclastic was erupted, there was some cause to suspect that it might have been very recently. There was evidence that the valley was within a Tycho ray, and the craters of the Central Cluster had been interpreted as secondaries from this event, and since their rims were dark this implied that the pyroclastic was younger. While confirming this would be the most significant result, the case was so weak that it was subordinated to determining the nature of the massifs and the subfloor.

The geologists were further favoured at Taurus-Littrow with a fan of light-toned material at the base of the South Massif. From overhead, it looked like an avalanche. If so, what had prompted it? As the slope was not beyond the angle of repose in the weak lunar gravity, the material must have been dislodged by a shock. Although the flanks of the mountain were typically clear of craters, there was a cluster of craters on the summit. The shock of their formation might have initiated the slide. It was a long-shot, but these craters were in-line between Tycho and the Central Cluster, and perhaps they were produced together. If so, then the exposure age of the material in the avalanche would match that excavated by the Central Cluster, and this would be a clear measure of the age of Tycho, the most prominent ray crater in the Southern Highlands. In order not to prejudge the issue, the putative avalanche fan was listed as the 'light mantle'.

As if this was not enough, there was a 75-metre-tall scarp across the valley floor, with the eastern end of the valley on the lower level. If the slope were to prove to be amenable, Cernan and Schmitt were to drive up onto the elevated side to reach boulders that had rolled down the South Massif where the light mantle had engulfed the crater Nansen. If it were not for the fact that the scarp continued up the flank of the North Massif, which suggested a thrust fault, it could readily be explained as the front of a lava flow from Mare Serenitatis. It was hoped that by viewing it from ground level Cernan and Schmitt would be able to offer some insight.

In order of priority, Apollo 17's geological objectives were therefore:

A depiction of the Taurus-Littrow landing site, showing the planned traverses: the first to some of the craters of the Central Cluster to the southeast; the second up the scarp to sample the base of the South Massif; and the third to the North Massif and the Sculptured Hills.

(1) to characterise the rim of the Serenitatis Basin by sampling boulders whose context could be related by tracks to specific outcrops on the North and South Massifs;

(2) to determine the nature of the subfloor unit by sampling the craters of the Central Cluster;

(3) to demonstrate that the Moon had been volcanically active in the recent past by sampling the dark-halo craters Shorty and Van Serg;

(4) to sample the light mantle;

(5) to investigate how the Sculptured Hills related to the basin rim; and

(6) to investigate the nature of the scarp.

Given such a rich variety of features, all potentially highly significant, Schmitt lobbied for a four-EVA mission, but the only way that sufficient consumables could have been carried to allow an additional day on the surface would have been to shed some of the scientific experiments. It had already been decided not to fly packages (such as the seismometer) that could be said to have made their contribution to lunar science, in order to give new experiments the opportunity to make a discovery. The geophysicists were unwilling to yield to the geologists. Accordingly, three traverses were devised to address the various objectives. The first traverse was to be south to inspect some of the Central Cluster craters, but the route would be contingent upon setting up the ALSEP on time. The second would involve a long drive west, up onto the elevated side of the scarp to inspect the boulders associated with the light mantle and to sample Shorty. The third would be north to investigate boulders on the North Massif and to sample the Sculptured Hills. It was an *extremely* ambitious program.

ROUTINE LANDING

"Okay, Houston, the Challenger has landed," Cernan reported matter-of-factly, with a hint of homage to Neil Armstrong.

"That's super," acknowledged Gordon Fullerton, a rookie astronaut.

Challenger had started off a little high at PDI, but the PGNS had readily corrected this discrepancy, and thereafter the descent had been textbook. Having caught sight of the South Massif early on, Cernan was comfortable with his line of approach, and when the LM pitched at 7,000 feet and he was presented with a view of the valley he knew just where he was and where he was heading; it was just like the simulator. He simply selected the flattest-looking spot, and set down. He was 230 metres east and 60 metres north of the designated point, but in view of the fact that they were to explore the 10-km-wide valley from end to end, this was of no consequence. In fact, the descent was so *nominal* that there was 2 minutes of propellant remaining in the tanks – sufficient, as Schmitt would later joke, to have conducted a hovering tour of their traverse routes.

"Houston," Cernan called a moment later, "you can tell America that Challenger is at Taurus-Littrow." This reference to 'America' was to CMP Ron Evans, who was about to start a comprehensive program of observations of his own.

FIRST IMPRESSIONS

"Look at that rock out there!" exclaimed Schmitt.

In addition to the litter of cobbles, there was a 3-metre-tall block some 200 metres directly ahead. It appeared to be intensely fractured, and had a mottled appearance. Lifting the focus of his attention, he could just make out the rim of Camelot, which was the largest crater in their immediate vicinity. Beyond that, partially lost in the zero-phase glare, was the brilliant line of the scarp, crossing from the South Massif on the left to the North Massif on the right. Far to the west was Family Mountain. In general the horizon matched what was sketched in the traverse planning documents, but the stark clarity of the surface detail was awesome. There were boulder tracks "all over" the South Massif, Schmitt reported, and when Cernan drew his attention to outcrops near the summit he launched into a description of how the "tan-grey" of the mountain's flank contrasted with the "blue-grey" of the exposure. The horizontal outcrop was just above the break in the slope where the 30-degree gradient reduced for the run up to the summit, so it was no wonder that the boulders had broken off and rolled down. As one of their main objectives was to sample a variety of such boulders, this was excellent news.

ALL IN ALL, A GOOD START

As had been intended for Apollo 16, but not achieved due to the late landing, the first EVA was to be undertaken as soon as Houston confirmed that Challenger was healthy, so when they were given the 'Stay' order Schmitt curtailed his site report to devote his attention to the post-landing checklist.

Four hours after the landing, Cernan was heading down the ladder. The traditional TV by the MESA camera had been cancelled on Apollo 16 due to a communications problem, and it had been decided not to bother with it this time.

"As I step onto the surface at Taurus-Littrow," Cernan called from the footpad, "I would like to dedicate the first step of Apollo 17 to all those who made it possible." Having taken his 'small step', he turned around and was struck by the beauty of the sight that filled his visor. "Oh, my golly! Unbelievable!"

The contingency sample had been deleted, since it was felt that a scoop of regolith and a few pebbles would not shed any light on such a geologically complex site, so Cernan began with a once-around inspection of the LM. As soon as Schmitt joined him, Cernan leaned forward to enable Schmitt to deploy the antenna on the top of his PLSS, and was amazed to see that his colleague's boots were already black with dust. "Your feet," he exclaimed, "you look like you just..."

"... walked on the Moon, huh!?" Schmitt finished the thought.

"There's sparklies in the soil, Jack," Cernan noted.

"I think that's little fragments of glass," Schmitt speculated. Then he turned his attention to nearby rocks. "That's a vesicular rock, and it's obviously not a breccia." He inspected a few more. "*That's* like a breccia. *This* is something else again." He peered at one of the vesicular rocks. "It looks like very light-coloured porphyry of some kind." He meant there were tiny phenocrysts in the crystalline matrix. "About 10 to 15 per cent vesicles." He scanned his vicinity, trying to identify the rocks through their coating of dust to determine the dominant population – most were a pinkish, coarsely textured crystalline material. His colleagues in the Backroom, as yet unable to see for themselves, listened carefully and tried to visualise the scene.

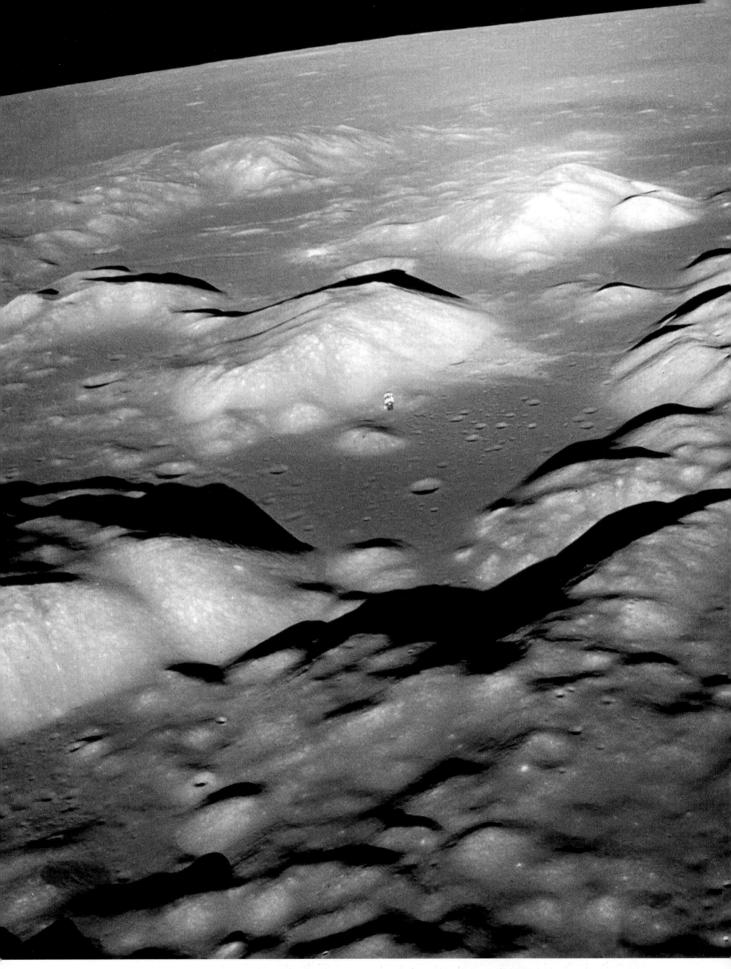

The beautiful valley of Taurus-Littrow. Contrast this confined landing site with the open plain selected for Apollo 11.
The Apollo 17 mothership, America, can be seen in the centre of the frame, set against the base of the South Massif.

Astronaut Bob Parker, Apollo 17 mission scientist.

As always, the ALSEP was to be deployed on the first day. As the most distant sampling objective of the first traverse was only 2 km away, it had been decided to set up the ALSEP first. However, Schmitt was particularly concerned that any delay involving the ALSEP would shorten the first traverse as a consequence of tightening the walkback constraint.

Because it would be the key item in determining the surface activity plan, the first task was the Rover. If this failed to deploy, or was unable to be used, they would be reduced to traverses on foot. Cernan inspected the package. Unlike previously, the walking hinges had not popped. "The Rover looks pretty good, Bob," he told Robert Parker, the mission scientist who was to serve as the CapCom during the traverses.

Cernan took the lanyard that controlled the deployment mechanism, and Schmitt took the one that would draw the vehicle away from Challenger.

As Schmitt walked backwards with his lanyard, he studied the granularity of the surface. "I think it's safe to say that this surface was not formed yesterday." He was looking for an indication of a fine layer of recently deposited pyroclastic ash which might form the dark mantle. "There's a classic regolith." By this, he meant that the surface was well gardened, which meant that it was a very old surface. This led him to conclude that

whatever the dark mantle material might be, it had been mixed into the regolith. As had his predecessors, Cernan had learned to make basic geological observations, but Schmitt was a professional geologist and therefore able to *interpret* what he saw. In fact, his observation that there were fragments of glass in the regolith would prove to be the key to understanding the nature of the dark mantle.

As Cernan played out his lanyard in hand-over-hand fashion, trailing it across the ground, the dust turned his forearms black. "I look like I have been out for a week, already!"

The Rover's deployment was uneventful. They took just 15 minutes to lower and configure it, which was precisely as allotted in the plan. While Schmitt took a pan of the landing site, Cernan gave the ultimate off-road vehicle a test drive. "Challenger's baby is on the roll," he called delightedly.

Although Schmitt had not yet unpacked his sampling tools, he decided to pick up a rock to inspect it more closely. On trying to bend to reach for it, he lost his balance and fell. "I just got my initiation to getting very dirty," he chuckled as he scrambled to his feet. He had not managed to retrieve the rock, but he had certainly gained a close look at it! "The basic light-coloured rock type in the area looks very much like the gabbro in the mare basalt suite." By this he meant samples recovered from mare sites. Although it was undoubtedly a pyroxene-enriched plagioclase-based rock, it was too coarsely grained for an extruded basalt flow. Gabbro is chemically identical to basalt, but is able to grow larger crystals as it slowly cools underground. The fact that this rock was vesicular meant that it was hypabyssal – i.e. it had solidified near the surface. He decided to call it 'intermediate gabbro'. "I haven't learnt to pick up rocks yet," he admitted in embarrassment.

Once Cernan had parked the Rover, Schmitt retrieved the pallet of sampling tools from the MESA. Even as he worked, he mulled over what he had seen so far. "I've seen an awful lot of rocks that look just like those pyroxene gabbros. The pyroxene is iridescent; it looks like clino-pyroxene."

"Hey, Jack," Cernan called as he squinted through the optical sight to align the TV antenna. "You owe yourself 30 seconds to look up at the Earth."

"Agh," Schmitt dismissed, "you see one Earth, you've seen them all." He had spent much of the long cislunar coast monitoring terrestrial weather patterns, and his mind was now focused on his new environment.

Challenger, with the North Massif in the background.

The quality of the TV proved to be better than ever. "I guess you'll believe we're here now, huh?" Cernan joked as the camera came to life and made its first pan of the beautiful valley in the mountains. He could not have guessed that within a few years cynics would be arguing that the entire program of lunar exploration had been faked.

The first surface experiment was to calibrate the Lunar Traverse Gravimeter. He took readings with the instrument both on and off the Rover in order to verify that it could be used while onboard. Readings were to be taken at every sampling station to determine the variation in the 'local gravity' across the valley floor, to investigate the nature of the subfloor and the underlying pre-Serenitatis basement. Readings were to be taken at the start and at the end of each traverse in order to verify the instrument's calibration. The procedure required only that a button be pushed to initiate the cycle and then again at least 3 minutes later to produce the result. The only proviso was that the instrument not be disturbed while it was active.

Before making a start on the ALSEP, there was the matter of the flag, and it was fitting that his particular flag had been on display in Mission Control throughout the Apollo program. Schmitt joked that they should plant it on the summit of the North Massif.

"This has got to be one of the proudest moments of my life," Cernan observed.

The ceremony over, Schmitt ran to inspect a 20-metre crater behind the LM, and, now more comfortable in his suit, dipped to one knee to reach for his first rock,[1]* a "slightly tabular" piece of the intermediate gabbro material. His curiosity satisfied, he mounted the two ALSEP packages on the carrying bar and set off west to find a flat site.

Unfortunately, as Cernan shuffled around the Rover the handle of the hammer in his shin pocket caught on the extension of the right-rear fender. This was a perfect replay of how John Young had damaged his vehicle.

Even as Schmitt manoeuvred around the small craters with his bulky load, he was surveying the territory west of the landing site. "Some of the rocks that I believe to be gabbros have a texture not unlike a welded tuff," he noted. He was not saying that they *were* consolidated pyroclastic, just using the textural comparison as a shorthand communication with his colleagues. "They've got a mottled appearance that I have not figured out." As it happened, the 3-metre block that he had described during his initial view from the window, was exactly where the plan called for him to place the ALSEP, so he veered 30 metres north and put

his load down on a spot that looked as if it would accommodate the instruments. "The rocks still seem to be the pinkish-grey gabbro out here." In other words, there was no obvious variety, and the landing site may well be typical of the Central Cluster.

Having made a crude repair to the torn fender using duct tape, Cernan joined Schmitt at the ALSEP site. He was to drill the holes for the Heat-Flow Experiment while Schmitt deployed all the other instruments. This particular experiment had had more than its share of ill luck. It had been carried on Apollo 13, but had not made it to the lunar surface. At Hadley-Apennine, Dave Scott had not been able to emplace the sensors as deep in the regolith as hoped, and the veracity of the surprisingly high results was open to doubt. And at Descartes-Cayley, John Young had tripped on the cable and torn it loose even before the instrument was fully emplaced, which was all the more frustrating since Charlie Duke had been able to drill the first hole without any difficulty.

Cernan had no problems drilling the holes for the sensors, and was soon ready to take a 3.2-metre core to document the regolith in which the sensors were operating. The tube went into the ground fairly easily, but when he tried to extract it he found that it was stuck fast. "It didn't feel like this stuff was *that* hard," he complained. His efforts were not helped by the fact that as he pumped the jack the treadle dug itself into the soft dirt. He dipped to his knees with each throw of the jack's lever.

After he had set up the other instruments, Schmitt went to assist Cernan with the core. In an effort to increase the leverage on the jack, Schmitt literally threw himself onto the lever, and then sent the rack of tools flying when he kicked out with his feet to recover his balance. After initially slow progress, the tube suddenly began to ease out, so Schmitt left Cernan, now rested, to finish off, and went to sample a few more rocks. Extracting the core had cost 20 minutes. Before he could move on, however, Cernan had to slide the Neutron Probe into the hole. This was to measure how the flux and energy of neutrons varied with depth in the regolith. To shield the sensor from the RTG, which emitted neutrons, he had drilled in a shallow crater in the 'shade' of a rock. When cores from earlier missions were studied, it had been found

---

* Information pertaining to each sample is given on page 369.

In taking 'tourist' shots of the flag raising ceremony, Gene Cernan held the camera low and aimed high over Jack Schmitt's shoulder in order to illustrate that even on the lunar surface one is never far from the 'Good Earth'.

that radioactive isotopes were *created* in the regolith by the absorption of cosmic-ray neutrons. Charlie Duke had noted that the hole of his deep core had not collapsed as he withdrew the tube, and the Neutron Probe was an opportunistic use of the hole this time to determine the flux and energy spectrum of the neutrons in the regolith. The aim was to calibrate a dating method that used the abundances of the isotopes to measure the rate at which the regolith was being 'turned over' by micrometeoroid gardening. The Neutron Probe was to be recovered and returned to Earth for analysis. The core showed the regolith at this point to be a series of distinct layers, alternating between fine dust and fragments of rock.

One of the new instruments deployed by Schmitt was the Lunar Seismic Profiler, for which he laid out four geophones in a 'Y'-shape near the 3-metre block, which was promptly named 'Geophone Rock'. "All of these big boulders are the same," he observed. The gabbro was seemingly ubiquitous. The fact that there were such large pieces of it laying around strengthened the case for it being the predominant type of rock in the subfloor. "It is possibly upwards of 50 per cent plagioclase, rather than the 30 per cent of the mare." This inference was based on the fact that the rock was brighter than the mare basalt he had seen in the laboratory. However, as he would discover when he examined a bit of it in the LM, the mottling he had observed was due to micrometeoroids shattering the crystalline plagioclase, thereby enhancing its reflectivity, at which time he would reduce his estimate back toward the norm. Until then, however, he worked under the misapprehension that the magma that flooded the valley was unusually rich in plagioclase.

"One block," said Schmitt, meaning Geophone Rock, "has very sharply defined parallel parting planes." He was directly addressing his colleagues in the Backroom. The rock was intriguingly fractured. One advantage of being able to study a boulder was that it was possible to observe textural variation on a scale that was impossible to infer from studying small rocks. "I think there's foliation paralleling the parting," he said hesitantly. A 'lineation' is the result of the alignment of elongated minerals under directional stress; 'foliation' is the alignment of tabular minerals, and it makes a planar variation in the rock; and a 'parting' refers to how a rock breaks, which is usually parallel to a foliation plane. It meant that the magma had been stressed as it cooled, in all likelihood because it was flowing. "The parting planes go through the whole boulder." On

searching for, but not finding, any alignment of the crystals, he revised his assessment. "The layering, or the foliation, or the parting – whichever – is the result of variation in vesicle concentrations." It was not easy, but he peered into the cavities. "The vesicles are spherical, but they have fairly rough outlines; it looks as if there has been some recrystallisation." This supported his belief that the magma had cooled slowly, because the minerals in the gas which made the vesicles had recrystallised on the cavity walls. There had not been sufficient condensation to turn the voids into irregular vugs. To complement his visual inspection, he bagged a piece[2] of the rock.

As Cernan stripped down the core tube, Parker announced that as they were half-an-hour behind, it had been decided to designate the crater Steno as the objective of their first traverse.

"I don't care if we're 30 minutes late," Cernan replied. He was content to have been able to deploy the ALSEP without too many problems.

DASH SOUTH

Parker refined the traverse objectives. "We're planning on going to the west side of Steno – to the boulder field." A number of boulders had been spotted on overhead imagery right on the rim of this 600-metre crater, so these could safely be assumed to have been dug from the deepest point. If there had been time, they were to have skirted Steno's rim on the way to Emory, a similar crater a kilometre further south, where the dark mantle appeared to peter out and there was the prospect of sampling a variation in the regolith that would indicate the contact.

"I guess we ought to press on," said Schmitt. Nevertheless, it was nearly half an hour before they were finally ready.

To start with, due to uncertainty in precisely where they had landed, they were not sure where they were, and it wasn't until they encountered the eastern rim of Trident, and recognised the distinctive septum where the walls of the component craters met, that they got their bearings. Having this reference, Cernan swung right to head for Steno, some 750 metres away.

As they ran up a slight ridge, Parker warned that there would be "very large boulders" laying around.

"There certainly are," Schmitt laughed. But because their context was ill-defined these were useless for investigating the *structure* of the subfloor.

A view of the deployment site of the Apollo Lunar Surface Experiment Package.

The flag in the foreground marks the location of one of the seismic sensors. Jack Schmitt named the large nearby boulder the 'Geophone Rock'.

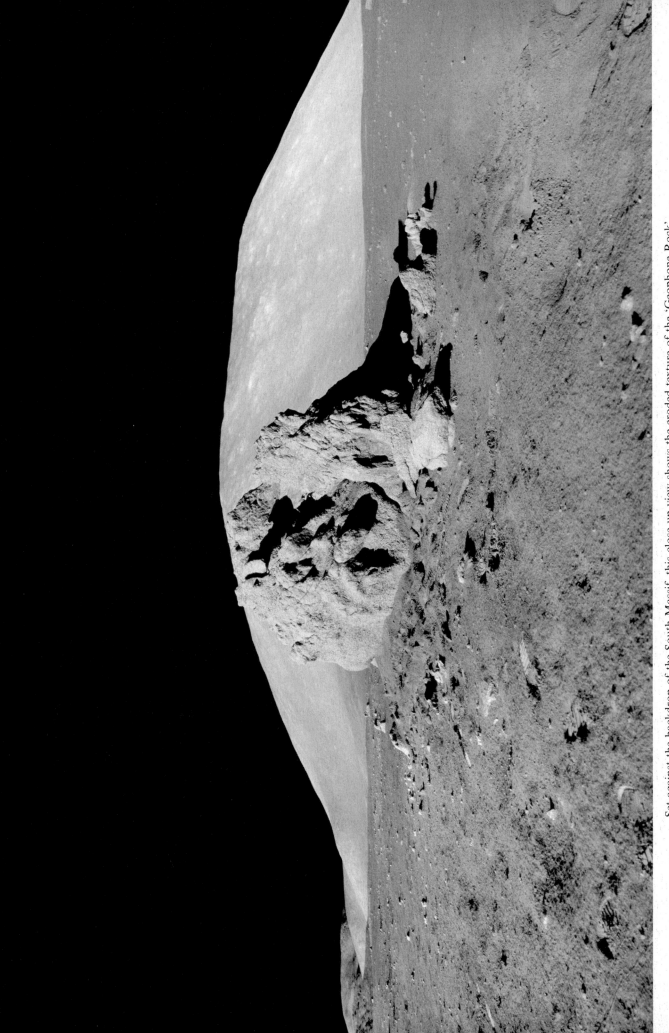

Set against the backdrop of the South Massif, this close-up view shows the eroded texture of the 'Geophone Rock'.

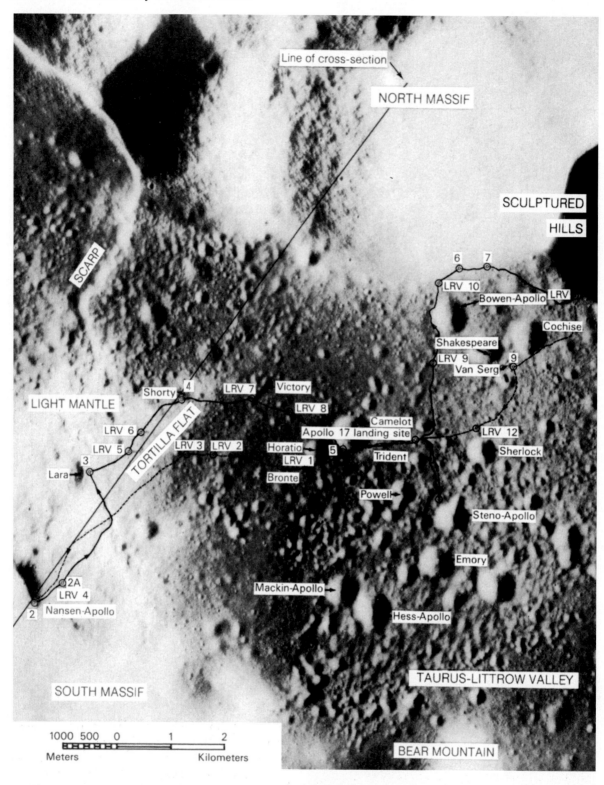

The actual Rover traverses and sampling stops made by Gene Cernan and Jack Schmitt. Although the large crater near Station 6 is specified as 'Bowen-Apollo', the crew referred to it as 'Henry the Navigator'. Courtesy the Lunar and Planetary Institute and Cambridge University Press.

"We think you're just about there," Parker prompted several minutes later. The Backroom had erred in determining the landing site, and Parker believed them to be significantly further south than they actually were.

"It sure doesn't look familiar," Schmitt replied, "as far as Steno's concerned." All the craters that dominated the overhead view were ellipses that were very difficult to discern; it was amazing how different the terrain looked at ground level. His concern was that they would waste time seeking a specific position. Their objective was to sample a crater with a blocky rim. "I think they can locate us if we work that boulder field right there," he suggested, indicating a 20-metre crater. Cernan parked alongside it. Once they had figured out where Challenger was, they would be able to identify this position from the Rover's 'nav' data – it would later be established that Station 1 was 150 metres from the northwestern rim of Steno.

SOMEWHERE NEAR STENO'S CRATER

After dismounting the Rover, Cernan turned to study Emory, which was visible on a rise to the south. It was heavily littered with rock, and would have been an excellent site for a full-length station, but if they had pressed on they would have had less than 15 minutes on-site, which would not have done it justice.

Schmitt emplaced a seismic charge for the Lunar Seismic Profiling Experiment, and then went to the crater. "All the blocks on the rim, look like the vesicular rocks at the LM." He went to a half-metre boulder on the southwestern rim. "There's the parting plane. We'll try and get a sample along it." This similarity in the rocks at the two sites suggested lateral homogeneity in the subfloor. "Get your hammer," he told Cernan, "we're gonna need it." In the sampling, Cernan was content to play the role of field assistant to Dr Schmitt the geologist.

Parker told them that they had 30 minutes, maximum. The first priority was to be sampling the boulders. If there was time after a comprehensive sample with the rake, they were to take a core sample to determine the vertical structure of the regolith. If they had managed to find Steno, and had had the time, they would have undertaken a radial sample to investigate the structure of the subfloor.

Having set up the Rover's TV and started the gravimeter, Cernan joined Schmitt. "You got one picked out?" he enquired, eagerly wielding the hammer.

Schmitt indicated for Cernan to chip at the parting plane in one of the rocks. "We have got to figure out why they have foliation in them."

Cernan manoeuvred down into the wall of the crater in order to gain access to the northeastern corner of the rock, struck an overlapping fracture and detached a small chip. While Schmitt went to retrieve the fragment[3] using his scoop, Cernan delivered a series of heavy blows to a point where it looked as if the rock would be weak. "The whole thing is going to fracture off here," he predicted, but it failed to yield and he gave up. He had better luck on the far side of the parting plane, when a piece the size of his fist came away. "A whole big slab!" He leaned against the rock and bent to collect his prize,[4] which was so angular that its jagged edge cut through the thin plastic sample bag, so he dumped it straight into Schmitt's SCB in order not to waste time.

"Look at the dark minerals in there," Cernan said, upon inspecting the scar on the boulder.

"That may be ilmenite," Schmitt ventured. "Or fresh pyroxene," he offered as an alternative. The sparkling recrystallised minerals in the vugs implied slow cooling.

To round out the sample, Schmitt took a scoop of soil off the boulder's fillet, then set off around the crater's rim in search of variety. Cernan, following, saw a block out from the southern rim that had an interesting pattern of vesicles. Schmitt took a look. There was a transition from coarse to fine vesicles. "That's what I'm after!"

"Let's see if I can get that contact," said Cernan, pulling his hammer out of his shin pocket.

"Get a piece of both," Schmitt directed, indicating that Cernan should sample both sides of the contact.

"Do you guys see any 2-metre boulders around?" Parker broke in. The Backroom had had Fendell make a TV pan, and were puzzled that they could not see the large blocks that were evident on the overhead imagery.

"We're not where you think we are," Schmitt insisted. "We're not sure where we are."

As it was awkward for Cernan to reach this half-metre high block, and there was no convenient slope to exploit, Schmitt took the hammer and secured a sample from each side of the contact – one side[5] was coarsely vesicular and the other[6] was finely vesicular. "It's a *small* chip," Schmitt apologised as he bagged the finer sample, "but it will 'tell the story', I think."

"I'm going to take a close-up stereo of that

The blocky crater located "somewhere near Steno" that was selected for Station 1.

Some of the rocks that were sampled on the rim of the crater at Station 1.

The Rover parked at Station 1, with the South Massif in the background.

A westward-looking view of Jack Schmitt using the rake at Station 1. The box with the antenna (left) is one of the charges that were to be detonated to facilitate seismic profiling of the valley floor after the astronauts had left the Moon.

contact," Cernan said. The pictures would complement the rock samples by expanding the context.

"When you get done with that boulder," Parker interjected, "we would like you to move on to a rake sample."

"Get a close-up," Schmitt agreed, "I'll get started on the rake." He went to rake a clear patch about 6 metres northeast of the Rover.

"We'd like to have you guys driving in 10 minutes," Parker prompted.

"Nag, nag, nag!" Schmitt complained light-heartedly.

After taking his close-ups, Cernan grabbed a small rock[7] from nearby to augment their haul, then took the site pan.

Schmitt's rake did not penetrate the regolith more than 3 centimetres, but the soil was thick with rock fragments and he was able to fill two bags.[8] The fact that most of the fragments were vesicular indicated that they were chips off local boulders.

"Now we need a kilo of soil," Parker reminded. The soil would complement the pebbles, and also provide a point of reference for working out the variation of the regolith across the floor of valley as part of the investigation of the dark mantle. In this respect, the loss of a sample from Emory's eastern rim, at the fringe of the dark mantle, was particularly unfortunate.

"Have we got time for the core?" Cernan asked Parker as he wrapped up the site pan.

"Negative!"

As soon as they were on the Rover, Cernan turned and raced north at top speed. They had not reached Emory, and not having had the satisfaction of viewing Steno they would never know whether its walls were lined with bedrock outcrops, but they had not done too badly; not too badly at all.

BACK HOME

The long ride back enabled Schmitt to reflect upon what he'd seen since venturing out. "My impression," he offered, "is that the dark mantle is indistinguishable from the regolith derived from the rocks." It certainly was not a thin veneer of pyroclastic overlying a mature regolith. The dark stuff was mixed in, but what was it? "The more I look at this dark dust, the more it *doesn't* seem like the kind of thing you would expect to have been derived from the underlying bedrock. It just seems much too fine-grained. You don't have the

impression you're getting the size distribution you'd expect by having all the blocks around." He meant that if the dark material was off the rocks, there would be chunks of it. But it was all *fine* material, mixed with a soil that *was* derived from the rocks. "I think there's two populations." The mantle *had* to be an additive. The tops of the rocks were perfectly clean, so whatever it might be, the dark dust had been added a long time ago.

At Trident, Cernan swung west. An ominous shadow implied that the damaged fender had collapsed and the wheel was throwing up a rooster tail.

As they approached the LM, Cernan paused to let Schmitt get off where they had left the transmitter for the Surface Electrical Properties Experiment, then he lined up and drove a series of dog-legs in order to mark a north–south and east–west cross on which they were to string the 35-metre-long arms of the antenna. On the remaining traverses, a receiver on the Rover was to record the experiment's radio signals, and the data would later be processed to provide insight into the structure of the subfloor to a depth of a kilometre or more. It was a simple experiment that could beneficially have been deployed at Hadley-Apennine and at Descartes-Cayley.

"This dust isn't going to be fun tomorrow," Cernan pointed out upon finding that there was so much dust on the covers of the Rover's batteries that he could write his name in it with his fingertip.

Walking back to the LM, Schmitt lifted a football-sized rock[9] as an opportunistic sample. It was a coarsely grained vesicular gabbro, and had one very flat face that suggested it had split off a parting plane of a partially buried boulder nearby.

Once they were back inside Challenger, Parker invited their interim conclusions regarding the dark mantle. Schmitt summarised his deliberations. "I do *not* have an intuitive feeling that it has been derived from the boulders, because those are fairly light. It could be that it is derived from some *other* material that blanketed the area." It was certainly not the recent pyroclastic that had been hoped. "I don't think we have that answer yet."

Cernan offered that the boulders probably represented the subfloor unit, because some of the rock in the walls of the few craters that they had managed to peer into as they drove by had seemed to be *in situ*; but it was not possible to be sure they were outcrops of bedrock. Parker, not wishing to be too interpretive, accepted that they had almost certainly been sampling "the *top* of the subfloor".

Schmitt suggested that if Camelot Crater failed

Jack Schmitt tends the Surface Electrical Properties Experiment's transmitter, east of Challenger.

to resolve the matter of the subfloor, they might revise the final traverse. The plan was to conclude this by sampling a large crater east of the landing site. Perhaps they should instead make a high-speed run for Emory.

As Schmitt signed off for the evening, he suggested that the Backroom consider the possibility that the dark mantle was an ancient pyroclastic layer that had been laid down directly onto the subfloor, and hence had been mixed into the regolith at the very start. Although this did not stimulate a reaction, and the issue would remain 'open' until the soil samples were analysed, Schmitt had just solved the mystery of the dark mantle.

OUT AGAIN

Gordon Fullerton woke up Cernan and Schmitt by playing Wagner's 'Ride' at full volume, then launched straight into the procedure that had been devised to overcome the loss of the fender. The support crew had devised a plan for taping together some laminated maps in order to create a rigid flap that could be fastened to the remains of the fender using tiny clamps taken off the cabin lamps. The trick would be to ensure that there was no air trapped between the maps as they were taped together, or else it would open the seam. John Young had even suited up to verify that the job could be done by gloved fingers.

Upon taking over, Bob Parker relayed the results of the Backroom's cogitations. It was thought *likely* that they had sampled the subfloor at Station 1, but because they had not had an opportunity to inspect the stratigraphy of the walls and floor of Steno, it was *possible* that they only sampled material from a late lava flow that was not representative of the true subfloor beneath.

On the forthcoming excursion, therefore, it was crucial that they reach the rim of Camelot, which was comparable to Steno, both to ensure that they sampled material from the deepest point of excavation and to photograph any layering exposed in its wall. Although sampling the subfloor at Camelot was a high priority, the walkback constraint required that it be the last stop on their itinerary. They were to visit it on the way back from sampling the South Massif, the light mantle at its base, the scarp and a dark-halo crater, *each* of which would be at the limit of their operating radius at the time they were present – so there was no scope for slippage in the schedule. In

The *ad hoc* substitute for the Rover's damaged fender.

order to provide time to fix the fender, Parker announced that it had been decided to delete the originally planned soil-mechanics trench on the light mantle at the base of the scarp and the radial sample at the dark-halo crater, both of which were now subordinated to Camelot. If the nature of the subfloor had been able to be resolved at Steno, then Camelot would have been sacrificed instead.

On the topic of the subfloor, Schmitt reported on his own overnight deliberations. He had examined a sample of his 'intermediate gabbro' using a hand-lens, and found that its light tone was caused by halos around the zap-pits. In the harsh sunlight, this had led him to overestimate the amount of plagioclase, which he revised down from 50 per cent to "30 to 40 per cent". He concluded that the subfloor was "probably just a good gabbro; a clino-pyroxene gabbro with a fair amount of ilmenite". Picking up on his early suspicion that the regolith contained a lot of tiny fragments of glass, he proposed that if the pyroclastic was derived from the same magma reservoir as the extrusion that made the subfloor then it, too, would be rich in ilmenite, and therefore "darker than usual".

"Okay, Houston. Apollo 17 is ready to go to

work," Cernan called as he stepped onto the surface once more. "What a nice day."

"There's not a cloud in the sky," Schmitt agreed deadpan. "Except on the Earth!"

Cernan was impressed by the Earth's rotation. To a terrestrial observer, the Moon tracks across the sky as the Earth rotates beneath it, but it maintains the same face towards the viewer. To a lunar observer, the Earth remains in the same place in the sky and turns on its axis. The Earth had been overhead at the equatorial sites; and Hadley-Apennine, although northerly, was too far west for the Earth to be visible in the west-facing windows of the LM. Taurus-Littrow was the only landing site to be both at high latitude and east of the meridian, and Cernan had been able to watch the Earth in comfort. Furthermore, not only was the disk four times wider than the Moon's to a terrestrial observer, it was also alive with colour. In fact, the previous day he had held the camera low and aimed it obliquely upwards to include the Earth in the frame with Schmitt and the flag.

With the make-shift fender attached, Cernan went for a test drive. "Jack, how's the rooster tail look?"

"Looks like a good fix," Schmitt confirmed.

DRIVING WEST

About 250 metres beyond the ALSEP, they paused to enable Schmitt to drop off a seismic charge. Such 'Rover stops' had been devised to save the time that it would have taken to unstrap and disembark in order to lift a rock or to scoop up soil using a long-handled tool, but as he leaned out to set the charge he noticed that his scoop was missing. The limited visibility of the suit prevented him from searching the footwell, so he dismounted to search for it. If it had fallen off the vehicle then it would have been irretrievable, but fortunately it was still onboard.

"As you go by Camelot," Parker called as they got underway again, "you might keep an eye out for blocks along the rim." The Backroom wanted to ensure that the high priority assigned to this crater was justified. A specific site on the southern rim had been chosen on the basis of what appeared on the overhead imagery to be a field of rocks. If this were to prove not to be the case, then the drive-by might identify an alternative site for sampling on the return trip.

"There's Camelot," Schmitt exclaimed as they crested a slight rise and the crater appeared before them.

"Woowee!" added Cernan. "*That* is a 600-metre crater."

"It's very blocky," Schmitt confirmed.

"Man, are there blocks there!" Cernan echoed.

In fact, the field of boulders extended far outside the rim, suggesting an excellent prospect for there being interesting sights within the crater. Schmitt estimated about 30 per cent coverage for rocks in the 2- to 4-metre size range, and the fact that most of them appeared to be the seemingly ubiquitous gabbro supported the case for this being the subfloor.

On entering a broad depression beyond Camelot that was so shallow that it never cast a real shadow, and therefore had not been recognised by the mappers, the block coverage decreased to at most 1 per cent. His route clear, Cernan ran the Rover "full bore". Schmitt noted that the small craters were different. The fresh ones had lots of regolith breccia, but no rock, which implied that the regolith was free of fragmented blocks. In this area, the valley floor was essentially flat, although not level, because it formed a very gentle slope.

"We can see the light mantle," Cernan told Parker. This was several kilometres ahead, extending out across the plain in a broad fan. They were to sample its fringe for comparison with material from the base of the massif. It was prominent from a distance, but would they be able to establish the moment that they ran onto it?

"There's Lara," Schmitt added. This crater was on the slope of the 75-metre-tall scarp, and they were to sample just below it.

"And Hole-In-The-Wall," Schmitt continued. This feature marked a joint between two sections of the scarp, and the plan was to use it as a route up onto the elevated level. In fact, the scarp was not the obstacle that had been feared, since the slope was smooth and there was no evidence of outcropping.

The 400-metre crater Horatio loomed to the right, and was strikingly different to Camelot. "The blocks don't go to the rim," Schmitt pointed out. The debris was so completely contained within the crater that even on the rim the coverage was no more than 1 per cent, so Cernan ran right along it to give Schmitt an opportunity to study the interior. There was a transition in the slope 100 metres or so inside the rim, within which it fell away steeply. In fact, the litter was not merely inside the rim, it was also confined by this bench. Schmitt was impressed by the 10-metre blocks, but what really got his attention was the stratigraphy in the walls. The outcrops implied the presence of a 20–30-metre-thick layer of regolith on the subfloor. Data from the seismic charges, and from the SEPE, which was active at this point, would confirm this observation.

As Horatio fell behind, Schmitt returned to studying the variation. The terrain that they had entered after Camelot persisted. There was evidently a much greater block coverage within the Central Cluster, where they had made their first traverse. It was very different out here. As the diameter of a crater was related to its depth, the depth of the regolith could be estimated from the extent to which craters of different sizes had excavated bedrock.

"Bob, we're seeing craters as much as 20–30 metres in diameter *without* blocky rims," Schmitt reported.

Shortly thereafter, they spotted a 10-metre secondary that had "a big mass of rock in the bottom". This prompted Schmitt to call a 'Rover stop' in order to sample the spray of rock on the southwestern rim that was clearly the shattered remains of the projectile. Cernan paused on the rim, and Schmitt used his long-handled scoop to collect one of the "very irregular and jagged rocks", which proved to be a breccia.[10] Cernan then ran around the crater to enable Schmitt to take some 'locator' pictures, after which they resumed the westerly drive.

A little further on, the block coverage suddenly increased and an ellipse opened up directly ahead. "My God!" cried Cernan, as he swung hard right to detour around the northern rim of the 200-metre crater.

"It looks like Brontë has got the subfloor," Schmitt pointed out. The presence of "egg-sized" vesicles indicated that this magma had solidified near the surface. There were many blocks on the rim, but they were tiny compared to those within Horatio. As soon as they had passed beyond the crater, everything returned to normal. Apart from in the immediate vicinity of large craters, the valley floor outside the Central Cluster was remarkably clear.

THE LIGHT MANTLE

"We're only 100 metres from the light mantle," Cernan informed Parker. But then as they closed in on it, the boundary became less distinct. "You can't see the contact, but you know you're coming to something lighter."

Schmitt reported that the walls of the craters were now "much brighter", which suggested that the light mantle, at least out on its fringe, was a veneer of fines. "We ought to sample the rim of one of these craters," Schmitt prompted. Cernan paused on the rim of a 10-metre crater to enable this to be done.

As they drove across a ray of light mantle that projected across the valley about a kilometre east of the main fan, the glare was dazzling, and Cernan had to slow down because the obstacles were invisible until he was almost upon them. The valley floor between the ray and the fan was named Tortilla Flats. As they crossed it, Schmitt noted that the few blocks that were present were the familiar gabbro. The plan called for a sample of the dark mantle of the Flats, so Cernan halted and Schmitt scooped a rock[11] and then some soil.[12] Meanwhile, Cernan noticed a change in texture where the scarp ran onto the flank of the North Massif. Specifically, the 'wrinkled' texture of the mountain was absent westward of the scarp. "It looks like the scarp *overlays* the massif," he ventured.

"The scarp is smoother, less cratered and less lineated," Schmitt agreed. "It's not just the slope, it's the material on the west side of the scarp." It looked as though the scarp had 'ridden up' on the flank of the mountain.

After leaving the Tortilla Flats, they drove onto the fan of the light mantle, which spanned both levels of the scarp.

"Hole-In-The-Wall, Bob," Cernan called, "is a very long, very subtle, very gentle slope." They should have no difficulty in driving up it.

Schmitt's attention was directed a kilometre or so to the north, where they were to undertake Station 3. The sequence of events was evident: Lara had hit the scarp and was later mantled by the fan.

"Bob, I'm starting up," Cernan announced, as he made a straight run at the scarp. "I don't even think the Rover knows it's going uphill!"

However, the scarp proved to be rather rougher than it had appeared, being somewhat undulatory, and Cernan had to slow to 5 kph and switch to 'tacking' back and forth. "The Rover is really working now," he reported, "but we're almost there!"

Upon finally reaching the crest, he turned southwest for Nansen, which was to be the site for Station 2.

SAMPLING THE TALUS

The South Massif filled their visors. It had not been the height of Mount Hadley Delta that had so impressed Scott and Irwin, but its *bulk*, which had been all the more amazing because it stood alone. Having served as Irwin's back-up, Schmitt had followed that mission in detail. Gazing up at the South Massif, he appreciated how his predecessors had felt.

Earlier, he had described the flank as tan-grey and the outcrops above the break in the slope as blue-grey. Now it was apparent that the outcrops contained two types of rock: light-tan-to-white and blue-grey. "There's a lot of boulder tracks coming down from the blue-grey outcrop," he reported. They did not have time to detour very far, and they could not drive far up the slope, but if he could find an accessible boulder with a track that could be traced up to a specific part of the outcrop then it would be worth making that their objective. Parker agreed that this was "a nice idea", but only if there was a boulder in the immediate vicinity of Nansen.

The nature of the light mantle changed as they approached the mountain. The rock coverage increased, and the dominant type changed to a light-toned rock that had a mottled pattern which suggested they were breccias.

On approaching the South Massif, they discovered that there was a broad trench at its base. "Look at Nansen!" Schmitt exclaimed. It was in the trench, and had been partially overrun by the landslide. "My goodness gracious." To his delight, a number of tracks ran down into the crater. They decided to sample a scattering of more readily accessible large boulders in the talus just east of Nansen's rim.

Even although the make-shift fender had served its purpose well, during the 7-km drive the Rover had become thickly coated with dust, and it took several minutes to brush this off the electronics boxes and to work through the other chores preparatory to sampling.

"We'll never get started," Schmitt moaned. As soon as he could, he made his way 30 metres up-slope to inspect the nearer of two pleasingly large boulders. "The blue-grey rocks are breccias," he confirmed. The block was only a metre high, but it was twice that across. It was composed of several different units, and was fractured and foliated. The friable matrix was so thoroughly eroded that the more indurated clasts projected in large knobbly mounds. Some of the clasts were elongated parallel to the plane of the layering. The matrix itself was relatively free of fragments. The zap-pits had faint halos. "They might be vitric or glassy breccias," he pronounced. The light clasts were fine-grained, and he ventured that they might

be "shattered anorthosite". The broad fillet on the up-hill side indicated that the boulder had been present for a long time. It was not clear whether it was partially buried or resting on the surface, but if it had slithered into position it had not ploughed itself a furrow because there was no ridge on the down-slope side.

As Cernan joined him, Schmitt boldly stated the rock's context. "I'll bet you it's the same as the blue-grey rocks we saw higher up." It was a sure bet that in sampling this block they would be sampling the outcrop near the summit of the massif, which could be either the near-surface of the pre-Serenitatis crust or the material deposited by the 'base surge' from the newly excavated basin. Either way, this site was a lunar geologist's paradise.

"We ought to sample across that layering," Schmitt prompted, and Cernan obliged with the hammer, obtaining several samples[13] across the foliation.

Having collected both matrix-rich and fragment-rich material from the dark matrix, Cernan wanted to sample an inclusion. "How about this?" he offered, pointing to a large clast bulging out of the matrix.

"That's a good representative fragment," Schmitt agreed.

"It's a *football*-sized fragment," Cernan laughed, as a considerably larger piece of the clast broke free than he had been expecting.

"This is a light-matrix breccia," Schmitt informed Parker, referring to the clast.[14] It was so chunky that he could not close the metal tabs to seal the bag.

To the Backroom's frustration, the Rover's seats blocked their view of the sample site. "Do you guys see any tracks coming down to these boulders?" Parker asked.

"Unfortunately, no," Schmitt replied. The tracks that he had been studying during the drive ran down into Nansen, but he was sure that all the blocks were from the same source. He scooped the fillet, skimmed the regolith, and dug out a subsurface soil sample to establish the rock's current context.

"We'd like to sample a variety of blocks," Parker prompted, lest they spend the entire 40 minutes assigned to Station 2 working on this one boulder.

"We're going after a lighter block now," Schmitt pointed out.

In fact, Cernan was already heading to an even larger boulder, 10 metres further up-slope. It was a tough climb. "Hey, Jack, don't come up here unless you bring the rake."

A northward-looking view of the Rover at Station 2, taken from a position on the lower slope of the South Massif. The boulders almost certainly rolled down from outcrops further up the mountain, beyond the reach of the astronauts, and as such were the primary sampling objective.

The first large rock to be sampled at Station 2 is visible on the left in the previous illustration.

"We'll want to rake away from that," Schmitt warned, "because it's probably shedding." A rake was supposed to provide a representative collection of the pebbles in the regolith, and little purpose would be served by sampling too close to a friable rock that was shedding fragments. Upon reaching his commander, he jammed the rake in the ground as if it were a ski-stick, and went to inspect their new objective.

"Hey, that's a different rock," Schmitt announced. It was a *green*-grey breccia. Parker's desire for variety was being fulfilled. It had about the same footprint as the first block, but was over twice as tall. It was also smoother, more rounded,

more uniform and less weathered, and although it had several irregular joints it did not have such clear fractures.

"It's got a flakey fracture pattern all over it," Cernan pointed out.

"This is crystalline," Schmitt announced. "There are fragments, I think." Then he laughed upon realising that he had been inspecting a half-metre angular inclusion. "As a matter of fact, here's a big fragment of what looks like a porphyry caught up in it." A porphyritic rock is a fine-grained crystalline mass with phenocrysts. In this case, however, later analysis would show that the 'phenocrysts' were the same as the

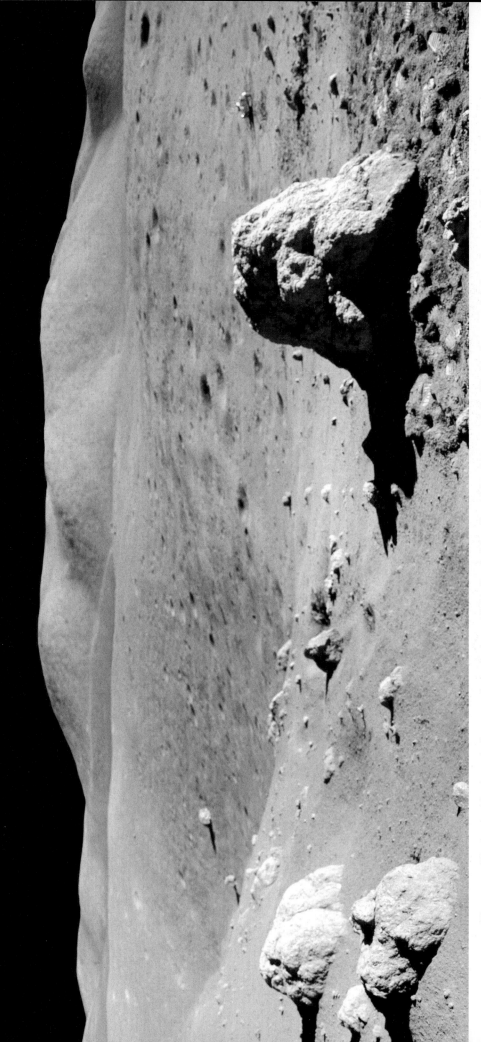

The crater Nansen resides in a shallow trench at the base of the South Massif, and has been partially filled in by material that slumped off the mountain.

The second large rock sampled at Station 2 is in the foreground in the previous illustration.

ground-mass; the crystalline rock had been so badly shocked that most of it had been reduced to powder, and as it was reconsolidated the remaining crystalline fragments were caught up in it. This was a testament to the rock's violent history prior to being shattered and included in the breccia. Cernan hammered off several fragments of the large inclusion.[15,16] "It's highly feldspathic, maybe an anorthositic gabbro," Schmitt ventured. "We need to get the host too," he prompted. Cernan obliged and released a chip[17] of the green-grey host. "The host *appears* to be crystalline," Schmitt observed, "but it may be a *re*crystallised rock." If the material had been subjected to melting, it could well have originated from near 'ground zero' in the creation of the Serenitatis Basin.

Accepting the potential of this site to the primary objective of the day's traverse, the Backroom had decided that if the astronauts wished, they could extend Station 2 by 10 minutes, but this would be at the expense of Shorty, the dark-halo crater which had already had its radial sample deleted. Parker passed on the invitation.

"We'll think about it," Schmitt replied noncommittally, as he continued to inspect the green matrix. "It looks like it has some olivine crystals in it."

"Bob, how much time *have* we got?" Cernan asked as he gave Schmitt two more pieces[18,19] of the matrix that he had just released.

"Twelve minutes, unless you take the extra ten," Parker replied.

"Let's take it," Cernan decided.

It was time to rake the talus, so while Schmitt prepared the site, Cernan took a pan from above the large boulder. "Bob, I'd say we can meet our walkback constraint." By this, he meant that there was no need to be overly enthusiastic in preserving the 'margin'.

"Can you see the LM?" Parker asked suspiciously.

This rock at Station 2 attracted attention for having a bright inclusion. In this view the gleaming patch marks where a fragment of the clast had been hammered off.

"Oh, no," Cernan replied automatically. The view was extremely limited, because although the North Massif poked over the edge of the trench most of the valley was hidden. "I won't stretch the walkback constraint," he assured, realising that Houston might think that by being on the high side of the scarp they were especially exposed to navigational difficulties. Deciding to quit while he was ahead, he went to assist Schmitt to pour the raked pebbles[20] into a bag.

"Do you see any more 'different' blocks up there?" Parker asked.

"We haven't had a chance to look around!" Schmitt retorted.

"We want a rake in the light mantle," Parker persisted, "and you might as well get that by the Rover." In his interface function he had to relay recommendations from the Backroom, but his sympathies were with the crew, especially Schmitt, a fellow scientist-astronaut. He offered a little leeway. "You might look around for a couple of documented samples up there." If there were no distinctive boulders, the smaller rocks would be worthwhile.

In fact, they had already moved down-slope to return to a block that had attracted Schmitt's attention on the way up. It was only half-a-metre wide, but owing to a pair of mutually orthogonal planar fractures it was very boxy. It was a grey breccia with a fascinating inclusion. "Let's get that big clast," Schmitt directed upon reaching their target. Cernan delivered several hammer blows to the matrix in an effort to loosen the clast, but to little effect. "Pretty hard, isn't it, Schmitt observed unnecessarily. As Cernan continued to pound at the matrix, the rock moved slightly. It was evidently not very firmly situated on the slope. Having made no progress with the matrix, he turned his attention to the clast, and readily gained several fragments. While Schmitt inspected a fragment of the clast,[21] Cernan finally released a piece of the matrix, which, owing to the energy of the blow, shot off at speed.

"See it?" Cernan asked.

"See it!?" Schmitt taunted sarcastically. "You hit me with it!!" It had ricocheted off his arm and rolled several metres down-slope. But it was such a large chip[22] that he had no trouble in locating it.

The blue-grey matrix was finely grained and poor in fragments.

Cernan decided to exploit the rock's instability, and roll it. As he had worked on it from downslope, he moved above it and gave it a good shove with his boot. "Look out, Jack!"

Schmitt, now re-examining the clast fragment, leapt clear, but the block stopped before reaching him and he casually resumed his study. "It has green, fairly rounded crystals in a fine-grained white-to-pinkish-tan matrix," he observed. It was familiar, and yet mysterious. "It looks like olivine in *something*."

In fact, this clast was 'olivine in olivine', in the same sense as the earlier samples in which the crystalline rock had been so shocked that it was reduced to a chemically homogeneous but texturally heterogeneous mixture. Analysis would reveal it to be a doubly valuable sample, in that not only was it almost *pure* olivine, making it dunite, but at 4.5 ($\pm$0.1) billion years, it was also the *oldest* dated sample returned from the Moon. There had apparently been considerable chemical differentiation at a very early stage in the Moon's history, and this may well have been a piece of a mantle-intrusion into the crust. It had barely survived the formation of the Serenitatis Basin, had been deposited on top of the massif, had become outcropped, and then, some considerable time later, had rolled down the hill.

The second rake was meant to establish any variation between the regolith on the lower slope of the massif and the light mantle in the trench at its base. Interestingly, whereas Schmitt had easily secured a full bag on the slope, he struck out on the flat, some 25 metres north of the Rover. "There just aren't any rocks!" he pointed out. It was just clods of regolith breccia. Evidently, in the trench the light mantle was all fines. This made sense, in that the dusty material could be expected to migrate down the slope and accumulate in the trench. The fan that spread out across the valley, however, must have been part of the avalanche. With luck, the soil samples taken on the drive in, and those that were to be taken on the way out, would help to resolve the nature of that flow. "One-scoop Schmitt, they call me," he chuckled, as he poured a rake of soil straight into the bag without spilling a drop.

"We ought to start moving," Cernan acknowledged. In fact, he now had no option because they were hard up against the walkback constraint.

"Let's go," Schmitt agreed.

"Beautiful station, guys," Parker commended. The main objective of the traverse seemed to have been achieved, and there was every prospect of resolving the doubts about the subfloor at Camelot. They had stolen some time from Shorty, but this had been well spent. In a change to the plan, the SEPE receiver was turned off because the dust had penetrated its thermal insulation and the telemetry indicated that it was running hot.

BACK DOWN THE SCARP

"More or less follow our tracks until we're over the big hump," Schmitt offered. He meant the scarp. "I've got Station 3 pretty well spotted."

"My golly!" Cernan exclaimed, as they drove out of the trench and he got his first view back east from a vantage point above the scarp. "Look at that valley!"

Now that they were on their way, Parker gave them the latest change to the plan. The scheduled 'Rover stop' away from the base of the massif to scoop a sample of the light mantle had been augmented by a gravimeter reading. This extra time was to be taken from some as-yet undetermined activity. A battle over experiment priorities had been fought in the Backroom and, as Parker put it, "the gravity people won".

Cernan obligingly halted at a 2-metre crater about half a kilometre from the South Massif. As the gravimeter reading would take only a few minutes he announced that he would remain on the Rover, but Parker insisted that he dismount to preclude his presence on the vehicle from perturbing the delicate instrument – as a dubious result would be worthless. So he retrieved the 500-mm Hasselblad in order to document the outcrops on the South Massif. This was to have been done at Station 2, but the view up-slope there had been so foreshortened that he had deferred the task until he had a better line of sight, and this was as good an opportunity as he was likely to get.

Schmitt, meanwhile, had sampled both the regolith and the glass in the crater. His assessment of the light mantle was that none of the craters above the scarp had dug up bedrock. Like this one, they were littered with clods, but that was to be expected of any high-energy strike. And being on the light mantle, their rims were brightened as a result of having penetrated a layer of lighter material several centimetres below the surface. The older-looking craters appeared to have 'faded', but that was because the light mantle itself was actually veneered with a thin deposit of grey dust. As he returned to the Rover, he was astounded to see a "yellow-brown rock", so he took prior-to-

Gene Cernan at the Station 2 rake site on the light mantle.

An uphill view from the Station 2 light-mantle rake site. The route of the sampling activities among the boulders is evident from the boot prints.

sampling pictures and moved in to pick it up – only to see the bright colour suddenly disappear ... and then return as he withdrew! "That really fooled me," he laughed. It was the Sun reflecting off the golden mirror of the TV camera's thermal shield.

"Boy, that's a sight, isn't it," Schmitt exclaimed as they finally reached the scarp.

"Spectacular," Cernan agreed. It was difficult to believe that they were the only people on the *entire* planet. On the Tranquillitatis plain, Armstrong and Aldrin had had no horizon features, and they had referred to the "magnificent desolation" of the lunar surface; but this was a strikingly beautiful site. Having perfected the Apollo system at such great cost, it would have made sense to make a thorough exploration of the Moon, but here he and Schmitt were, the final crew, dashing about in an effort to address as many scientific objectives as possible in a single traverse. "I still don't know where the LM is," he innocently taunted the 'worriers' back home.

Looking down over its crest, it was evident that the scarp was not a linear feature. Its face was a set of juxtaposed lobes of different radii and elevations.

"Okay," Cernan warned as he started down the undulatory slope, "hold on."

"It likes to spin when you turn going downhill," Schmitt muttered, referring to the Rover's handling characteristics. In the weak gravity, the Rover's wheels had so little grip that, as Scott and Irwin had discovered, it readily spun end for end.

"Watch it!!" Schmitt extorted as Cernan dodged an obstacle.

"I bet my heart rate indicates the terrain we're going over," Cernan chuckled, knowing that the flight surgeon would be monitoring his biomedical sensors. Upon reaching the crest of the final terrace, he chose a clear route and let the Rover run straight down, with the result that they shot out across the plain at a tremendous rate.

"We're off!" Schmitt reported, slightly breathlessly.

"Look at the hill we came down," Cernan laughed as he turned around and started north along the base of the scarp, towards their next objective.

"I'd rather not," said Schmitt. He launched into a succinct assessment of the light mantle for Parker's benefit. The surface texture was really the same everywhere – on the low side of the scarp, on the slope, and up on top – and there was no variation in either the crater morphology or rock

coverage. But there were still open issues. "The one thing we don't have a handle on yet," he noted, "is the fragments mixed with the mantle." They really needed to get a rake sample somewhere out here. It was a pity that the Station 3 trench had been deleted even before they had started the day. Once he was on-site, he would reassess that decision. Houston did not really 'control' the surface activity; it was a support system.

The navigation to Station 3 was fairly relaxed, as it was not particularly important that they find a specific spot. All they required was a small crater that had excavated the base of the scarp, and when Schmitt had surveyed the area below Lara on the run in he had spotted two likely candidates.

TRIALS AND TRIBULATIONS

The first 10-metre crater had the attraction of having excavated several half-metre-sized blocks, so Cernan parked about 15 metres away from its rim. As they worked through their chores, Parker updated them on the plan, which had been pared back. Furthermore, to minimise the time spent at Station 3, the Backroom wanted them to work separately, with Cernan taking a core while Schmitt took the pan and sampled rocks. But Schmitt had his own plan, and made an exploratory scraping of the rim of the crater. "There's quite a 'marbling' of light and dark soil," he noted, referring to the cloddy nature of the subsurface material. It was the same sequence as above the scarp. "I'm going to sample the soils," he announced, having decided to reinstate the cancelled trench. He found solo-sampling frustrating, but eventually managed to bag the surficial material, the layer beneath, and each of the cloddy zones in the trench.

Meanwhile, having started the gravimeter, Cernan assembled a double-core tube and moved 10 metres east of the Rover in order to be clear of the inverted regolith in the crater's ejecta blanket. By recording the layering of the light mantle to a depth of almost a metre, the core would yield the vital view into the third dimension which would complement the areal coverage of the wide-ranging traverse. To his delight, the tube was readily hammered into the ground. Upon pulling the tube out almost as easily as it went in, he observed that the end was packed with fragments of rock and clods of dark regolith. He swung it around and showed it to the TV camera to enable the Backroom to see for themselves. To preserve the sample in a pristine state, the lower tube was stored in a vacuum can.

"Jack," Parker called. "When you finish the trench, you might look at one or two of those blocks."

In a circuit of the rim, Schmitt selected 15 rock fragments, of which 12 proved to be breccias, with four of these being of the blue-grey type that had almost certainly come from the outcrop on the South Massif, and whose presence tied the rocky avalanche to the fines of the light mantle. As with soil sampling, collecting rocks on his own was taxing. After scooping up a rock, he leant the tool against his leg to free his hand in order to bag the sample, but the scoop fell, and to retrieve it he dropped momentarily to one knee and blindly felt for the handle. After this had happened several times, he devised the technique of placing his foot on the right-angled head of the scoop in order to elevate the handle high enough for him to reach it. His frustration was compounded by the fact that the chest-mounted dispenser for his sample bags became detached. Worse still, having put a spare SCB on the crater's rim, he inadvertently kicked this over, and after dropping down onto both knees in order to retrieve the bagged-samples which had spilled out he flailed wildly as he struggled to regain his footing. After watching these antics, Parker suggested that the crater be named Ballet.

In fact, while the TV was following Schmitt's ordeal, Cernan was also suffering problems. To hold the core tube as he hammered it, he had used the handle from the rake, and was now having trouble reconnecting it. Dust was the culprit. It was nasty stuff, and although the powdery dust looked as if it would act as a lubricant, it was actually highly abrasive. It had not only clogged the quick-lock fastener of the rake handle, it had also disabled the catch of the Rover's tailgate.

Station 3 had taken its toll on Parker's nerves, too. Once again, what was to have been a brief stop – set by the time needed to take a core sample – had become an exercise in watching the clock, and in this case fully half an hour had elapsed, with the result that when Cernan stowed the rake and asked what was next, the reply was: "Nothing! Get on the Rover, and leave!"

Despite Parker's imperative, Cernan chose to linger to take telephoto pictures of where the scarp had encroached on the flank of the North Massif, and then, since he had the camera out, he swung around to take another sequence of the outcrops on the summit of the South Massif in order to facilitate stereoscopic analysis.

"We assume you guys are ready to go now," Parker observed sardonically after waiting out the mission commander.

"Yessir," Cernan agreed. Sensing Parker's frustration, Cernan made a joke that drew on an incident in training. "Bob, they were all taken with the lens cap *off*."

"Splendid," Parker allowed.

The next stop was to be at Shorty, the dark-halo crater located on the edge of the light mantle fan. Although slightly closer to the LM, this was really 2 km around the arc that was defined by the diminishing walkback constraint, which meant that Station 4 was going to have to be a flying visit. The pressure on the timeline would not lift until they made the long drive back to Camelot.

HOLY GRAIL?

Angling away from the scarp, Schmitt spotted a 3-metre block sitting all on its own on the light mantle, so Cernan detoured to let him take a closer look. In terms of its tone and texture, it was a blue-grey breccia just like the talus at the base of the South Massif; either it was ejecta blasted off the mountain, or it had somehow rolled all the way out.

Upon encountering a 15-metre crater with a blocky rim and an unusually deep pit that was a striking exception on the otherwise bland light mantle, Cernan opted for a 'Rover stop' to let Schmitt sample it. "It's a very, very fragmental crater," Cernan informed Parker. "The ejecta coverage is about 50 per cent, and it's all small angular fragments." The samples[23] were regolith breccia. "A 30-second stop," Cernan noted delightedly as he moved off again.

The plan was to scoop a sample of the light mantle just short of the dark-halo crater, but Parker relayed the Backroom's cancellation of this for the reason that the crater they had just sampled had "ticked that box".

"Bob!" Schmitt retorted, "I thought the purpose was to sample the light mantle?" The crater had been sampled because it was *atypical*, which meant that it would be useless for recording the variability of the fines in the light mantle itself.

"They're anxious to get to Station 4, I guess," Parker replied meekly.

"Well, how about it, Gene?" Schmitt prompted.

"I think we've got to do it," Cernan agreed, after all it would cost them only a few seconds, and he drew to a halt to enable Schmitt to collect the sample as planned.

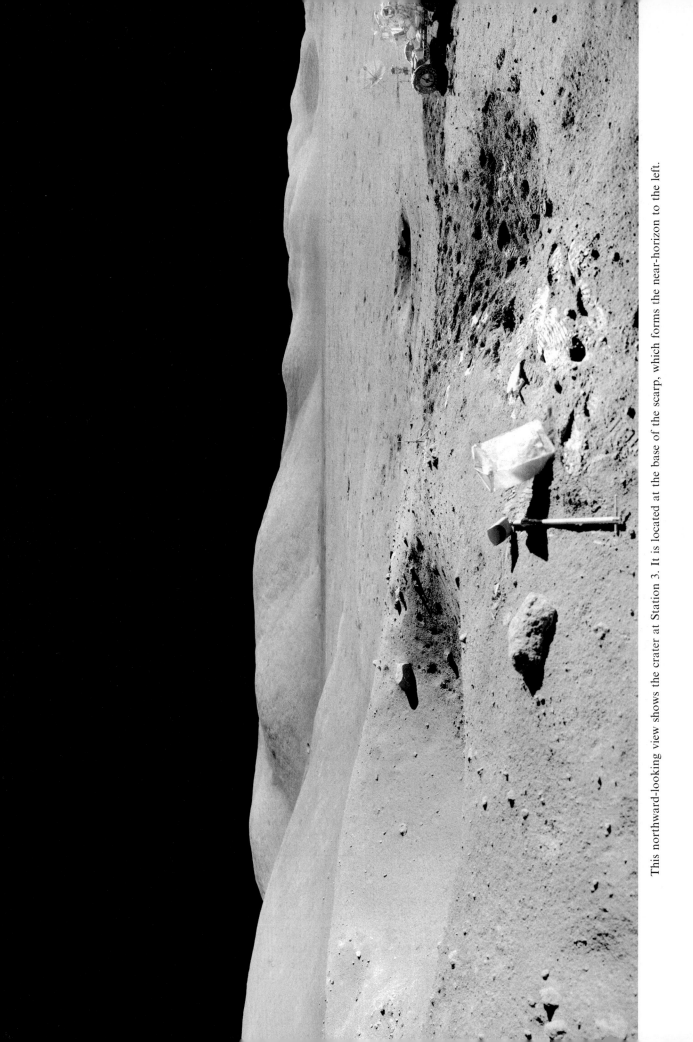

This northward-looking view shows the crater at Station 3. It is located at the base of the scarp, which forms the near-horizon to the left.

"Something's dark out there," Schmitt reported a few minutes later. There was a dark patch on a mound, which was evidently the raised rim of the dark-halo crater, directly ahead.

"That's Shorty, I think," Cernan said. He was navigating using the Rover's 'nav', and so was not surprised to find their objective at the 12 o'clock position.

"My goodness," exclaimed Schmitt as the 100-metre-diameter pit opened up immediately ahead. "It's different!"

Cernan parked near a 2-metre block that was on the southwestern rim. "This is an impressive one," he promised Parker as he went to aim the TV antenna. "Wait until you see the bottom of it!"

While Cernan looked after the chores, Schmitt made his preliminary site report: "Shorty's clearly a darker-rimmed crater. The inner wall, except the western side, is blocky. The floor is hummocky. The central mound is *very* blocky, and jagged." He offered an interpretation: "The mounds look like slump-masses off the side, but I'm not sure; they have a bench-like appearance."

Parker interrupted to say that all they would need would be a pan of the crater's interior and a few samples off the rim. The original intention to perform a full radial sample had long-since been abandoned. This site would have to pay the price for all the time 'lost' earlier.

"We've got a large boulder of intensely fractured rock right on the rim," Schmitt continued, ignoring the interruption. He went to take a closer look at this rock. "It looks like a fairly vesicular version of our gabbro." As he stepped away from the boulder, he saw a patch of colour where his boots had made an indentation. "Hey!" he exclaimed. "There's *orange* soil!!"

"Well don't move it until I see it," Cernan replied rather sceptically, recalling that Schmitt had previously been fooled by a reflection off the Rover.

"It's all over!!" Schmitt persisted excitedly. It was brightest where his boots had broken the surface, but as he looked around he saw there was a 1-metre-wide strip of the stuff running along the rim. "I have *got* to dig a trench, Houston."

"I guess we'd better work fast," Parker replied, making his first contribution since Schmitt had made the discovery.

Rather than attempt solo-sampling, Schmitt chose to wait until Cernan was finished at the Rover, and in the meantime he took a pan.

"How can there be orange soil on the Moon?" Cernan asked rhetorically, upon joining the geologist at the exposure, and then supplied a possible answer: "It's been oxidised!?"

"It *looks* like an oxidised desert soil," Schmitt agreed. "If ever there was ..." he began, but then he stopped and laughed. "I'm not going to say it." But then he laughed again and finished his interpretive remark. "If ever there was something that looked like fumarolic alteration, this is it." Was this evidence of 'recent' volcanism?

"I've trenched across the orange," Schmitt reported. "There's light-grey material on either side." There was a distinct contact on either side of the orange strip.

"That's incredible!" Cernan said enthusiastically. This patch of orange soil could well be the Holy Grail that had prompted the selection of the Taurus-Littrow site for this final mission.

Schmitt cut away the eastern wall of his radial trench in order to illuminate the vertical western wall sufficiently to show any subtle changes of hue that might indicate differences in chemical composition, and since his camera was loaded with monochrome film whereas Cernan's had colour, he asked Cernan to take a down-Sun picture.

"Do you want any of this in a can, Bob?" Schmitt asked Parker.

"They're debating that right now," Parker replied. The Backroom had burst into hearty applause when Schmitt reported orange soil, and they were considering how best to sample it. In particular, they were wondering whether it would be best to put it into a SESC can in order to preserve any volatiles that might still be present if the crater really were a recently active fumarole.

While his colleagues debated, Schmitt took a scoop of material from the centre of the trench.[24] Extending the trench away from Shorty, into the surrounding regolith, he discovered a steep contact which indicated that the orange material was not simply a thin veneer; it was an integral part of the rim. This meant that it was from the deepest point of excavation. After taking a scoop from the outer contact[25] he added the inner contact,[26] to transect the deposit with a miniature radial survey.

After helping with the bagging, Cernan set off to fetch the SESC from the Rover, but at that moment Parker relayed the Backroom's decision: the can would not be required; instead they were to take a double-length core sample.

"Do you want it in the orange?" Schmitt asked.

"We can put cores in grey soil anytime," Parker retorted sarcastically.

In fact, Schmitt was concerned that if the orange material ran as deep as its near-vertical subsurface contact hinted, then it might be better

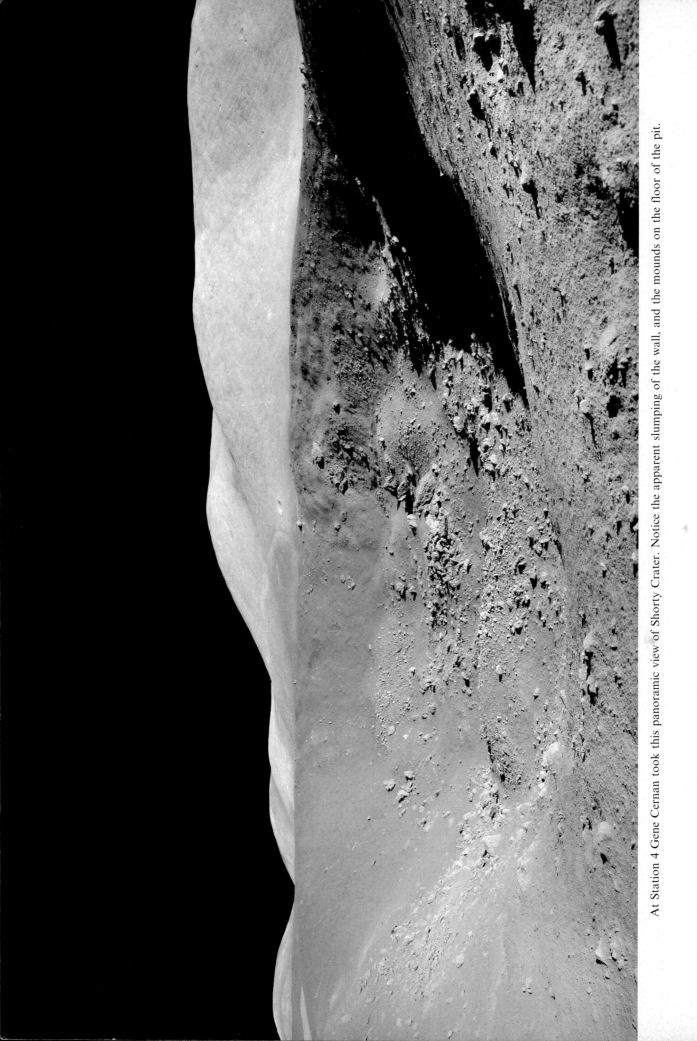

At Station 4 Gene Cernan took this panoramic view of Shorty Crater. Notice the apparent slumping of the wall, and the mounds on the floor of the pit.

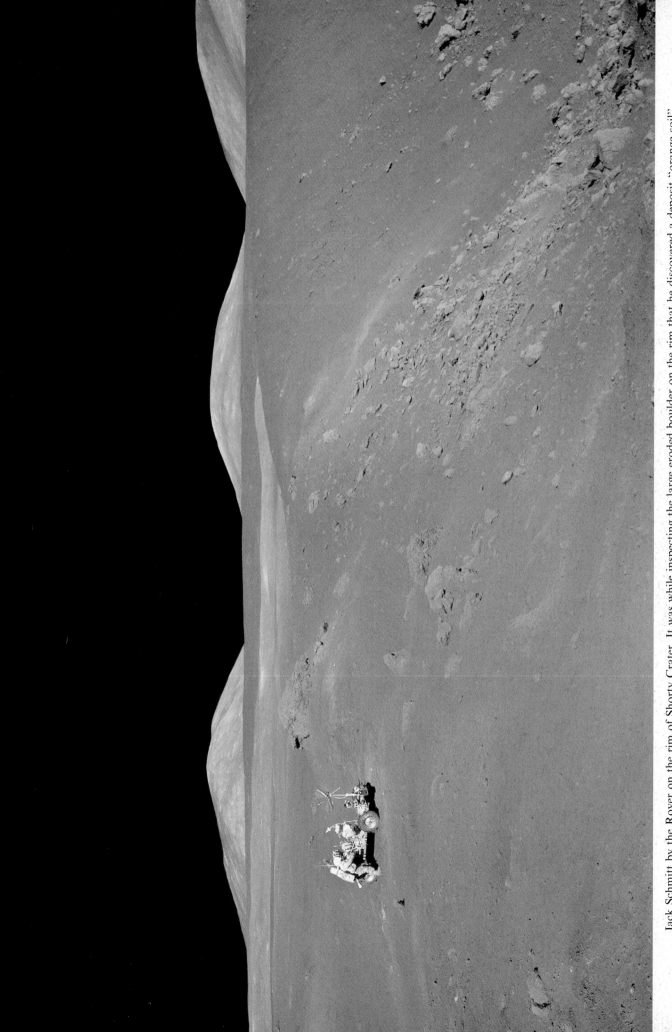

Jack Schmitt by the Rover on the rim of Shorty Crater. It was while inspecting the large eroded boulder on the rim that he discovered a deposit "orange soil".

to angle the tube to ensure that it crossed that contact and recorded the chemical variation that was implied by the transition from crimson in the centre, through orange to yellow toward the edge. While he waited for Cernan to assemble the core tubes, he scraped a shallow trench down the flank on each side of the visible deposit to test whether it was masked, but it was strictly confined.

As Cernan hammered the tube into the orange deposit against a surprising degree of resistance, Parker warned that there was no option of an extension. It was not a matter of priorities, so further sampling at Shorty could not be 'bought' at the cost of sampling the subfloor at Camelot; the walkback constraint was inviolable. They had only 20 minutes remaining on-site. Having had such a hard time driving in the core, Cernan and Schmitt both took a firm grip of the tube in the expectation of having difficulty in pulling it out, but it yielded easily, and Cernan stood back to let Schmitt finish the job.

"Even the core tube's red," Cernan reported, upon seeing that the fine-grained material had coated the metal tube. As the rest of the tube emerged, however, it turned dark. The orange deposit was evidently quite shallow.

"God, it's black!" Schmitt exclaimed. This subsurface layer was the darkest material that he had ever seen.

"Fantastic!" Parker said. A few seconds later he reminded them that they had still to sample the boulder.

While Cernan took the core to the Rover, Schmitt inspected the block. It was so eroded that he was able to gain a piece[27] using his fingers. On his way to the Rover, he lifted a fist-sized chip[28] from near the trench that had probably been shed by the boulder. "If ever I saw a classic alteration halo around a volcanic crater," he mused, "this is it."

For the second pan, Cernan moved 40 metres east, a change in perspective that offered him a view of the western wall of the pit, north of where they had sampled. "The mantle on the inside of the west rim turns from grey to a very dark material," he said, commenting on the extent of the dark mantle. But then he saw something else. "And there's lots of orange stuff that goes down, radially, into the pit." In fact, the extent of the orange 'outcrop' correlated with Schmitt's earlier observation that the western wall was considerably less blocky. As he finished the pan, he lamented that there was not time to sample the really dark material on the western rim.

"We need you rolling in 7 minutes," Parker insisted. In all likelihood, the traverse would finish 'on time', and then there would be plenty of time for the chores in the vicinity of the Challenger. That was the terrible irony of the walkback constraint – it preserved a safety margin until it was not needed, and then everyone relaxed and resented the lost opportunities. But the astronauts were literally living on what they carried on their backs, and it would be irresponsible to risk the Rover failing while they were beyond the range from which they could walk home in the time available. As an item of good news, Parker said that the SEPE receiver had now cooled down and so could be reactivated.

"We're moving, Houston," Cernan reported, as he ran down off Shorty's flank to start the long drive east to Camelot.

"Thirty-seven seconds *early*," Parker congratulated.

Brief though Station 4 was, it delivered its share of clues to the early history of the valley. An analysis of the 'orange soil' showed it not to be the result of chemical alteration by volcanic gases, but to be composed of tiny glass spherules. Although at first it was frustrating that having selected Shorty specifically for having a dark halo there had not been time to sample this material, this was almost certainly represented by the jet-black material at the bottom of the core tube. Significantly, the orange and the black materials proved to be chemically identical. In fact, the black material was the devitrified (crystallised) form of the orange glass. As most people had believed, Shorty was an impact crater. There was a fire fountain, but it was contemporaneous with the volcanism that made the subfloor billions of years ago. On emerging from the fire fountain, if the ilmenite-rich magma droplets had cooled instantly they had vitrified, and if they had cooled slowly they had crystallised. The glassy spherules were orange because of the iron-to-titanium ratio (just as those found at Hadley-Apennine were green because of their high magnesium content). The extremely vesicular character of the boulder on the crater's rim implied that it was from the very top of the subfloor. It was theorised that after the main slab of the subfloor had formed a solid crust, this had received a thick coat of pyroclastic, with the orange and black materials forming distinct layers. Before this material could be gardened it was covered by a relatively thin lava flow, almost certainly extruded from the vent that issued the fire fountain, and although this was gardened its presence protected the pyroclastics beneath. But the impact that made Shorty punched through the veneer and excavated the deposits. Some of this compacted material was built into the rim, just as if it were solid rock, and the rest was scattered on the flank. The mystery of the 'dark halo' was resolved.

The lava flow which had protected the pyroclastics that were excavated by Shorty was evidently a local feature, because elsewhere in the valley the products of the fire fountain were left exposed and were gardened, which is why the dark mantle was indistinguishable from the regolith.

CAMELOT'S BOULDERS

After two 'Rover stops' to scoop soil,[29,30] they drove around Horatio's northern rim, crossed the isthmus between it and Camelot, and parked on the eastern fringe of Camelot's boulder field.

"You've got 25 minutes," Parker announced.

While Cernan handled the chores, Schmitt worked his way through the boulders, which he estimated to be 30 per cent coverage, on a tour of inspection. When the TV came alive, Fendell sought out Schmitt and followed his progress.

"This is certainly a uniform rock type," Schmitt reported after he had examined a dozen or so of the metre-sized rocks, all of which were the familiar gabbro. Most of the boulders seemed to be partially buried. The few which were sitting fully exposed appeared to have been tossed about by smaller impacts on the rim. Peering across Camelot, he noted that the remainder of the rim was clean. Recalling Horatio's thick overburden of regolith, he wondered whether the regolith was thinner on Camelot's southern rim. However, not having seen Steno, he could not say which was atypical.

It was 10 minutes before Cernan was able to join Schmitt in the boulder field, but he got straight to work with his hammer and broke loose two typical fragments from a rock chosen at random.[31]

"I wish we'd started on that rock," Schmitt said, "because we're going to run out of time." During his tour of inspection he had found a block of diabase gabbro with fine vesicles arranged in distinct layers, and he wanted to sample it for its variety.

"Did you have one picked out?" Cernan apologised.

As if to reinforce the point, Parker interrupted to say that they had 10 minutes left.

Cernan gave the diabase boulder a series of heavy blows, but to no effect. Taking over the hammer, Schmitt released a slabby fragment[32] from a different location with a single strike.

"Beautiful call!" Cernan said in praise of his colleague's assessment of the most likely fracture plane.

"That comes from 15 years as a hammer bearer," Schmitt chuckled, referring to his training as a geologist. "You learn where to hit rocks."

When Parker prompted them to try to find a rock that had dust on its upper surface, Schmitt selected a flat-topped boulder which, although fully 3 metres wide, was one-tenth of that off the ground and was coated by a 1-cm-thick layer of dust. The rock-top sample had been placed on the agenda in the expectation that the dark mantle would be a recent deposit, and that to scrape such material from the top of a rock would collect a near-pristine sample of the pyroclastic. In fact, it looked to Schmitt as if the dust was soil that had been thrown up onto the rock by a nearby impact, and although it was now clear that they had misinterpreted the nature of the dark mantle, it would nevertheless be productive to compare the material on top of the rock with that from the side, so he scooped up a full bag.[33] When he said he was going to take a soil sample for context. Parker interjected to say that the rake had been cancelled, and that if they were going to sample the regolith within the boulder field they ought to do it in as "open" an area as possible.

"Let's do it here," Schmitt insisted, sticking to his plan to make a comparison with the fines on the rock. While Schmitt sampled a shallow trench[34] alongside the boulder, Cernan snatched a small piece of rock[35] that had been resting on its tabular surface. When they were finished, Schmitt emerged from the boulder field and ran back to the Rover. Cernan lingered to take another pan.

"We'd like you to leave *immediately*," Parker insisted. "If not sooner!"

On returning, Cernan switched from his usual lope to a gallop. "Hippity-hoppety, over hill and dale, hippety-hoppety along," he sang to tease his impatient colleague. It had been a fine day's exploration, and he was not going to let any 'clock watcher' ruin his mood.

"My impression," Schmitt informed Parker as he packed up, "is that Camelot does not seem to be mantled to the degree that Horatio is." By 'mantle' he meant regolith.

The site for Station 5 had been well chosen. The overhead imagery had suggested that there would be blocks, but the scale of the boulder field was a surprise. It was a shame that time had been so short. However, because the boulders were on the

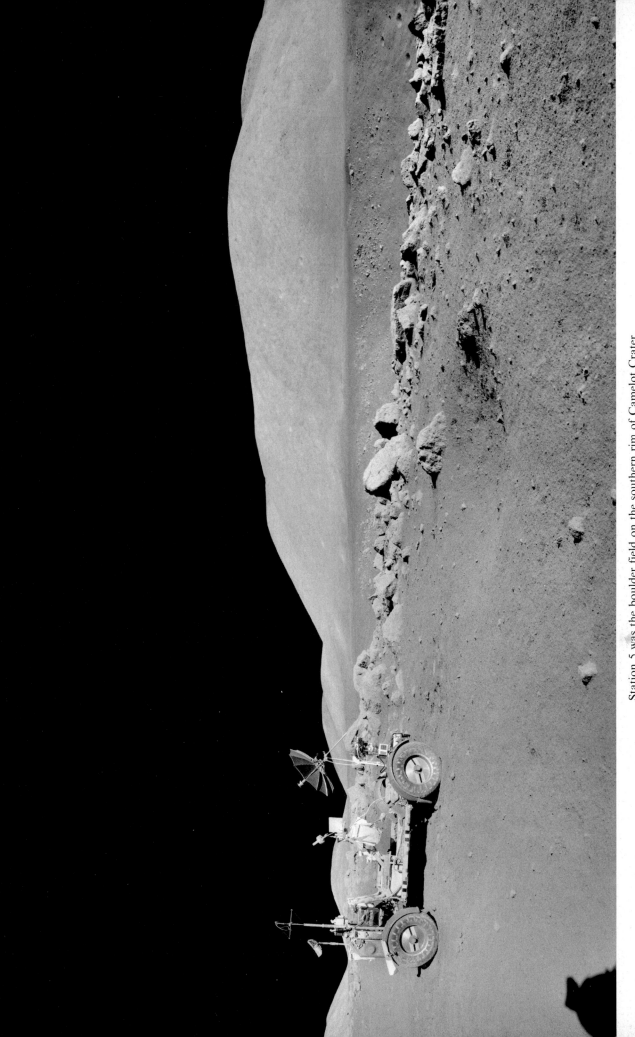

Station 5 was the boulder field on the southern rim of Camelot Crater.

A northward-looking view across Camelot Crater.

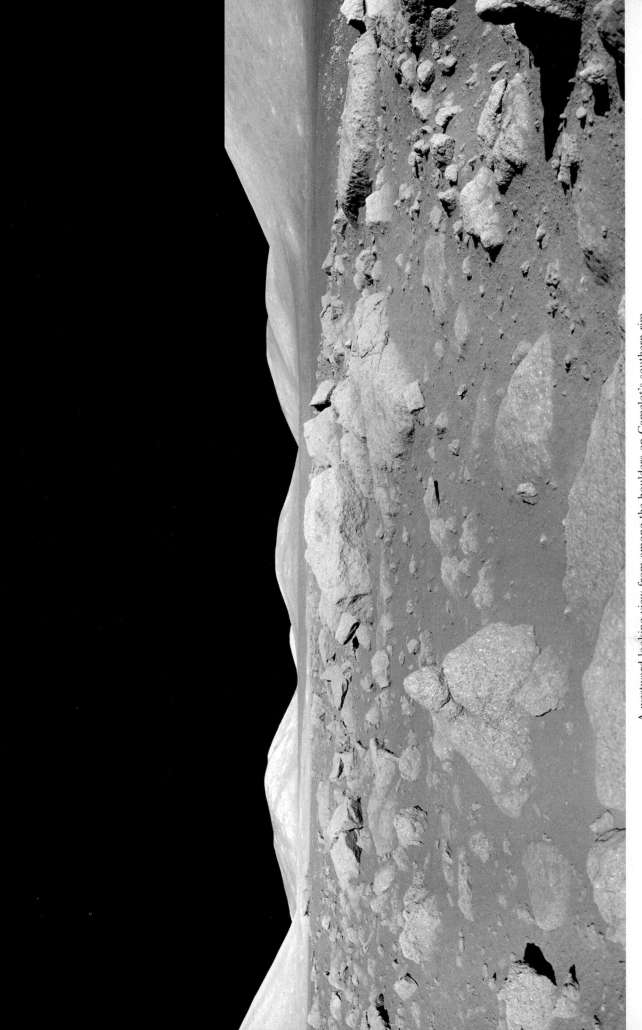

A westward-looking view from among the boulders on Camelot's southern rim.

Sampling the 'diabase rock' in the boulder field on Camelot's southern rim.

Jack Schmitt sampling alongside the 'flat rock' in the boulder field on Camelot's southern rim.

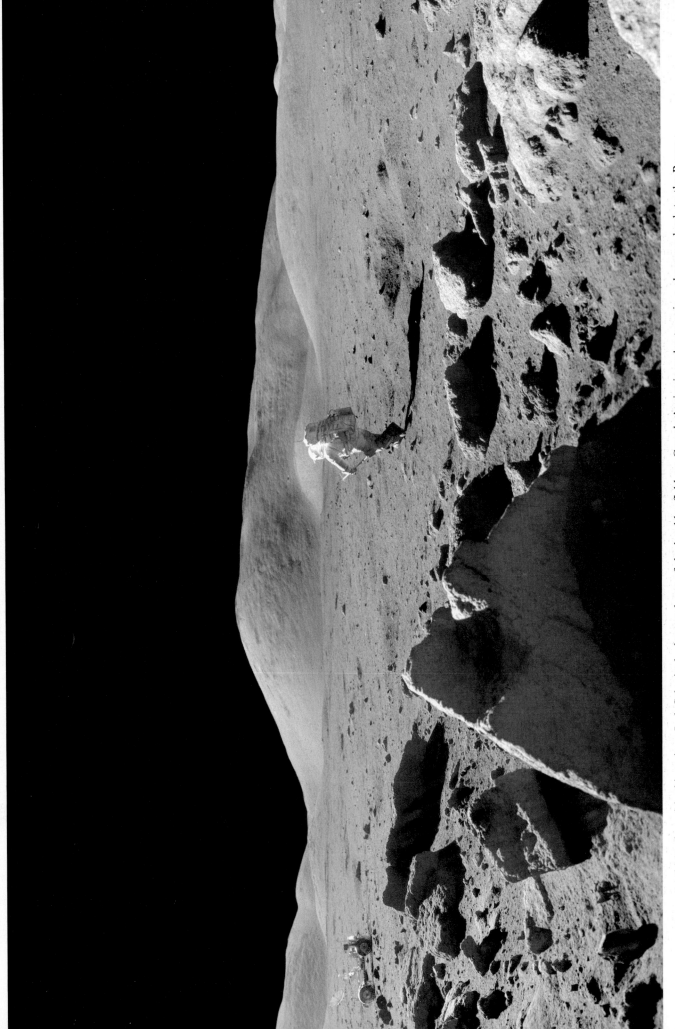

In this eastward-looking view, Jack Schmitt had moved out of the boulder field on Camelot's rim in order to gain a clear run back to the Rover.

rim it was safe to assume that they were from the deepest part of the excavation, which, in the case of a crater the size of Camelot, might be from a depth of 180 metres. This context left no doubt that the valley floor was gabbro. Furthermore, the similarity of the rocks across the valley indicated the subfloor to be extremely homogeneous. And, of course, now that its nature had been established, there was no need to refer to it neutrally as the 'subfloor unit'; it was an extrusion of ilmenite-rich pyroxene basalt. The absence of the crust of the flow, which would have been finer-grained as a result of having cooled rapidly, was explicable if this had been gardened to form the thick regolith.

By the time that Cernan and Schmitt arrived back at the landing site, their traverse had covered a total distance of 20 km. When they re-entered Challenger they had been out for 7.6 hours. In terms of distance and duration, this penultimate excursion marked the high point of Apollo lunar exploration.

THE GEOLOGY OF TRACY'S ROCK

The aim of the final traverse was to explore the northeastern corner of the valley. The North Massif was to be compared with its southern partner, and then with the domical Sculptured Hills. In essence, this chain of stations formed an arc 4 km from the landing site, so the walkback constraint was less significant and this outing was expected to be fairly relaxed in comparison to the previous day's hectic pace. In fact, as the nature of the subfloor had been settled, the final station, at Sherlock Crater, was down-graded to offer a margin to enable an earlier station to be extended if this should be requested by the astronauts. Given the discovery of what appeared to be fumarolic activity at Shorty, it was hoped that the dark crater Van Serg would prove to be similar. This traverse was not meant to be the longest lasting, nor the farthest roving, but it promised to be a fitting finale to the Apollo program, and, in the event, it strikingly demonstrated the value of sending a professional geologist to explore the Moon.

While Cernan prepared the Rover, Schmitt undertook chores at the SEPE site, and as he did so he found a piece[36] of what appeared to be the elusive fine-grained basalt that would have formed the crust of the subfloor, but in the event this turned out to be a clast-poor dark-matrix breccia.

The outbound leg was due north, and Cernan relished the benign illumination.

Looking up at the North Massif, Schmitt noted that the boulders seemed to have come from lines of outcrops. Whereas on the South Massif the outcrops were above the break in the slope, in this case they were on the lower flank. The tracks did not run straight down, but ran cross-slope, often zigzagging, and in many cases were chains of indentations where a boulder had bounced. Several rocks were clearly house-sized. There were rocks high on the Sculptured Hills, but they had not rolled, which meant that the sampling opportunities there would be more limited.

"Coming up on Prince Henry the Navigator," Schmitt informed Parker, as a 600-metre crater opened up ahead. Sampling Henry was not on their agenda, but Cernan skirted its western rim to enable Schmitt to peer in. It had rocks exposed in its wall, but the litter did not start until about 30 metres down, which suggested the presence of a thick regolith, as at Horatio. Most of the rim was clear, but there were partially buried boulders on the northwestern flank that had been dug up by a 50-metre crater in Henry's flank. "They look gabbroic, fine-grained; a little greyer," Schmitt mused. In the event, it would prove unfortunate that they did not stop to sample one.

Immediately beyond Henry was 100-metre Locke. Although the rim was clear, the pit held a lot of rock. After clearing Locke's eastern rim, Cernan headed for a dark boulder at the base of the massif which, since it was to be a checkpoint, had gained the moniker of 'Turning Point Rock'.

At 6 metres tall, the boulder towered above the low-slung Rover. Cernan paused alongside it to enable Schmitt to scoop a soil sample.[37] Not only was the base filleted, but there was a thick cover of dust on its ledges.

"It looks like a breccia," Schmitt speculated upon spotting mottles in the tan-grey surface. "It's very coarsely vesicular." But he wasn't sure, since the mottled texture might be the result of zap-pits brightening the dark gabbro. It was conceivable that it was an enormous piece of the subfloor, but the source crater was not obvious. It was more likely to have come from the mountain. It was not a listed objective, however, so they did not disembark to chip off a sample. Given the absence of pressure on the timeline, this was unfortunate, as this rock would later be recognised as a significant piece in the puzzle of the massif's origin.

After the tour of inspection, Cernan, swung east, to run cross-slope up the North Massif to Station 6. There was no indication of a contact between the valley floor and the mountain and,

Gene Cernan, shortly prior to setting off for the third traverse.

indeed, the coverage of cobbles remained much the same, but the flank was littered with blocks up to several metres in size.

Schmitt's attention was fixed on their primary objective: a massive boulder which had left a prominent track in tumbling down the massif. On drawing near, it became apparent that the final impact had broken the boulder into several pieces. The Rover had taken the drive up the mountain in its stride, but Schmitt, sitting on the downhill side, was very aware that they were on a 20-degree slope. He joked that it might be wise to jam rocks under the wheels, but Cernan parked side-on about 20 metres west of the southernmost fragment, and the vehicle did not display any tendency to slide. As they disembarked, it became evident that this would not be the easiest of sites to work. Cernan dealt with the chores while Schmitt made an initial assessment of their objective.

There was a 3-metre-wide gap between the two main blocks, which Schmitt chose to call the northern and southern boulders, even although each had split into several pieces. The train of fragments was 25 metres in length, so this was the biggest rock they had encountered. Schmitt trudged up to the top of the train, so that he would be able to conduct his inspection while moving downhill, which would be easier. "It's a coarsely vesicular finely crystalline rock," he announced after inspecting the texture. "Probably anorthositic gabbro." He moved in for a closer study. A dark glassy melt in the zap-pits would indicate a significant proportion of mafic minerals. A lower mafic content would correspond to gabbroic anorthosite. "It's anorthositic gabbro," he confirmed upon finding a patch of dark glass. Further down the boulder he found some prominent vesicles. "The vesicles are flattened," he noted. The cavities were about 20 cm long, but only one-third as wide, which was an indication that the rock had been under directional stress while in a molten state. "There's a strong foliation in the rock," he continued.

"Outstanding," Parker replied, borrowing Tony England's favourite response.

"The foliation doesn't run all the way through the rock," Schmitt pointed out as he worked down the boulder. The inhomogeneity made the rock interesting. It held a surprise, too: "There's a large inclusion of non-vesicular rock within the vesicular rock." It seemed that a piece of rock had been caught in the molten material. Finding a xenolith in an igneous rock was a geologist's delight, because this could have been ripped off the side of the feed pipe as the magma forced its way up

through the crust. On the other hand, it might just be a chunk of rock that had been tossed in by an impact after the lava had been extruded and before it had solidified. It certainly did not look like the type of rock that would be expected to occur deep underground. "It looks mineralogically like those light-tan breccias from the South Massif."

His initial visual inspection of the northern boulder finished, Schmitt moved off to the west to take the site pan, and in so doing documented the distribution of the boulders in the train.

Meanwhile, Cernan had scraped a flat spot for the gravimeter, which would not work on the vehicle on such a steep slope. This done, he looked up to find Schmitt, and saw that he had gone further uphill to take a soil sample[38] from the trench which the rock had scraped as it slithered to a halt. "Think how it would have been if you'd been standing there when the boulder came by!"

"I'd rather not think about it," Schmitt replied.

"Did you pick a good spot?" Cernan asked, meaning had Schmitt selected the first part of the boulder to sample.

"Gene, if you hit fragments off in there," Schmitt replied, indicating the shadowed western side of the rock where he had made his inspection, "it's going to be awfully hard to find them." It would be better to sample the illuminated eastern face. "Let's go photograph the Sun side." As they passed through the gap between the two rocks, Schmitt opportunistically took a scoop of soil[39] from the base of the southern one, in case it was in permanent shadow. When they turned right on the far side, the TV lost sight of them.

Cernan delivered a dozen hammer blows to a shoulder-height protrusion on the southern boulder, but it was not the most convenient angle and he made no progress. "I've got to find a corner that I can get at." As he sought a better target, he spotted a scar marking where a large fragment had broken off. "It looks like somebody's been chipping up there!"

"Like there's been a geologist here before us," Schmitt laughed. There was a chunk of rock on the ground at the foot of the boulder that it was fair to assume had spalled off. "I think we can just lift that." This rock[40] was so large that he had to slide it carefully into the SCB.

"The more I look at the southern half of this boulder, the more heterogeneous in texture it looks," Schmitt said. "It's either a recrystallised breccia or an anorthositic magma that captured an awful lot of inclusions." That is, it was either an impact-melt breccia, or it was a fragment-rich igneous rock. After a moment of mulling the

A view of 'Turning Point Rock' on the lower slope of the North Massif, taken by Jack Schmitt as the Rover drove by.

matter over, he laid his bet on the table. "I guess I prefer the latter, because of the extreme vesicularity." Standing back to gain an overall impression of the inclusions, some of which were the size of a grapefruit, he realised that they were well rounded, which meant that they had been through the mill prior to becoming caught up in this matrix; they had not been freshly shattered. They clearly had a story of their own to tell. "A few of them are *very* light-coloured." He attacked a clast using the hammer, but to no effect. On seeing "a finer-grained vesicular rock" on the ground at his feet, he collected this[41] as a consolation.

At this point, Parker suggested that instead of dividing the 80 minutes assigned to the North Massif between two stations, they might spend it all at this site; but if they did, then they would have to take care to identify the variety of rock types.

"Okay," Schmitt replied ambiguously as he slogged uphill in order to inspect the illuminated side of the northern boulder. Cernan followed, content to accommodate whatever recommendation the geologist might make concerning the extension. "This looks *different*; it's greyer," Schmitt reported, on reaching the top of the train. "It's a blue-grey rock – crystalline, I believe. The inclusions are more sharply defined." At this point his training came to the fore, and he saw what was *absent* – there were no vesicles in this matrix. Having toured the entire boulder, he put the pieces together. There were basically two rock types in it. There was a light-tan rock at the southern end and a blue-grey rock at the northern end, but, intriguingly, there was some of the tan rock at the southern end of the northern fragment. Although the block had been split by the shock of coming to rest after rolling down the hill, it had not broken at the contact, which was evidently an immensely strong bond. Finally, he reckoned he had the 'big picture': "This blue-grey rock is included in, or at least it's in contact with, that very vesicular anorthositic gabbro."

At this point, Schmitt poked one of the large white inclusions in the blue-grey rock and, to his surprise, it broke free.[42] "It looks a lot like that 'sugary' textured rock I sampled." He meant the final boulder they had sampled in the talus at the base of the South Massif, the one with the dunite clast.

Exploiting the friability of the inclusions, Cernan tore another one[43] from the blue-grey rock: "That's a different kind; more beat-up."

"It looks like a piece of breccia that got caught up in this thing," Schmitt mused, after he had inspected the fragment that Cernan had obtained.

"The whole thing is obviously a breccia," Cernan ventured rather boldly.

"I'm not sure it's 'obviously a breccia'," Schmitt cautioned. "I think it might be an igneous rock with breccia inclusions."

"But look at all these things," Cernan persisted. By not dwelling on the subtleties, his 'big picture' involved large clasts in a fine-grained breccia matrix. "Let's get the host." He hammered the blue-grey material, without result. "I may have been a little optimistic."

"Do you have a feeling that these are different rocks?" Parker interjected. He had not picked up on the fact that the contact between the types of rock was within the northern component of the boulder, and was wondering whether their juxtaposition on the hillside was simply by chance.

"They're all one boulder," Schmitt insisted.

"And you have it pretty well photo-documented, right?"

"We're working on it," Schmitt promised.

Having persisted with the hammer during this exchange, Cernan finally gained a piece[44] of the blue-grey material, and then secured another light-toned inclusion[45] as a bonus.

As Schmitt studied the fragment of the blue-grey host, he revised his assessment. "This looks like a partially recrystallised fragment-rich breccia." This implied that the blue-grey rock, or at least this bit of it, had once been severely heated. Might this have been as a result of its proximity to the tan-grey gabbro? "I'm going to look at that contact," he announced as he shuffled down-slope.

"I want to do some more documentation," Cernan said. On climbing back to the top of the train, he sampled[46] the thick layer of dust on the broad ledge at the northern tip of the boulder. In doing so, he left an imprint. It did not occur to him until much later that he had neglected an excellent opportunity for lunar graffiti, and he regretted not writing the name of his daughter on the ledge. After he related this story the Station 6 boulder acquired the nickname of 'Tracy's Rock'.

"Jack?" Cernan prompted, upon realising that he had not heard anything from his colleague for some minutes.

"Yeah, I hear you," Schmitt replied dismissively. The interface between the two types of rock had been difficult to identify, but once he realised what was seeing, he was astonished. The contact was a 1-metre-wide transition and, intriguingly, there were fragments *beyond* it. "Bob, there are inclusions of blue-grey in the anorthositic gabbro!"

"Positively outstanding!" Parker enthused.

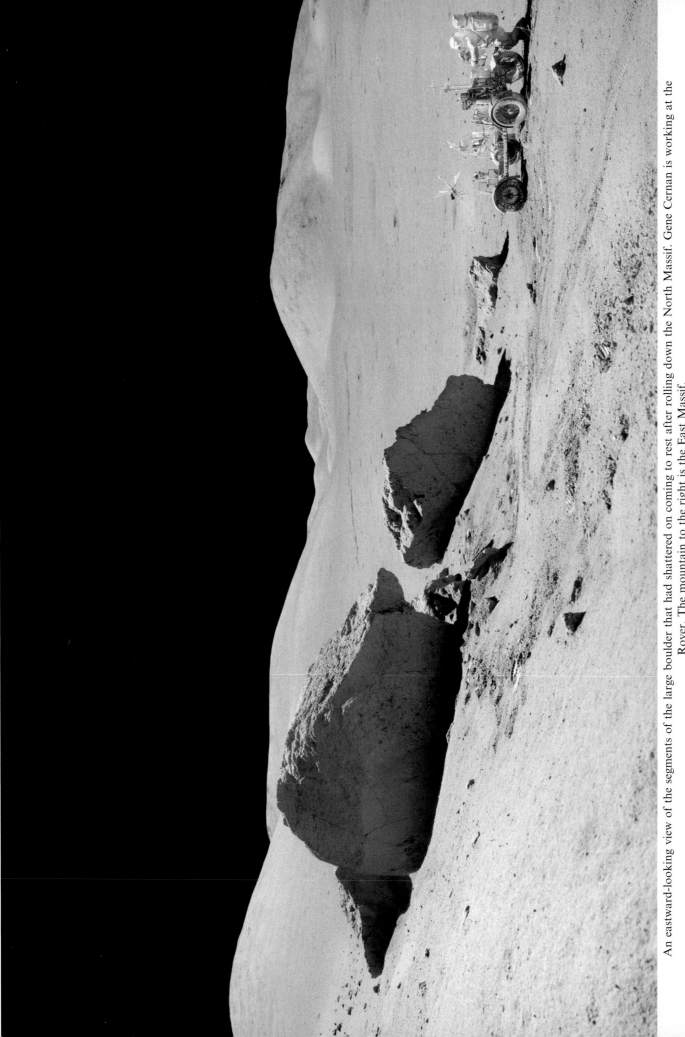

An eastward-looking view of the segments of the large boulder that had shattered on coming to rest after rolling down the North Massif. Gene Cernan is working at the Rover. The mountain to the right is the East Massif.

Jack Schmitt at the southern end of the main part of the split boulder at Station 6.

After climbing a little way up the North Massif, Gene Cernan took this view of the split boulder at Station 6. The large crater behind the rock is Henry the Navigator, and Challenger is visible on the bright streak just to the right of the top of the rock. It was only after they had departed that Cernan wished he had etched the name of his daughter, Tracy, in the thick accumulation of dust on the surface of the nearest fragment of the boulder.

Cernan was taken aback. "Are you saying this whole big blue-grey thing is an inclusion!?"

"Yessir," Schmitt confirmed. In fact, he had meant only that there were pieces of blue-grey rock in the gabbro, but if, as seemed likely, the breccia had been swept up by the magma, then Cernan's extrapolation was correct.

"And within the blue-grey, we've got all these other fragments?" Cernan mused.

"It's several generations of activity," Schmitt explained. This was a story in time as well as in space. "It looks like the gabbro picked up the fragmental breccia." It had taken him half an hour, but he had 'read' the rock. "Well, Bob. I think I've done the best I can." He speculatively hit one of the blue-grey inclusions in the transition zone, and a fist-sized lump[47] fell off; it was clearly more friable than the main part of the rock. "It seems to be pretty well metamorphosed." On making another pass over the transition zone, he discovered that there were tiny vesicles in the blue-grey matrix close to the tan-grey rock, which confirmed that the heating had been caused by the breccia coming into contact with what he had identified as a gabbro, and not with some other material at an earlier time.

As the laboratory analysis of the samples would demonstrate, Schmitt's story was accurate. The blue-grey rock *was* a breccia, but the tan-grey rock was *not* gabbro, it was impact-melt, which made it a fine-matrix breccia. But this misidentification was excusable, because it was difficult to tell an igneous rock from a fragment-poor melt in the laboratory, let alone 'in the field'. And, to be fair, Schmitt had considered and rejected this hypothesis. One thing was certain, however: if they had merely taken a chip off one end of the train and moved on, they would have completely missed the story that the rock had to tell. In retrospect, it was clear that both kinds of rock were also present in the talus at the base of the South Massif, but there the tan-grey rock had not been vesicular. The first vesicular variety they had seen was Turning Point Rock. Nevertheless, by sending a professional geologist, and granting him the time needed to really study the rock *in situ*, this station was the closest that Apollo came to reproducing the spirit of a terrestrial geological field trip. It was fortunate that this opportunity coincided with the sampling of this particular rock. The massif had been up-thrust essentially instantaneously by the shock of the impact which excavated the Serenitatis Basin. The boulder's track could be traced to an outcrop one-third of the way up the massif, which made its context known. The scale of the transition zone within the northern fragment of the boulder reflected the scale of the event that had produced it, and because the story of this rock was the story of the massif, it seemed likely that this extraordinary contact was formed at the time of the impact that excavated the Serenitatis Basin.

Noting that they were finished with the boulder, Parker requested a rake/soil and a core sample. As Schmitt went about the routine chore of raking[48] a small crater near the Rover, Cernan said that he was heading further uphill to take a pan from above the boulder. Although he was only 75 metres above the valley floor, and hence no higher than he had been on the scarp, he was stunned by the view, and awed by their achievement. "You know, Jack, when we finish, we'll have covered this valley from corner to corner!"

"That was the idea," Schmitt replied pointedly.

ANOTHER PIECE OF THE PUZZLE

There were boulders at their elevation to the east, several hundred metres away, but their route obliged them to angle down to the valley floor in order to 'cut the corner' on the way to the Sculptured Hills.

With the Rover parked cross-slope and with his seat on the lower side, Schmitt did not even attempt to embark. Instead, he asked Cernan to drive to a small crater about 100 metres in the direction of their next station, where the crater's wall would cancel the slope, and he would make his way there on foot, sampling as he went. Prior to setting off, he grabbed "a nice big one" that he called a gabbro but turned out to be a tan-grey breccia,[49] and a fragment[50] of "crushed anorthosite" that was similar to the white clasts in the large boulder. As he made his way to the crater he hammered a piece off a blue-grey breccia.[51]

After angling down for half a kilometre, it was decided to perform Station 7 near a 2-metre-tall boulder, and Cernan parked in another small crater to simplify access. In order to minimise the time at this site, the Backroom requested a sweep to gather fist-sized rocks with the minimum of documentation, which meant taking a pair of pans, one prior to sampling and the other afterwards. Although it would not be easy to locate individual rocks, their footprints would show the sampling route. As such, it was a compromise between 'grab' and 'documented' sampling.

While Cernan was aligning the TV antenna, Schmitt took the first pan and set about collecting rocks, rapidly filling two bags.[52] In contrast to his ordeal at Station 3 the previous day, he had no trouble in solo-sampling. His chores complete, Cernan added a rock[53] to the bag and then went to the boulder, which was 25 metres back along their tracks.

"It's one of those blue-grey rocks," Cernan reported, matter-of-factly. There was evidently an entire outcrop of the stuff somewhere up on the mountain. He moved around to the other side, taking pictures, and found an enormous clast. "It's got a light fragment that runs the *full height* of it, about a metre-and-a-half thick!" As he moved close to inspect the clast, he saw something that they had not encountered in the other rocks. "I'm not absolutely positive," he began, "but it sure looks like a dikelet in the inclusion." It looked as if the blue-grey material had been injected into a crack in the clast. Tracing the contact, he found several other such veins. Schmitt joined him, and confirmed that the inclusion (which looked to be intensely shocked anorthosite, but was later reclassified as an anorthositic matrix-rich breccia) had been intruded by the blue-grey material. The vein meant that the melt had been under pressure, which meant that this contact was formed underground, not as a surface-splash. They easily hammered off a bit of the clast[54] near one of the veins, but really had to work to get a chip of the material in the vein.[55]

When Parker broke in to say that they would have to be moving in 5 minutes, they ignored him and continued sampling. Schmitt had noticed that the intruding veins were somewhat darker than the primary blue-grey ground-mass, so they took a piece[56] of this for comparison and then, on spotting vesicles in the clast, they bagged some of this too.[57]

Having methodically sampled all the key points on the boulder, they set off back to the Rover, but on the way Cernan was seduced by a football-sized rock that was poking out of the ground and he dropped to his knees to dig it out.[58] After putting it on the Rover, he left Schmitt to tidy up and moved up-slope to take the second pan looking down on their tracks.

Although Station 7 had taken only 20 minutes, they had been working extremely efficiently and had augmented the requested suite of small rocks with an interesting boulder. The intruded clast had been particularly satisfying, because their immediate interpretation exploited their analysis of the transition-zone contact in the rock at Station 6. If they had not stayed at that rock until they had figured it out, they would not so readily have realised the significance of the dikelets. In sampling the valley of Taurus-Littrow "from corner to corner", they were developing an understanding of the processes that had created it – and that was what field geology was about. It was therefore with some satisfaction and much lighthearted banter that they drove down off the North Massif, two intrepid explorers at home in their strange environment, heading for their next objective.

THE SCULPTURED HILLS

Cutting the 'corner' heading for the Sculptured Hills, Cernan and Schmitt detoured south around SWP, a crater they had named for the Science Working Panel that set the program's scientific objectives. On its southeastern rim they found a 35-metre crater which, while very blocky, had not excavated bedrock; its 70 per cent coverage of litter was entirely of regolith breccia. On seeing that its rim was dark, they paused to enable Schmitt to use his long-handled scoop to lift a clod,[59] thereby unwittingly obtaining a sample that would later prove to be an interesting point of comparison.

At Station 8 they were to sample boulders whose context could be inferred. There were some boulders on the lower flank, but none had left a discernible track. It was therefore evident that they would have to rely upon pebbles to characterise the hills, which meant raking the rim of a small crater to sample whatever this had dug out of the regolith, and so they parked alongside a 20-metre crater.

As Cernan tended to the chores, he discovered that one of the clamps used to hold the make-shift fender in place was missing.

After inspecting one of the nearby cobbles, which he said resembled the subfloor material, Schmitt headed 50 metres up-slope to what passed for a 'big' rock. When Cernan pointed out that he had neglected to take the hammer, Schmitt promised to roll the rock down to the Rover for Cernan to work on. "It's a big chunk of shattered anorthosite," he offered. It was also evidently coated with glass. It was extremely well rounded. In fact, it was ellipsoidal, with a 60-cm major axis and a 30-cm minor axis. If it had rolled down the hill, the smooth profile explained why it had not left a track. But why had it stopped? It was not in a depression. "I'm going to roll it," he repeated.

"Go ahead," Cernan replied warily. Having just hammered a chip[60] off a small block near the crater, he turned to watch.

Jack Schmitt is working at the Rover at Station 7.

The objective of Station 7 was to sample the boulder on the right. The crater in the distance is Henry the Navigator, and Challenger sits on the bright streak beyond.

It took only a nudge with Schmitt's boot to start the rock rolling, but it stopped after a few cycles, which, given its shape, was surprising. "I'd roll on this slope," he said to the rock, "why don't you!?" The fun over, he sampled the soil[61] where the block had been, as this would indicate how long the rock had been in place.

On joining Schmitt, Cernan delivered a series of heavy hammer blows to the well-indurated rock, and finally secured a chip.[62] "That's pretty inside!" he exclaimed. "I haven't seen anything like *that* before."

"It's stained by the glass," Schmitt explained. This had clearly been hot. In fact, there were tiny vesicles in the rock just beneath the glass coating. "It's about 50–50 blue-grey plagioclase and a light-yellow-tan mineral that is probably ortho-pyroxene, and it's fairly coarsely crystalline." He sampled the glass[63] for completeness.

The boulder proved to be norite, a plutonic rock from deep within the early crust that was dated at 4.34 billion years old. Interesting though this rock was, the fact that it was almost certainly an interloper meant that it did not reveal anything about the Sculptured Hills.

Schmitt returned to the Rover to rake the rim of the crater, which proved to be thick with pebbles.[64] Cernan lingered to take a pan from alongside the rock, and then he too started back to the Rover. On the way he spotted fragments of a "pure white and very friable" rock in a small crater, and he took a piece[65] for the simple reason that it was different to the ubiquitous subfloor ejecta. However, the fact that it was concentrated in the pit suggested that the white rock was the remains of the projectile, which meant that it had been tossed in, and therefore this anorthositic matrix-rich breccia, which is what it proved to be, was not typical of the site either.

The objective of Station 8 was to determine whether the Sculptured Hills, which were morphologically distinct from the massifs, were compositionally similar, but in reality there was little that Cernan and Schmitt could do on such a bare slope, and, in any case, Parker was urging them to leave, so after barely half-an-hour they packed up and departed.

MYSTERIOUS VAN SERG

Back on the valley floor, Cernan ran the Rover up to full speed because, while their next objective was only 2 km away, they had to follow a winding route around the southern periphery of 600-metre Cochise which, with similarly-sized Shakespeare to the west and Gatsby to the south, hemmed in the 85-metre Van Serg which was to be Station 9. Unfortunately, with the make-shift fender flapping loose, it, as Schmitt put it, "started to rain".

On reaching Cochise, they saw prominent layering in its far wall. The stratigraphy had not been evident in overhead imagery. "I can see a contact within the subfloor," Schmitt informed Parker, "between albedo units, one of which is light tan-grey and the other is light blue-grey. It looks like it dips to the north at about 20 degrees." Because the blue-grey material was on top, he speculated that this might be a late flow resting on top of the subfloor. Another possibility, not considered until later, was that it was a deposit of blue-grey breccia which had either been tossed off, or slumped off, the nearby massif. "Cochise is like Horatio," Schmitt continued, "in that it has a blocky wall but a mantled rim." This was a sign that the regolith was thick. Exploiting the fact that the southern rim was clear, Cernan crossed the rim and drove around the upper part of the interior wall.

After leaving Cochise's southwestern rim, it was a straight run to Van Serg. It was soon apparent that this was in the centre of what Schmitt referred to as "an extreme block field". In fact, there were so many rocks of a size sufficient to scrape the underside of the Rover that Cernan decided not to try to find a way through, and he parked about a crater's diameter out from the southeastern rim.

While Cernan brushed the thick deposit of dust off the electronics boxes, Parker said they had 25 minutes, and the objective was to be a radial sample from the rim of the crater to the Rover.

"I'll go on up to the rim, and see what we've got," Schmitt said. As he meandered through the litter of rock he sang, "Tip-toe through the tulips." On reaching the rim, he was amazed by the mess. Not only were there blocks poking through the shallow walls, there was a central mound. Although most of the rocks were thick with dust, he was impressed by their "shocked" appearance. A bench about half-way down the northern wall implied a shallow regolith, which was surprising because the material on Cochise's rim had seemed thick. He instantly dismissed the notion that Van Serg was a volcanic crater, since it was self-evidently the product of an impact. What was puzzling was that it appeared to have dug up such a large quantity of rock. This was in marked contrast to the clear rims of the larger craters nearby. The fractured blocks certainly looked like gabbro. If the regolith was sufficiently thin for a

In this westward-looking view at Station 8, the North Massif is in the background and Gene Cernan is setting up the Lunar Traverse Gravimeter Experiment on the bland lower slope of one of the Sculptured Hills.

crater the size of Van Serg to have excavated bedrock, why wasn't Cochise heavily littered? It was implicit in his logic that Van Serg was the result of a primary impact.

"Holey Smoley!!" Cernan exclaimed, as he joined Schmitt and got his first view of the pit. Getting straight to work, he went to a flat rock with "interesting patterns". The surface of the rock "flaked" when Schmitt scraped it using his scoop. Up to this point, he had assumed it to be gabbro, so it's extreme friability came as a surprise. With remarkable ease, Cernan pushed his tongs into the material and prised loose a 25-cm-long slab.[66] "That's a breccia," he announced on spotting small white specks in the matrix.

"These might be pieces of the projectile," Schmitt said, considering the possibility that Van Serg might be a low-energy secondary impact. He moved inside the rim to find a less-shattered block that might be solid rock in order to determine the nature of whatever bedrock this impact had excavated. "A lot of these blocks, particularly the more fractured ones, are a grey-matrix breccia," he noted. There did not appear to be any bedrock as such. "I'm not sure I understand what's happened here."

When Parker said they had only 10 minutes left, Schmitt let his frustration show. "Hey, Bob, we ought to find out what the rock is, here!" He was in no mood to be hustled just to meet the timeline.

Cernan, meanwhile, having remained on the rim, had collected a grapefruit-sized fragment[67] with a thick coat of glass that told a story of a violent impact. Then, upon realising that his boots were scraping up small fragments of glass, he suggested that they sample it. As Schmitt made his way back, he saw a "cow-pie bomb" composed of several splashes of fairly viscous glass that had solidified without flattening out, so they switched their attention to this agglutinate.[68] Next they lifted a nearby solid-looking dusty rock[69] that was in their earlier documentation, but this proved to be a blue-grey breccia. Parker suggested that they start their radial sample, and this time he was not chewed out for hinting that they start back.

First, however, Cernan and Schmitt moved apart to take pans for stereo mapping of the distribution of blocks in and around the crater. Unfortunately, Cernan ran out of film part way through. Leaving Schmitt to do the radial by solo-sampling, Cernan returned to the Rover to use the 500-mm Hasselblad to document the boulder tracks on the North Massif.

No sooner had Schmitt taken his first sample[70] for the radial than Parker told him the Backroom had decided that the ejecta was so dense that he should abandon the task. "I think that's a smart move," Schmitt admitted. "I don't think it would tell you much."

Parker then announced that Station 10 had been reinstated, so they were to make best-speed to Sherlock.

As Schmitt waited alongside the Rover for Cernan to take the gravimeter reading, he scraped an exploratory trench and, to his surprise, exposed some "very light-grey material" at a depth of 15 cm, so he sampled this layering.[71] In doing so, it occurred to him that if they had landed at this point and this had been his first sample, then it would have led him to conclude that the dark mantle was a distinct veneer. But having already explored the valley floor extensively, he knew that this was not the case; the light-toned subsurface material was another oddity to be considered in assessing this specific site. The discovery of this material had caught the interest of the Backroom, and Parker announced that it had been decided to delete Station 10 after all, and they should take a double-core in order to probe the stratification of the regolith at Van Serg.

"It's too rocky," Schmitt complained.

"Let's give it a try," Cernan decided.

"You're not even going to debate the issue!?" Schmitt enquired.

"Nope," Cernan confirmed. Debating the chances of striking a rock could easily take longer than it would take to find out by trying. If the tube stopped short, he'd go home with whatever he was able to extract.

"I hope you're right," Schmitt acceded.

In the event, after an easy start, the tube did stop, but a few hefty hammer blows were sufficient to clear the obstacle and then it went all the way. Even better, it came out easily.

"It looks like you proved me wrong!" Schmitt admitted. His uncertainty regarding the nature of the ejecta had prompted him to believe that the regolith would be laced with rocks, as indeed it almost certainly was, except that they were merely clods and Cernan had been able to drive the tube through them. Even as Parker was hustling them back onto the Rover, Schmitt grabbed a "light coloured rock" that had the single virtue of appearing to be well-indurated.[72] As they packed up, he restated his puzzlement: "The thing that amazes me, is that there's no subfloor around here."

Parker reported that the flight director wanted

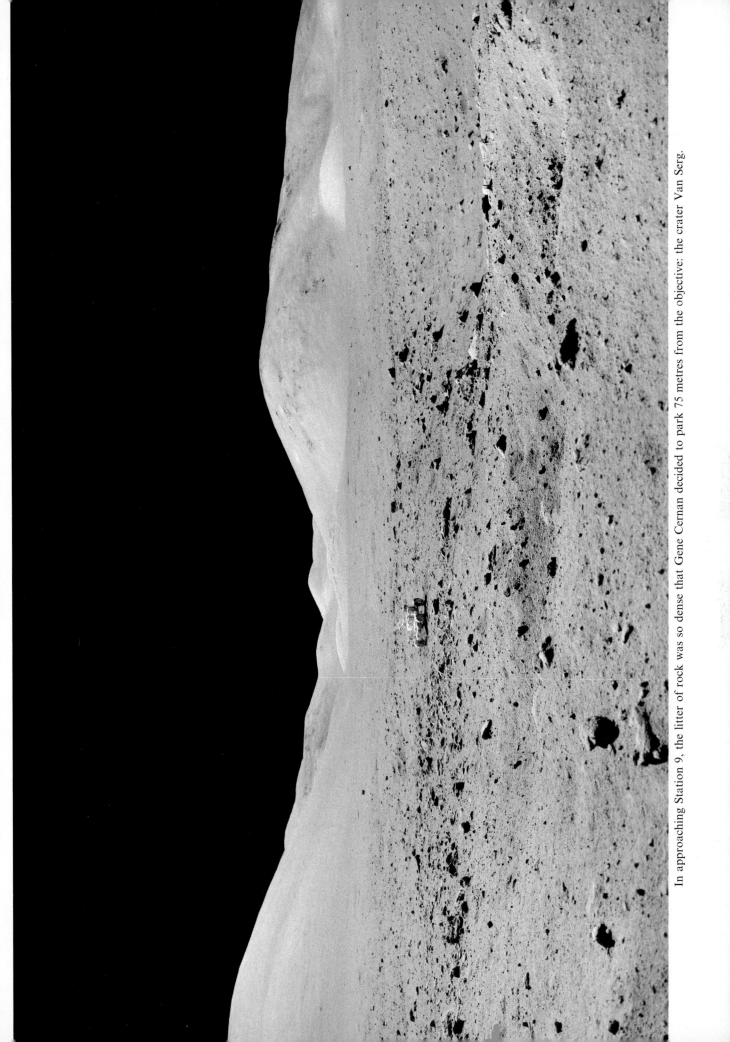

In approaching Station 9, the litter of rock was so dense that Gene Cernan decided to park 75 metres from the objective: the crater Van Serg.

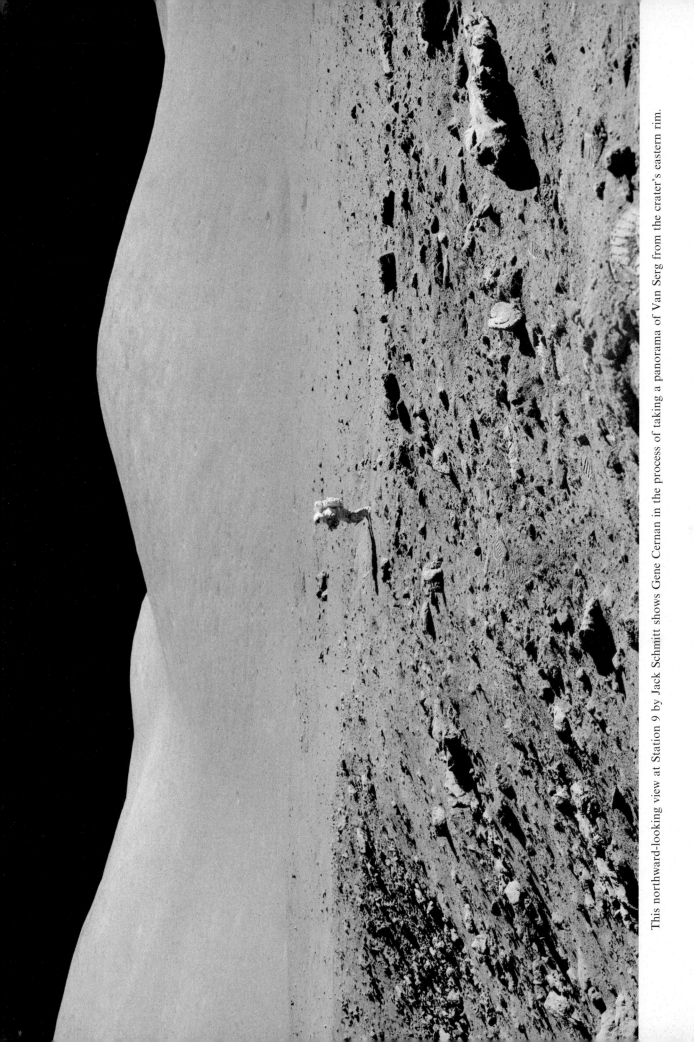

This northward-looking view at Station 9 by Jack Schmitt shows Gene Cernan in the process of taking a panorama of Van Serg from the crater's eastern rim.

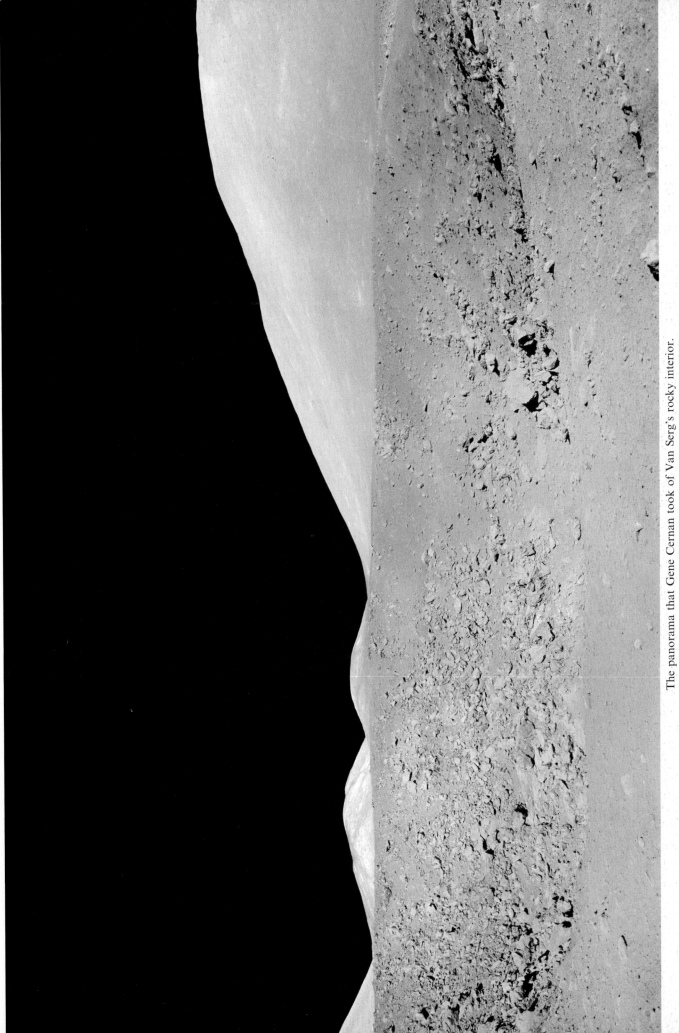

The panorama that Gene Cernan took of Van Serg's rocky interior.

them back at the LM as soon as possible in order to ensure that they would have sufficient time to finish the surface activities without encroaching on the safety margin.

As soon as they exited Van Serg's ejecta and the rock coverage fell to 1 per cent, Schmitt noted that they were back in the domain of the subfloor, and upon seeing a particularly large boulder Cernan made a close drive-by to enable Schmitt to check. "Yep," he confirmed. "Vesicular subfloor."

Schmitt didn't figure out Van Serg until he was on his way back to Earth. It really was a primary impact crater, but it had struck a particularly thick patch of regolith and had not reached the subfloor. However, the intense shock of such a large impact had so consolidated the soil as to create angular fragments which, despite their size and rocky appearance, were really no more than clods. The bench that had suggested a thin regolith and therefore a lot of excavated bedrock, was due to the way in which the pressure varied around the 'ground zero' point, and most of the glass would have been formed within this ring, producing the bench. Dave Scott and Jim Irwin would have seen it immediately as a scaled-up counterpart of their Station 9 crater. If Schmitt had remembered the cloddy crater near SWP, this might have put him on the right track. But the rock-like clods had thoroughly deceived him.

The lay of the land obliged Cernan to stick close to the planned route, which ran along Sherlock's northern rim, so he rode along the crest in order to confirm that the boulders which they were to have sampled for the subfloor really were there. Parker took advantage of the fact that they were at the Station 10 site to ask for a 'Rover stop' to collect a scoop of soil; Schmitt obliged, and even added a rock[73] for good measure.

After Sherlock, it was a straight run home, and when Cernan drew up at the SEPE antenna to check the calibration on the Rover's 'nav', Schmitt hopped off to fetch a large rock that he had noticed while deploying the antenna on the first day. It was a very-fine-grained chunk of gabbro.[74] When he had first seen it, he had not known how rare the crust of the subfloor would be. In fact, at football-sized, this was the biggest rock returned from Taurus-Littrow. On going to stow the rock on the Rover, he found that the latch of the pallet, long-since clogged by dust, had slipped and let the tailgate swing open during the drive. An inventory revealed that although the rake and the scoop had fallen off, all the SCBs – which at this stage were all that really mattered – were still present.

MAN'S DESTINY

It took an hour and a half to pack everything away, but as Cernan prepared to follow Schmitt up the ladder he paused for a moment of reflection on the significance of what he was about to do.

"As I take Man's last step from the surface, I would like to just say what I believe history will record – that America's challenge of today has forged Man's destiny of tomorrow. And as we leave the Moon at Taurus-Littrow, we leave as we came and, God willing, as we shall return, with peace and hope for all Mankind." With that, he took the 'small step' onto Challenger's footpad. "Godspeed the crew of Apollo 17," he added as he climbed the ladder for the final time.

FINAL INSIGHTS

It is a characteristic of a terrestrial avalanche flow that the larger debris settles at the bottom of the deposit. The fines fill in the cavities and then accumulate on top to make a smooth fan. This seemed to match the light mantle at the base of the South Massif. The surface was all fines, the fragment size and their number increased with depth, and only the largest craters had dug up large rocks. The solar wind 'exposure age' of the light mantle indicated that it was deposited about 100 million years ago. Following the reasoning set out prior to the mission, this was tentatively interpreted as dating the crater Tycho – the impact must have been a spectacular sight but only the dinosaurs were present to observe it.

When the seismic charges that had been placed around the valley floor were fired, they revealed the subfloor to be 2 km thick. This great depth explained why all the debris from the Central Cluster was subfloor material – the pre-Serenitatis basement and the overlying intra-massif were far beyond their reach. With the exception of the immediate area of Van Serg, the fragments in the regolith on the dark valley floor were derived from the basaltic subfloor. The fragments in the light mantle were tan-grey and blue-grey breccias from the South Massif. The SEPE yielded data only for part of the second excursion, but it showed that the regolith west of the landing site varied in thickness between 20 and 40 metres, and this was consistent with both the exposure of blocks on Camelot's rim and the thick overburden at Horatio. Van Serg was distinctive in having excavated only dark matrix-rich regolith breccia

Gene Cernan in the process of taking a double core sample alongside the Rover at Station 9.

Gene Cernan carries the double core tube to the Rover at Station 9.

Jack Schmitt at the Rover after the final traverse.

The Rover parked at the VIP site to document Challenger's departure.

which lacked basaltic fragments. The fact that a crater of that size had not reached the subfloor showed the regolith was particularly thick at that location. It is probable that the overlapping ejecta from the large craters that surround Van Serg had built up a thick deposit of regolith. Van Serg had been selected for sampling because it looked fresh, and so it is – it was excavated a mere 4 million years ago. In propagating through the regolith, the shockwave caused partial melting, which produced

HERE MAN COMPLETED HIS FIRST
EXPLORATIONS OF THE MOON
DECEMBER 1972, A.D.
MAY THE SPIRIT OF PEACE IN WHICH WE CAME
BE REFLECTED IN THE LIVES OF ALL MANKIND

EUGENE A. CERNAN
ASTRONAUT

RONALD E. EVANS
ASTRONAUT

HARRISON H. SCHMITT
ASTRONAUT

RICHARD NIXON
PRESIDENT, UNITED STATES OF AMERICA

A depiction of the commemorative plaque on the main strut of Challenger's landing gear.

a glassy agglutinate that contained the debris of mineral fragments. In contrast, the glass at Shorty, and indeed in the regolith of the valley floor, was a homogeneous basaltic composition and was present in the form of microscopic beads. Traces of sulphur, zinc, lead and other volatiles on the surface of the beads proved that they derived from a gas-rich magma issued by fire fountains. It was not a recent pyroclastic, however, because at 3.64 billion years, it was almost as old as the subfloor: 3.72 billion years.

Upon being informed of the discovery of orange soil at Shorty, Evans had looked but seen no sign of it from orbit. However, a little further west around the rim of the Serenitatis Basin, in the vicinity of Sulpicius Gallus, he noted a subtle but extensive "orange hue". On rejoining him, Schmitt confirmed this observation, and mapped it.

The premise for focusing on the dark mantle was the contemporary empiricism that mare material was dark when extruded and gradually lightened with exposure to micrometeoroids. The dark lava flows around the southeastern rim of the Serenitatis Basin were therefore taken to be somewhat younger than the lighter material in the middle. No one knew how much younger, of course, but that was the reason for sampling it.

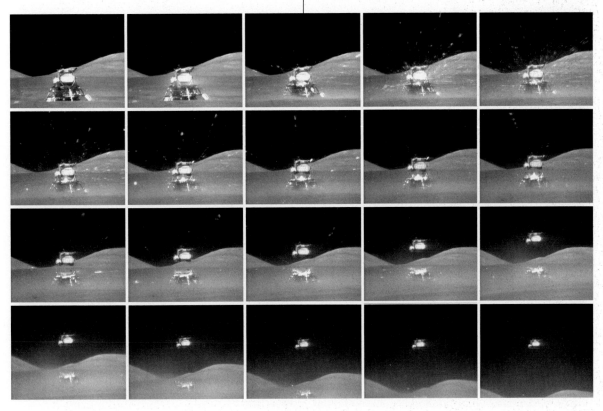

Challenger's lift off concluded the first phase of Mankind's exploration of the lunar surface.

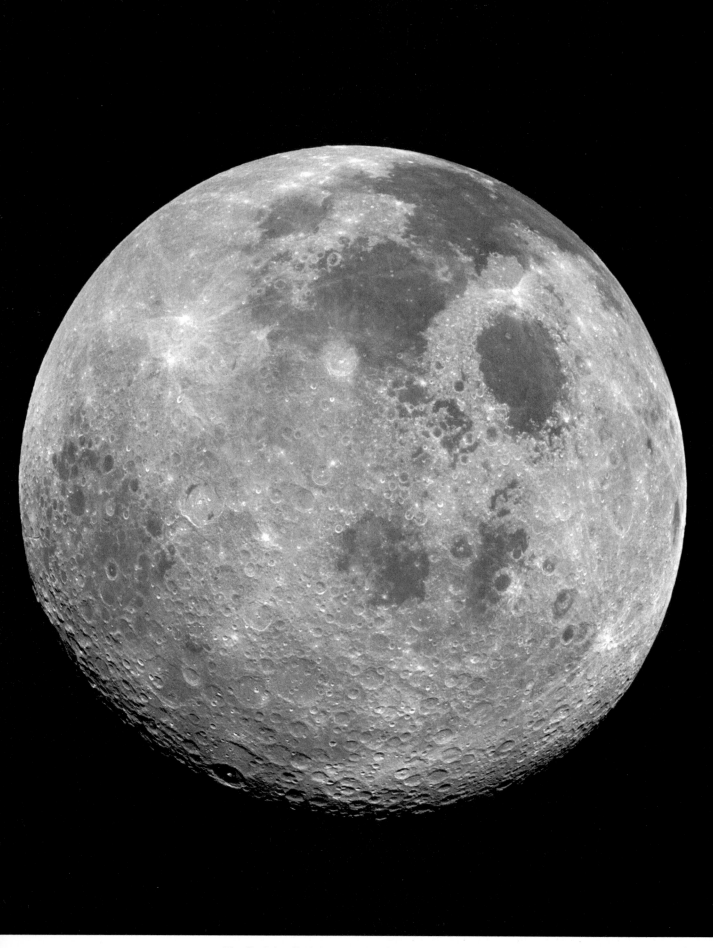

The final Apollo lunar crew heads home from the Moon.

Sample 71577, a 0.21-kg 'grab sample' from Station 1.

Sample 72355, a 0.37-kg chip from the second boulder at Station 2.

Sample 72435, a 0.16-kg chip of pure olivine from the clast in the third boulder at Station 2.

Sample 72275, a 3.64-kg chip from the first boulder at Station 2.

Actually, prior to the Apollo 17 mission, George McGill had noted that in some places the lighter mare material actually cut across the dark mantle, but the community at large had not been receptive.

After the discovery that the dark mantle was ancient, interest in finding 'recent' lunar volcanism waned, but this also reflected the fact that there would be no further opportunities to seek the ground truth required to test hypotheses. There are, in fact, several convincing candidates for Copernican Era volcanism. The Flamsteed Ring in

Sample 74255, a 0.74-kg fragment torn from the boulder at Station 4.

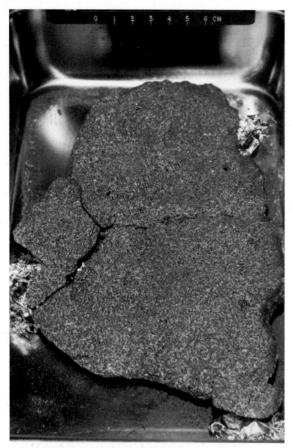

Sample 75055, a 0.95-kg slab chipped off the foliated diabase boulder at Station 5.

Sample 70215, an 8.11-kg very finely grained gabbro that Jack Schmitt collected near Challenger for the reason that it was atypical of the valley floor unit.

Oceanus Procellarum, where Surveyor 1 landed, was considered for Apollo 12 but Pete Conrad and Al Bean visited Surveyor 3, which was more conveniently located but less geologically significant. A site never considered for Apollo, is the 20-km crater Lichenberg in the northwestern Procellarum. It has rays, and therefore cannot be very old, and yet it is partially flooded. And then, of

Samples of the 'split boulder' at Station 6: 2.82-kg 76015 (top); 0.41-kg 76255 (centre); and 0.67-kg 76315 (bottom).

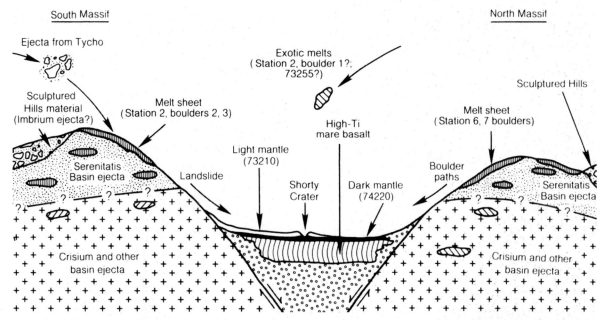

A schematic southwest to northeast geological cross-section through Apollo 17's landing site (see the earlier traverse map) showing the complex boundaries between the older highland (ejecta layers making up the North and South Massifs) and the younger mare basalt flows that underlie the valley (modified after Wolfe *et al.*, 1981). The massifs consist of a top thick layer of ejecta from the Serenitatis Basin, which is inferred to overlie an even thicker layer of complex ejecta from earlier basin-forming events (such as the nearby Crisium Basin). The valley developed as a down-dropped wedge (graben) between two fractures (faults) that may have formed at the time of the Serenitatis impact event. It was then filled, first with fragmental debris from the massifs, then by basalt lava flows and even younger pyroclastic dark-mantle deposits. More recent geological activity includes landslides and boulder falls from the higher massifs onto the valley floor. Numbers refer to specific collected samples that are representative of the various units inferred to be present. Courtesy the Lunar and Planetary Institute and Cambridge University Press.

course, there are the Marius Hills in Procellarum. If Apollo 13 had not aborted, Apollo 14 would have sampled the dark mantle west of Littrow, and the discovery that this was ancient would have meant that the final Apollo mission would not have been sent to Taurus-Littrow; but then we would not have had the pleasure of exploring such a beautiful valley.

8

Luna revival

The Soviet Luna program can conveniently be divided into those missions prior to Apollo's lunar landing, and those contemporary with, and subsequent to, America's exploration of the Moon.

SAMPLERS

The first spacecraft to use the large new Luna bus was launched on 13 July 1969. Apollo 11 followed 3 days later. Luna 15 entered orbit of the Moon on 17 July, and Apollo joined it on 19 July. The next day Luna 15 adjusted its orbit to produce a 16-km perilune in the eastern equatorial zone. Tranquility Base was established several hours later. As Neil Armstrong and Buzz Aldrin tried in vain to sleep prior to lifting off, Luna 15 crashed attempting to land in Mare Crisium. Its mission had been to scoop a sample of soil and beat Apollo 11 home. It would have been an engineering triumph if it had worked, and if it could have been achieved several years earlier it would have been scientifically significant, but being done in parallel with a human mission that returned with 22 kg of lunar material, a scoop of soil that would have been effectively no better than Armstrong's contingency sample would only have served as a useful point of comparison.

In September 1970, Luna 16 flew a similar trajectory and landed 100 km east of the crater Webb in the northeastern part of Mare Foecunditatis. An arm with a drill rotated down to the surface. Once the drill had secured a 35-cm core sample, the arm transferred it to a capsule in the upper stage, which promptly lifted off. The 0.1-kg sample was mostly fines, but there was granular material towards the base. In terms of its titanium content, the basaltic regolith was intermediate between the 'old' Mare Tranquillitatis from Apollo 11 and the 'young' Oceanus Procellarum sampled by Apollo 12, making it more aluminous, and it was dated at 3.4 billion years old. After

A depiction of the Luna 16 ascent stage lifting off with its lunar sample in the spherical capsule on top. (Courtesy Lavochkin Association)

exchanging some of its own material for a very small sample of Luna 16's material, NASA found fragments of recrystallised shocked brecciated anorthosite, norite and troctolite (rock types collectively known as 'ANT') that were derived from impacts in the highlands.

Landing a large automated spacecraft was a risky venture, and in September 1971 Luna 18 crashed 120 km north of its predecessor, near the crater Apollonius in the mountains between Mare Foecunditatis and Mare Crisium. However, Luna

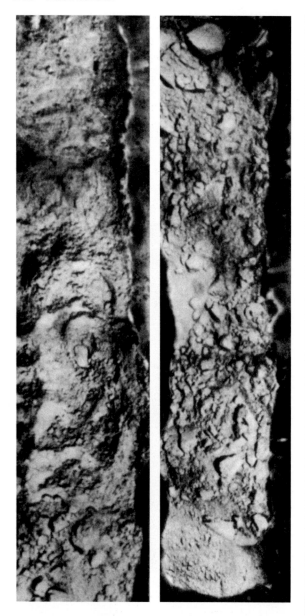

The majority of the 30-cm core sample retrieved by Luna 16, with the deepest point on the lower right.

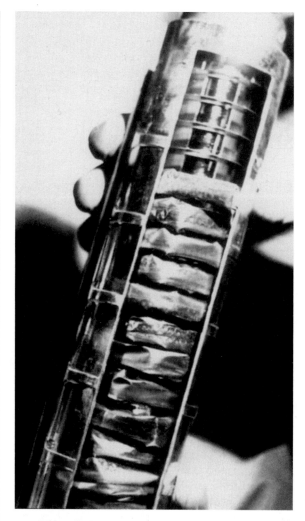

The coiled core sample retrieved by Luna 24.

20 had better luck in February 1972 and success-fully set down 2 km from the wreckage of Luna 18. Unfortunately, the drill encountered a layer of rock at a depth of 10 cm and managed to recover only 0.03 kg of material. It was nevertheless most welcome for being the first *in situ* sample of the lunar highlands. Whereas only tiny fragments of anorthosite had been found in the mare regolith, this was predominant in the Luna 20 sample, and, being calcic plagioclase, was why this sample had the greatest ratio of aluminium and calcium oxides yet observed.

Luna 23 set down in Mare Crisium in October 1974, but it was a controlled crash and the damaged arm was unable to recover a sample, so the ascent stage was not used. The last of the series, Luna 24, was sent in August 1976 in a final attempt to sample this site, and successfully landed 2 km from its predecessor. Its improved drill held its core in a flexible tube that was coiled up to enable a 1.6-metre sample to be stored in the return capsule. The 0.17-kg core was layered with powdery fines and a more granular component. Being highly feldspathic, this proved to have the lowest titanium ratio of any mare material. The basaltic fragments were dated at 3.3 billion years old, which implied that Mare Crisium is one of the freshest of the lunar maria.

ORBITERS

Like its predecessor, the new Luna bus was a multi-role craft, and Luna 19 was configured as an orbital mapper. It was launched in September

1971, and during the year that it operated it exploited its propellant capacity to manoeuvre to combine high-altitude regional mapping with low-altitude passes over a wide variety of sites. Launched in June 1974, Luna 22 flew a similar mission.

LUNOKHODS

An innovative use of the new bus was demonstrated in November 1970. Luna 17 set down in Mare Imbrium, near the Hercules Promontory, the western tip of the semi-circular rim of Sinus Iridum. Instead of an ascent stage, its payload was a *vehicle*. Once ramps had unfolded, this 8-wheeled remotely controlled 'Lunokhod' rolled down onto the surface. It had cameras like those used by the hard-landers for taking panoramic views, and a pair of TV cameras on the front to provide its operators with stereoscopic vision. It halted 20 metres from its bus, raised a flap to expose a solar panel to charge its battery, then 'buttoned up' for its first 14-day-long lunar night. It awoke with the return of the Sun, and then set off south, occasionally pausing to use a penetrometer to measure the bearing strength of the surface and an X-ray spectrometer to analyse the chemistry of the regolith, showing this to be basaltic. It parked for the night about 1.4 km from its lander, and then returned by a roundabout route the next day. Its mission was not over, however, and it set off north, this time never to return. It was still operating when Dave Scott and Jim Irwin explored Hadley-Apennine on the eastern side of the Imbrium Basin. By the time that it expired in October, it had travelled 10.5 km, returned 200 panoramas and analysed the regolith at 25 sites.

A few weeks after Apollo 17 left the Moon, Luna 21 placed the second Lunokhod inside the crater Le Monnier on the eastern shore of Mare Serenitatis. It promptly set off south, heading for the mountains of the Taurus Range. As it approached the foothills, the X-ray spectrometer found that the ratio of iron in the regolith declined, which made sense. In April, it encountered a 300-metre-wide rille, and ran along the edge of this for several kilometres, dodging boulders. In addition to its predecessor's instruments, it had a magnetometer, which showed variations in the strength of the field around the boulders. This data supplemented that from the portable instruments used by the Apollo crews. Unfortunately, Lunokhod 2 broke down in May. By that time it was 37 km from its landing site. If it had survived as long as its predecessor, the vehicle could conceivably have reached the valley which Gene Cernan and Jack Schmitt had explored, some 150 km further south.

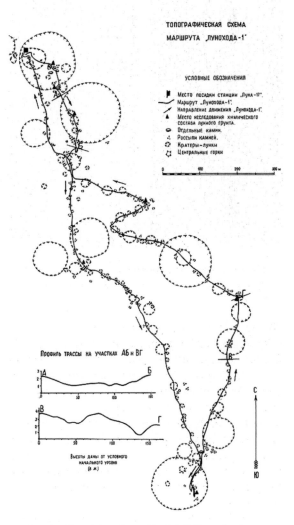

A depiction of Lunokhod 1 on its landing stage.

For its first drive, Lunokhod 1 followed a loop.

A view by Lunar Orbiter 4 of the crater Le Monnier breached by Mare Serenitatis, annotated with the route taken by Lunokhod 2 along the southern rim of the crater, and the location of the Apollo 17 landing site.

MISSED OPPORTUNITIES

Although the Luna sample-return mission hinted at a capability to collect samples from across the Moon without incurring the cost inherent in human spaceflight, the spacecraft was actually limited. The hard-landers had been sent to sites at 60 degrees west of the meridian because this allowed a straight-in vertical descent, eliminating the need to deal with lateral velocities. As the Lunokhod deliveries showed, because this new bus began its powered descent from orbit it could land in either hemisphere and away from the equator. However, even though the sample-returners could have landed anywhere, only near-equatorial sites 60 degrees east of the meridian allowed the return stage to perform a simple vertical ascent. This spacecraft was incapable of returning samples from the sites visited by the Apollo missions. Although it would have been able to land instruments to undertake *in situ* studies at widely distributed sites, and offered the advantage that its payload capacity exceeded the mass of the entire Surveyor spacecraft, the program was terminated in 1976.

9

Apollo in context

GROUND TRUTH

The Apollo Lunar Surface Experiment Package (ALSEP) was a selection of modular instruments that could be mixed and matched to study the various themes in lunar science. Generally, later flights carried larger packages, and much of the first EVA was required to deploy them. Some apparatus was carried by several flights in order to form 'networks' that could make coordinated measurements. Later flights carried instruments designed to follow up earlier results, so, overall, the program was well integrated. Several independent themes were addressed in parallel. The main focus of attention was the state of the Moon, but cislunar space was also studied, and several other projects were pursued on an *ad hoc* basis.

For the first landing, it was decided to carry only two independent instruments, as the Early Apollo Scientific Experiment Package (EASEP), but each later mission had an ALSEP which incorporated a Central Station to provide communications and to distribute power from a SNAP-27 Radioisotope Thermal Generator (RTG).

Passive Seismic Experiment (PSE)
The solar-powered seismometer which was left at Tranquility Base as part of the EASEP included a detector to measure the dust accumulation on and radiation damage to its solar cells, and an isotope heater to keep the electronics warm during the 14-day-long lunar night. Despite operating temperatures that exceeded the planned maximum of 30°, the instrument functioned normally through the

Instruments used by Apollo on the Moon

Apollo	11	12	13	14	15	16	17	Times
PSE	Y	Y	Y	Y	Y	Y	–	6
ASE	–	–	–	Y	–	Y	–	2
LSPE	–	–	–	–	–	–	Y	1
LTGE	–	–	–	–	–	–	Y	1
SEPE	–	–	–	–	–	–	Y	1
HFE	–	–	Y	–	Y	Y	Y	4
LNPE	–	–	–	–	–	–	Y	1
LSM	–	Y	–	–	Y	Y	–	3
LPM	–	–	–	Y	–	Y	–	2
CCGE	–	Y	Y	Y	Y	–	–	4
LACE	–	–	–	–	–	–	Y	1
LEAM	–	–	–	–	–	–	Y	1
SWCE	Y	Y	Y	Y	Y	Y	–	6
SWS	–	Y	–	–	Y	–	–	2
SIDE	–	Y	–	Y	Y	–	–	3
CPLEE	–	–	Y	Y	–	–	–	2
CRDE	–	–	–	–	–	Y	–	1
LSCRE	–	–	–	–	–	–	Y	1
LRRR	Y	–	–	Y	Y	–	–	3
UVC	–	–	–	–	–	Y	–	1
LSG	–	–	–	–	–	–	Y	1

period of maximum heating around local noon. With the output from the solar arrays in decline 5 hours prior to local sunset (on 3 August 1969) transmission was halted by command from Earth. It was reactivated early on the next lunar day, but the electronics had been damaged by the intense cold and the transmission was impaired. On 27 August, near local noon of the second day, the instrument ceased to accept commands, ending the experiment. Apollos 12, 14, 15 and 16 created a network of longer-term instruments in order to triangulate the sites of meteoroid strikes and moonquakes. By sending spent rocket stages and spacecraft crashing to the Moon, it was possible to refine the calibration of the network.

In comparison to Earth, the Moon is almost seismically inert. Most of the events that were detected would be lost in the general 'noise' of the continuously adjusting terrestrial crust. The largest event rated only magnitude 4 on the Richter scale. In all, until the network was switched off in September 1977, the largest meteoroid to strike the Moon was of the order of 5 tonnes in mass. This occurred on the Farside, and the seismic waves served to probe the Moon's internal structure.

The PSE network showed that the Moon has a crust, a mantle and a core, and so is a thermally differentiated body. The crust is plagioclase-rich and, on average, is three times thicker than that of Earth. The interface between the crust and the mantle is a sharp transition, as in the case of the Earth's Mohorovicic Discontinuity. The olivine and pyroxene mantle englobes a small core of less than 25 per cent of the Moon's radius – in the case of Earth the core is 50 per cent of the radius. The lunar crust is so deeply brecciated that it is very efficient at generating reflections, with the result that seismic energy is damped only very slowly, causing the Moon to 'ring like a bell'. A few internal events occur near the surface, but most originate at depths in the range 800–1,000 km, and are correlated with the tidal forces on the Moon resulting from its elliptical orbit of Earth. The damping of seismic energy below a depth of 1,000 km indicates that the rock is semi-molten. However, this inner mantle is 'warm' rather than 'hot', and too shallow for convection. Nevertheless, there may be isolated pockets of fluid in this zone, and the 'deep' seismic events may be due to movements of this magma in response to tidal forces.

Active Seismic Experiment (ASE)

The active seismic experiment probed the crust's upper kilometre at the landing site. It was flown only on Apollos 14 and 16, both of which sampled ejecta blankets set in highland terrain. For one part of the experiment, after a line of geophones had been emplaced, an astronaut placed a 'thumper' down at specific points along the line and fired a small charge (essentially a shotgun cartridge) to send a seismic signal into the ground. A mortar was also set up for use once the astronauts had returned to orbit, but Apollo 14's mortar was not fired, and only 3 of the 4 charges were fired using the Apollo 16 mortar. The low speed of the seismic transmission (at 100–300 metres per second, less than in the Earth's crust) indicated the crustal material in both cases to be brecciated, with the Fra Mauro Formation at the Apollo 14 site being 75 metres thick and the Cayley Formation at the Apollo 16 site at least 100 metres in thickness.

Lunar Seismic Profiling Experiment (LSPE)

As a follow-up to the Active Seismic Experiment's mortar, the Apollo 17 astronauts installed geophones at the ALSEP site and then emplaced 8 much larger charges at widely distributed points, and when these were fired after the crew had left the Moon the data showed the valley floor at Taurus-Littrow to be a slab of basalt at least 1 km thick.

Lunar Traverse Gravimeter Experiment (LTGE)

To probe the substructure of the Taurus-Littrow valley, the Apollo 17 astronauts took local gravity measurements on the valley floor and at various points around the base of the adjacent mountains. The variation of the field reflected the different

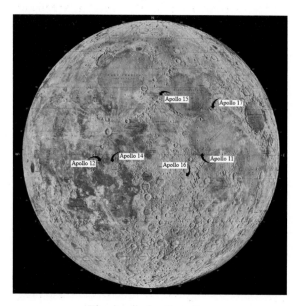

The Apollo landing sites.

densities of the massifs on the rim of the Serenitatis Basin and the material which later flooded the valley, and established that the lava flow that formed the valley floor was 2 km in thickness.

Surface Electrical Properties Experiment (SEPE)

Apollo 17's Rover was equipped to record radio signals broadcast by a transmitter that was laid on the ground at the landing site. It yielded information on the electrical conductivity of the subsurface to a depth of several kilometres. Its main function was to gain calibration data for the Lunar Sounder which was carried as part of the SIM for that mission, and which probed the CSM's ground track in a similar manner.

Heat-Flow Experiment (HFE)

One of the most important measurements that can be made of a planet is the rate at which it is losing heat to space. Once it had been established that the maria were lava extrusions, it was accepted that the Moon had a thermally differentiated interior. The HFE was to investigate the lunar 'heat engine'. It required strings of thermal sensors to be placed in holes drilled several metres into the regolith. It was first assigned to Apollo 13, but this mission did not reach the Moon. On Apollo 15, a design flaw in the drill stems made it impracticable to reach the intended depth through a thick slab of consolidated material. On Apollo 16 the experiment was lost when the cable was accidentally torn off the Central Station. Fortunately, on Apollo 17 everything went well.

The lunar surface is a harsh thermal environment, being baked when the Sun is in the sky and frozen when it is not. The HFE showed that this month-long cycle affects only the upper half-metre of regolith, since immediately below this the temperature is constant. At shallow depth below this, there is a build up of heat leaking from the interior. The heat-flow rate at Hadley-Apennine was surprisingly high, but similar results were obtained at Taurus-Littrow. However, as both of these sites are lava flows on basin rims, it is possible that the substructure is imparting a bias: that is, possibly as much as 10 per cent of the observed heat-flow is due to the fact that lava flows are efficient thermal conductors.

Lunar Neutron Probe Experiment (LNPE)

When the core samples from early Apollo missions were examined, it was found that radioactive isotopes were created in the regolith through the absorption of cosmic-ray neutrons. Since taking the 'deep core' for the Heat-Flow Experiment left a hole, for Apollo 17 it was decided that a sensor should be inserted into this hole in order to determine the flux and energy spectrum of neutrons at different depths in the regolith. This would serve to calibrate the technique by which the relative abundances of isotopes was used to infer the rate of regolith 'turn over' as a result of micrometeoroidal gardening.

The degree of regolith mixing by an impact is dependent on the size of the projectile. Although larger impacts have a greater effect and dig up material from greater depths, smaller impacts are more common, with the result that the shallower regolith is more intensively gardened. Cores revealed that while the turnover rate for the uppermost centimetre is of the order of a million years, it takes 1,000 times longer to recycle the top metre of regolith. The lunar surface is a remarkably inert place.

Lunar Surface Magnetometer (LSM)

Apollos 12, 15 and 16 deployed instruments to measure the Moon's magnetic field. These had three arms, each of which was to measure a specific Cartesian component of the local magnetic field. It was evident from the start that if the Moon possessed a dipole field, then this would be extremely weak. The data was to be compared to that from Explorer 35, which had been placed into lunar orbit in 1967 to study the lunar environment. This long-term database of the ambient fields enabled the Moon's own field to be isolated from the background fields associated with the solar wind and the presence of the Earth.

The dynamo of electrical currents flowing in the Earth's metallic core generates a dipole field. Although the Moon has a small core, this appears not to have produced a magnetic field. However, long-term measurements of the field at the lunar surface as the Moon entered and left the part of the Earth's magnetosphere that is blown 'down stream' by the solar wind allowed the electrical conductivity of the material within the Moon to be inferred – a process known as electromagnetic sounding. Electrical conductivity depends on both temperature and chemical composition, and while the data did not determine the internal constitution, it served to constrain models derived from other data. In particular, it placed a limit on the size of an iron core and on the temperature of the mantle. For an anorthositic crust over a mantle made primarily of the mafic silicates olivine and pyroxene (as suggested by the surface sampling), the temperature would reach $1,000°C$ (sufficient to induce melting) at a depth in the range 800–1,500 km (the actual depth depending on the assumed composition). This was consistent with the seismic data, which indicated a transition zone at a depth of 800–1,000 km.

Lunar Portable Magnetometer (LPM)

Following the discovery of a surprisingly strong magnetic field at the Apollo 12 site, it was decided to use a portable magnetometer to take readings at widely separated points on a given mission. The field varied between 43 and 103 gammas between the Fra Mauro landing site and the summit of the ridge a kilometre away, with the higher value being in the boulder field on Cone Crater's rim. For Apollo 16 the field varied from 121 to 313 gammas. (To put this into context, the highest observed value is two orders of magnitude lower than at the Earth's surface.) On the local scale, the field at the lunar surface is derived from remanent magnetism in the rocks, with the variation over short distances being caused by whether the orientations of the rocks cause the individual fields to cancel out or to reinforce.

Cold-Cathode Gauge Experiment (CCGE)

This was set up by Apollos 12, 14 and 15 to measure the pressure of any gas at the lunar surface. Although it demonstrated that there was indeed an 'atmosphere', it was extremely tenuous (the pressure being fully 14 orders of magnitude lower than at sea level on Earth) and the total amount of gas was no more than had been emitted by the LM in making its powered descent. The instrument was so sensitive that it could detect the presence of the astronauts by the water that was released from the sublimators of their PLSS backpacks.

Lunar Atmosphere Composition Experiment (LACE)

This mass spectrometer was set up by Apollo 17 to determine the composition of the gas at the lunar surface reported by the CCGE. It found the majority of the gas to be hydrogen, helium and neon. However, significant amounts of argon were detected at times of enhanced seismic activity. Since argon is the decay product of radioactive potassium, this could indicate ongoing venting from crustal faults. On the other hand, another instrument (SIDE) noted a correlation with impact rates which suggested that some gas was due to meteoroids vaporising regolith. Small quantities of methane, carbon dioxide, ammonia and water were occasionally detected, possibly as a result of

cometary fragment impacts. Owing to the Moon's low escape velocity, all but the heaviest gases will soon escape, and once ionised by solar ultraviolet they will be 'swept up' by the magnetic fields of the solar wind and carried away.

Lunar Ejecta and Meteorites Experiment (LEAM)

This experiment was set up by Apollo 17 to identify the nature of the small particles that strike the lunar surface. It used a grid of plates to detect the trajectory and energy of microscopic impactors. Although it detected debris flying on near-horizontal low-energy trajectories from nearby meteoroid impacts, it did not detect the expected rain of cometary dust, nor any interstellar grains streaming through the Solar System.

Solar Wind Composition Experiment (SWCE)

Set up by all missions except Apollo 17, this metal-foil sheet was erected facing the Sun in order to trap solar wind ions. Because it was returned to Earth for analysis, the determination of the chemical composition of the solar wind was more accurate than could have been achieved by a detector that was left behind and reported its data by radio, but the exposure time was necessarily very short.

Solar Wind Spectrometer (SWS)

This spectrometer was set up by Apollos 12 and 15 to record the energy density and temporal behaviour of the solar wind. Some 95 per cent of the solar wind is protons and free electrons, these being the products of ionising hydrogen. The detector could determine the direction of the particles as well as their energies. At 'night' and when the Moon was in the 'tail' of the Earth's magnetosphere, the solar wind flux fell to zero. Although ongoing monitoring would yield information on the variation of the solar wind over the 11-year solar cycle, the instrument was switched off after only half a cycle.

Suprathermal Ion Detector Experiment (SIDE)

This instrument was deployed by Apollos 12, 14 and 15 to detect ions with energies of less than 50 electron volts, corresponding to the *slowest* ions in the solar wind and gases at the lunar surface which had been ionised by solar ultraviolet. A correlation with the data from the seismometers suggested that at least some of the gas was regolith that had been vaporised by impacts.

Charged Particle Lunar Environment Experiment (CPLEE)

Deployed by Apollo 14, this instrument supplemented SIDE by detecting ions in the 50–50,000 electron volts energy range.

Cosmic-Ray Detector Experiment (CRDE) and Lunar Surface Cosmic-Ray Experiment (LSCRE)

Set up by Apollos 16 and 17, these two instruments detected ions in the energy range 100,000–150,000,000 electron volts, corresponding primarily to high-energy cosmic rays rather than the solar wind. They took the form of 'plates' that were exposed and returned to Earth for laboratory analysis of the 'tracks' produced by relativistic ions.

Laser Ranging Retro-Reflector (LRRR)

A small laser reflector was deployed by Apollo 11 as part of the EASEP, and larger units were left by Apollos 14 and 15. The corner-cubed reflectors could return a laser to its source across a range of angles of incidence. Although the signal strength was greatly diminished, by measuring the transit time of pulses it was possible to track the Moon with unprecedented accuracy. By taking simultaneous measurements from a variety of sites it was possible to determine the Moon's motion in three dimensions. Once the orbit was accurately known, it became possible to 'subtract' this in order to measure secondary motions, such as the tidal flexure of the lunar crust due to Earth's gravity during its perigee passage, and the diurnal flexure of the terrestrial surface as a result of lunar gravity. Long-term data established that the Moon is retreating from the Earth at 3.8 cm per year. The fine detail yielded information on the distribution of mass within the Moon, showing that there is a core of 20 per cent of the lunar radius. Once the motion of the Moon was understood, it was possible to isolate terrestrial contributions so as to reveal variations in the Earth's rotation rate and the precession of its axis. Indeed, in the longer term, it became possible to directly measure the rates at which the continents are travelling as a result of plate tectonics. Although the rest of the ALSEP network has long-since been switched off, these inert instruments (with a similar unit on Lunokhod 2) continue to facilitate geophysical research.

Far-UV Camera/Spectrograph (UVC)

Apollo 16 established the first astronomical observatory on the Moon by installing an ultraviolet camera. This comprised a 75-mm telescope and a spectrometer operating in the range 500–1,600 angstroms, a part of the electromagnetic spectrum that cannot penetrate the Earth's atmosphere. From a location in the LM's shadow, it ran through a preprogrammed list of targets. A total of 178 frames were exposed to investigate a variety of astronomical objects, the Earth and the lunar horizon.

Lunar Surface Gravimeter (LSG)

If any experiment sent to the Moon had the potential to produce a Nobel Prize, it was this gravimeter. Although Einstein's theory of relativity predicted the existence of gravity waves, they had proved difficult to confirm. This experiment was intended to work in conjunction with a counterpart on Earth to 'feel' the passage of such waves through the Solar System. Any signal detected by just one instrument could not be a gravity wave, but any signal that was detected by both could *only* be a gravity wave. It was taken to the Moon by Apollo 17, but the detector's working mass did not fully release from its carriage restraint, preventing it from attaining the sensitivity required to make the planned observation. However, it proved to be a capable seismometer, and returned useful data of this nature.

REMOTE SENSING

Early in the planning of the lunar phase of the Apollo program, it was envisaged that after the historic first landing there would be several 'H'-class missions and then a series of 'I'-missions on which, instead of the LM, there would be a module loaded with cameras and other instruments to survey the entire Moon from polar orbit. Then there would be a series of extended surface missions, in some cases involving robotic vehicles. When it became clear that the production line for Saturn V launch vehicles was to be closed, NASA decided to discard the 'H'-class format as soon as it could in order to use the remaining launchers for 'J'-missions on which (in addition to the LM ferrying a Rover and spending longer on the lunar surface) the main spacecraft would carry a suite of remote-sensing instruments. The Scientific Instrument Module (SIM) that was installed in a vacant bay of the SM provided a taste of what the 'I'-missions would have been able to achieve, although the operational constraints imposed by the landing sites precluded a polar orbit. The instruments carried by Apollos 15 and 16 included:

- a panoramic camera
- a mapping camera

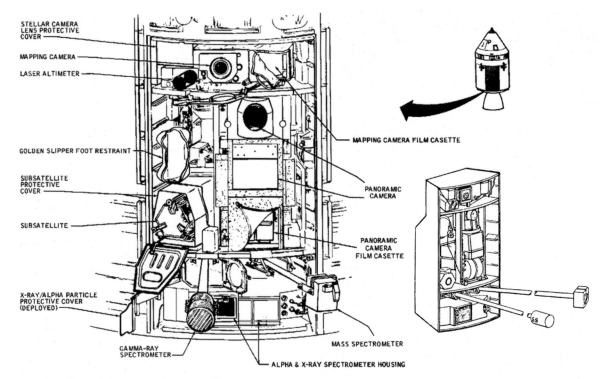

STELLAR CAMERA
LENS PROTECTIVE
COVER

MAPPING CAMERA

LASER ALTIMETER

GOLDEN SLIPPER FOOT RESTRAINT

SUBSATELLITE
PROTECTIVE
COVER

SUBSATELLITE

X-RAY/ALPHA PARTICLE
PROTECTIVE COVER
(DEPLOYED)

GAMMA-RAY
SPECTROMETER

ALPHA & X-RAY SPECTROMETER HOUSING

MAPPING CAMERA FILM CASETTE

PANORAMIC
CAMERA

PANORAMIC
CAMERA
FILM CASETTE

MASS SPECTROMETER

The first of the 'J'-missions, Apollo 15 had an unused bay in the Service Module loaded with remote-sensing instruments – the Scientific Instrument Module (SIM). The same instrument suite was flown on Apollo 16, but Apollo 17 used a different suite.

- a laser altimeter
- an X-ray fluorescence spectrometer
- a gamma-ray spectrometer
- an alpha particle spectrometer
- a mass spectrometer.

The mapping camera had a 75-degree field of view and could image an area 175 × 175 km from the CSM's 110-km mapping altitude. Although the panoramic camera's field of view extended only 11 degrees ahead, it looked out 55 degrees to each side, and each frame captured a strip that spanned 28 × 334 km. The laser altimeter was boresighted on the mapping camera and could measure the range with a resolution of 1 metre. The panoramic camera had sufficient film for 1,650 fames, and the mapping camera held 3,600 frames. The film cassettes were to be retrieved by the CMP making a spacewalk during the trans-earth coast.

The fact that the Moon has virtually no atmosphere opened a window for sensing its mineralogy from orbit. Solar X-rays stimulate the atoms in the surficial regolith to fluoresce, and because each element has its own energy spectrum it was possible to determine the concentrations of the most important elements in the crust: aluminium, silicon and magnesium. However, because the surface can fluoresce only when it is illuminated, only the sunlit part of the spacecraft's ground track could be monitored. Although only a narrow swath could be surveyed, and there was considerable overlap between successive passes, the fact that the terminator travelled 12 degrees westward in 24 hours meant that during the week that each mission was in lunar orbit the track which was surveyed exceeded more than half of the Moon's circumference.

Because the lunar surface is exposed to space, it is irradiated by cosmic rays. This rain of charged particles comprises both lightweight solar wind and relativistic heavy nuclei from the interstellar realms. The nuclei in the regolith react to this stimulation by emitting gamma rays. By monitoring the gamma-ray spectrum, it proved possible to measure the concentrations of iron and titanium. Furthermore, because radioactive elements spontaneously emit in this part of the spectrum it was possible to chart the distribution of uranium, thorium and potassium. A gamma-ray spectrometer offered the advantage that it did not require solar illumination, and so could sample over the entire ground track.

It is important to note that although the characteristic X-ray and gamma-ray data pertained only to the surficial material, the fact that in

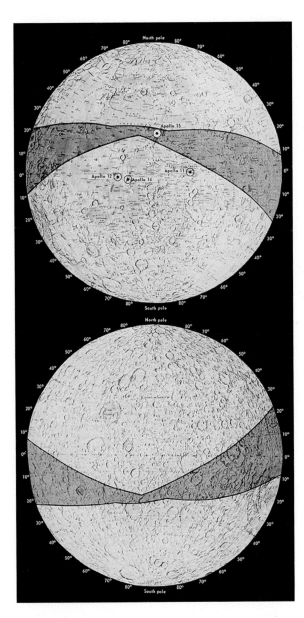

A chart showing the coverage of Apollo 15's ground track: the Nearside at the top and the Farside below.

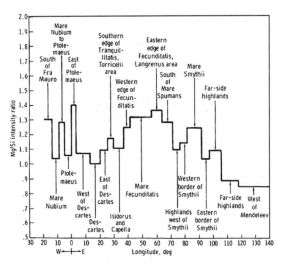

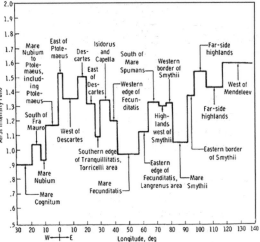

The variation of the magnesium-to-silicon (top) and aluminium-to-silicon ratios at various points along Apollo 15's ground track.

the main this was derived from the gardening of locally excavated rock meant that mineralogically distinctive ejecta around 'drill hole' craters could provide a view into the subsurface.

If gas was leaking from the lunar crust, the low escape velocity meant that all but the heaviest gases would dissipate. To investigate whether radon from the radioactive decay sequences associated with uranium and thorium might be concentrated in low-lying areas in the most radioactive KREEPy terrain, the SIM included a spectrometer to measure the alpha particles (helium nuclei) emitted by the decay of radon. As with the gamma-ray spectrometer, this instru-

ment was independent of insolation, enabling it to operate continuously.

Surface sampling and orbital work were synergistic in that laboratory analysis of the samples provided the ground truth required to calibrate the remote-sensing data and enable the elemental abundances to be interpreted in terms of *mineralogy*.

Silicon (Si) was chosen as the reference against which to measure aluminium (Al) and magnesium (Mg) because it is the most common element in the lunar crust. Not unexpectedly, the Al/Si and Mg/Si ratios indicated there to be a significant difference in composition between the maria and the highlands, but the discoveries were in the detail, since there were inhomogeneities in both types of terrain. The variation of the Al/Si ratio was particularly pronounced, varying by more than a factor of two from the highlands on the eastern limb (where it was highest) to the western maria (where it was lowest). As expected, this ratio correlated with albedo, confirming the visually obvious fact that the highlands are bright and the maria are dark. However, chemical composition is not the only factor to control albedo. A given feature might be bright because it was created recently – a 'ray' crater, for example. The inverse relationship between Al/Si and the gravitational gradient (inferred from radio-tracking) indicated that the lowest Al/Si ratio correlated with the strongest gravity, which confirmed that the 'mascons' were associated with the 'circular maria'. Also, the inferred density of the anomalous attractor was the same in each case. The data provided confirmation of earlier observations. For example, it had been noted that the 'circular maria' were brighter towards the red end of the spectrum and the others were brighter towards the blue end of the spectrum, and the X-ray fluorescence data showed that the lavas that rose through the floors of basins to flood them were distinctly more mafic than those which overran what appeared to have been 'open' territory. The gravity data was sufficiently detailed for some aspects of the substructure to be inferred. For example, the Marius Hills in Oceanus Procellarum were confirmed to have anomalously high gravity whereas the Apennine Bench in the Imbrium Basin did not – showing them to be different geological landforms. These 'reality checks' provided the confidence to interpret the data on a more general basis.

The natural radioactivity due to thorium, uranium and potassium in the crust was highest in Mare Imbrium and Oceanus Procellarum. The 'hottest' reading was at the southern extremity of the Fra Mauro Formation. It matched the level of radioactivity in the KREEPy material which was sampled by the Apollo 14 mission on this blanket of Imbrium ejecta. The Apennine Bench and the Aristarchus Plateau, which is in the western part of Procellarum, were also 'hot'. The eastern maria were intermediate. The highlands were generally 'cool', but there were 'warm spots' where basin ejecta had been deposited. In fact, one such area was the Descartes-Cayley where Apollo 16 landed, and this particular remote-sensing data supported the conclusion drawn from the surface sampling that the morphology of that area was impact-related rather than volcanic. The 'coldest' reading was in the highlands just beyond the eastern limb.

It had been hoped that charting the distribution of radon would reveal sites where the crust was outgassing, as a sign of current volcanism. The most significant result was over the Aristarchus

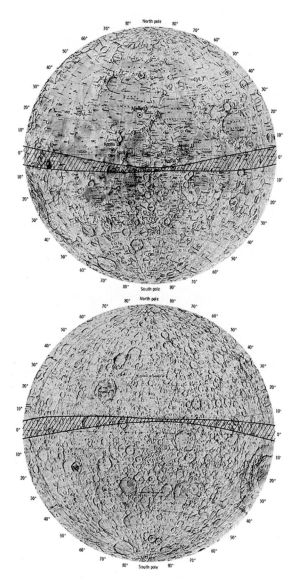

A chart showing the coverage of Apollo 16's ground track: the Nearside at the top and the Farside below.

Plateau. Other than this the data was ambiguous, largely because if radon was present then it was at a concentration that was right on the limit of the instrument's sensitivity.

As Apollo 16's ground track would complement that of Apollo 15, it was assigned the same SIM suite. However, because Apollo 15 had surveyed the landing site that was chosen for Apollo 17 there was considerable overlap of their ground tracks and it was decided to replace some of the instruments. For Apollo 17, the suite included:

- a panoramic camera
- a mapping camera

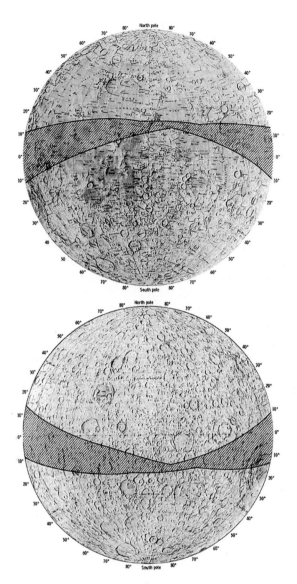

A chart showing the coverage of Apollo 17's ground track: the Nearside at the top and the Farside below.

- a laser altimeter
- a mass spectrometer
- an infrared scanning radiometer
- a radar sounder.

The lunar surface sounder was a crude imaging radar that profiled the topography and charted the substructure. The 3-frequency instrument measured the properties of radar pulses reflected from the surface. It was able to operate continuously because it supplied its own 'illumination'. Analysis of samples from earlier missions had shown that the electrical conductivity of the crustal rock was sufficient to allow the signal to penetrate about 1 km into the ground. The Surface Electrical Properties Experiment provided ground truth.

Since Apollo 17 had a similar ground track to Apollo 15, the radar data could be correlated with the surface mineralogy from that mission, which meant that the instruments which had provided it did not need to be carried.

The infrared scanning radiometer, which functioned as a passive thermal-imaging camera, was used primarily to study the rate at which the various crustal formations cooled in the lunar night. The Heat-Flow Experiment had established that the finely gardened regolith was an excellent insulator, and that the harsh thermal variability at the surface did not penetrate more than half a metre. The extent to which the surface absorbed solar energy depended on its albedo, the angle of illumination and the local slope. The re-emission of this energy after sunset varied with the physical properties of the material.

Although the low-viscosity mare lavas spread out to form vast open plains, they represent an accumulation of a succession of extrusions and are not totally flat. Also, different maria are at the different elevations. In particular, Mare Tranquillitatis has breached the southern rim of the Serenitatis Basin and the tectonic stresses resulting from the fact that it lies almost 1 km above the basin infill have produced a profusion of rilles in this area. The arcuate wrinkle ridges in the central part of Mare Serenitatis form a 'step' down to the central area, which lies 200 metres lower still. The gravity data showed that the mascon is confined to this lowest level, implying that the ridges resulted from mechanical failure following elastic flexure as the mare material settled in the basin, with the resulting densification forming the gravitational anomaly.

The mass spectrometer that analysed the gaseous environment at orbital altitude identified mostly gases derived from the spacecraft's attitude control thrusters.

Overall the area surveyed by the three SIM missions covered barely 20 per cent of the surface, so, despite Lunar Orbiter and Apollo, vast tracts near the lunar poles and on the Farside remained essentially *terra incognita*.

FOR A FEW DOLLARS MORE

Developing the technology to address Kennedy's challenge of "landing a Man on the Moon" was costly, but a large stock of Saturn V launch vehicles had been ordered as a precaution against early problems. In fact, the pace at which the goal was achieved in that incredible year of 1969

The Apollo missions' surface statistics

Mission	Class	Traverse (km)	Experiments (kg)	EVA Time (hr)	Lunar Samples (kg)
Apollo 11	'G'	0.25	102	2:24	21
Apollo 12	'H'	2.0	166	7:29	34
Apollo 14	'H'	3.3	209	9:23	43
Apollo 15	'J'	27.9	550	18:33	77
Apollo 16	'J'	27.0	563	20:12	94
Apollo 17	'J'	35.0	514	22:05	110
TOTAL		95.45	2104	80	379

enabled NASA to increase the capacity of its transportation system to fly advanced missions which were true scientific expeditions.

The three 'J'-class missions contributed 77 per cent of the equipment landed, 76 per cent of the surface EVA time, 74 per cent of the samples returned, and, with the benefit of Rovers, 94 per cent of the traverse distance. If Apollos 18 and 19 had matched Apollo 17, the ratios would have increased to 87 per cent of the experiments, surface time and samples and 96 per cent of the traverses. To have flown these missions would have added significantly to the scientific results for a trivial additional cost – after all, the rockets and the spacecraft had already been built. These magnificent vehicles are now adorning the parking lots of NASA facilities and languishing in museums.

STARTING OVER

Looking back, what can we say about Apollo? First, it achieved its specified political objective, set in 1961 by John F. Kennedy, of landing a man on the Moon before the decade was out; and also its unstated objective of beating the Soviet Union to the job. The transportation system was then devoted to exploration. It was admirably suited to expeditionary reconnaissance into *terra incognita*. But 'penny pinching' curtailed the program just as it became efficient, which had an effect on the potential scientific yield out of proportion to the number of flights. In 1968, shortly before Apollo's first manned test in Earth orbit, President Lyndon B. Johnson opined that America would "piss it all away" – and so it did. Just as Apollo had come out of nowhere and held centre-stage for a decade, it vanished from the public consciousness virtually without comment, as if it had never happened – indeed, almost as if it were an anachronism. In retrospect, it is apparent that the Apollo lunar missions were an element of twenty-first-century exploration which was somehow drawn forward in time.

So here we are at the start of the twenty-first century and NASA is slowly gearing up to resume the human exploration of the Moon, and this time to stay. Perhaps the first landing will be scheduled for 20 July 2019, precisely half a century after Neil Armstrong made the historic announcement that the Eagle had landed. But what will the next human say in preparing to step off the footpad onto the dusty regolith?

Spaceflight log

Craft	Launch	Mission
Luna 1	2 Jan. 1959	Intended impact, but made first lunar fly-by
Luna 2	12 Sep. 1959	First lunar impact
Luna 3	4 Oct. 1959	First circumlunar orbit, photographed the Farside
Vostok 1	12 Apr. 1961	First human spaceflight
Mercury	5 May 1961	First American suborbital flight
Ranger 1	23 Aug. 1961	Engineering test
Ranger 2	18 Nov. 1961	Engineering test
Ranger 3	26 Jan. 1962	Intended lunar impact, but missed
Mercury	20 Feb. 1962	America's first orbital mission
Ranger 4	23 Apr. 1962	Hit the Moon, but in a crippled state
Ranger 5	18 Oct. 1962	Intended lunar impact, but missed
Luna 4	2 Apr. 1963	Intended hard-landing, but made fly-by
Ranger 6	30 Jan. 1964	Lunar impact, but spacecraft malfunctioned
Zond 1	2 Apr. 1964	Intended Venus fly-by, but contact lost
Ranger 7	28 Jul. 1964	Lunar impact, first completely successful mission
Voskhod 1	12 Oct. 1964	3-man spectacular
Zond 2	30 Nov. 1964	Intended Mars fly-by, but contact lost
Ranger 8	17 Feb. 1965	Lunar impact, successful operation
Ranger 9	21 Mar. 1965	Lunar impact, successful operation
Voskhod 2	18 Mar. 1965	First spacewalk
Gemini 3	23 Mar. 1965	Testing the Gemini spacecraft
Luna 5	9 May 1965	Intended hard-landing, but crashed
Gemini 4	3 Jun. 1965	Spacewalk
Luna 6	8 Jun. 1965	Intended hard-landing, but made fly-by
Zond 3	18 Jul. 1965	Circumlunar orbit
Gemini 5	21 Aug. 1965	8-day test
Luna 7	4 Oct. 1965	Intended hard-landing, but crashed
Luna 8	3 Dec. 1965	Intended hard-landing, but crashed
Gemini 7	4 Dec. 1965	14-day test
Gemini 6	15 Dec. 1965	Rendezvous with Gemini 7
Luna 9	31 Jan. 1966	Successful hard-landing, first pictures from the surface
Gemini 8	16 Mar. 1966	First docking with Agena target
Luna 10	31 Mar. 1966	First to enter lunar orbit
Surveyor 1	30 May 1966	First soft-landing
Gemini 9	3 Jun. 1966	Spacewalk trial
Gemini 10	18 Jul. 1966	Spacewalk trial
Orbiter 1	10 Aug. 1966	Entered lunar orbit to survey landing sites
Luna 11	24 Aug. 1966	Entered lunar orbit
Gemini 11	12 Sep. 1966	Spacewalk trial
Surveyor 2	20 Sep. 1966	Intended soft-landing, but crashed
Luna 12	22 Oct. 1966	Entered lunar orbit
Orbiter 2	6 Nov. 1966	Entered lunar orbit to survey landing sites
Gemini 12	11 Nov. 1966	Spacewalk marathon
Luna 13	21 Dec. 1966	Successful hard-landing
Orbiter 3	5 Feb. 1967	Entered lunar orbit to survey landing sites
Surveyor 3	17 Apr. 1967	Soft-landed
Orbiter 4	4 May 1967	Entered lunar orbit for mapping
Surveyor 4	14 Jul. 1967	Intended soft-landing, but crashed
Explorer 35	19 Jul. 1967	Entered lunar orbit as interplanetary monitoring platform

Craft	Launch	Mission
Orbiter 5	1 Aug. 1967	Entered lunar orbit for mapping
Surveyor 5	8 Sep. 1967	Soft-landed
Surveyor 6	7 Nov. 1967	Soft-landed
Apollo 4	9 Nov. 1967	Test to crew-rate the Saturn 5 and Apollo CSM
Surveyor 7	7 Jan. 1968	Soft-landed
Apollo 5	22 Jan. 1968	Automated test of Apollo LM in Earth orbit
Zond 4	2 Mar. 1968	Test to crew-rate a spacecraft for circumlunar mission
Apollo 6	4 Apr. 1968	Test to crew-rate the Saturn 5 and Apollo CSM
Luna 14	7 Apr. 1968	Entered lunar orbit
Zond 5	14 Sep. 1968	Circumlunar test
Apollo 7	11 Oct. 1968	Test of Apollo CSM in Earth orbit
Zond 6	10 Nov. 1968	Circumlunar test
Apollo 8	21 Dec. 1968	First crewed spacecraft to enter lunar orbit
Apollo 9	3 Mar. 1969	Test of Apollo CSM/LM in Earth orbit
Apollo 10	18 May 1969	Full dress rehearsal of a lunar landing
Luna 15	13 Jul. 1969	Entered lunar orbit and then crashed attempting to land
Apollo 11	16 Jul. 1969	First crewed spacecraft to land on the Moon
Zond 7	8 Aug. 1969	Circumlunar test
Apollo 12	14 Nov. 1969	Lunar landing
Apollo 13	11 Apr. 1970	Intended lunar landing, but made circumlunar abort
Luna 16	12 Sep. 1970	First lunar sample-return
Zond 8	20 Oct. 1970	Circumlunar test
Luna 17	10 Nov. 1970	First lunokhod
Apollo 14	31 Jan. 1971	Lunar landing
Apollo 15	26 Jul. 1971	Lunar landing
Luna 18	2 Sep. 1971	Intended sample-return, crashed
Luna 19	28 Sep. 1971	Entered lunar orbit to conduct mapping
Luna 20	14 Feb. 1972	Sample-return, damaged during a rough landing
Apollo 16	16 Apr. 1972	Lunar landing
Apollo 17	7 Dec. 1972	Lunar landing
Luna 21	8 Jan. 1973	Second lunokhod
Explorer 49	10 Jun. 1973	Entered lunar orbit for radio astronomy
Explorer 50	26 Oct. 1973	Entered lunar orbit as interplanetary monitoring platform
Luna 22	29 May 1974	Entered lunar orbit to conduct mapping
Luna 23	28 Oct. 1974	Sample-return
Luna 24	9 Aug. 1976	Sample-return
Galileo	18 Oct. 1989	An odyssey through the Solar System to Jupiter
Clementine	25 Jan. 1994	Entered polar lunar orbit for multispectral mapping
Prospector	6 Jan. 1998	Entered polar lunar orbit for mineralogical survey

Note: This is a *selected* list of space missions. Although the focus is exploration of the Moon, some milestone missions have been included and other spacecraft which were lost in launch accidents or were deemed total failures have been omitted.

Lunar landing sites

Craft	Latitude		Longitude	
Luna 2	30	N	1	W
Ranger 4	15.5	S	130.7	W
Ranger 6	0.2	N	21.5	E
Ranger 7	10.7	S	20.7	W
Ranger 8	2.6	N	24.8	E
Ranger 9	12.9	S	2.4	W
Luna 5	31	S	8	E
Luna 7	9	N	40	W
Luna 8	9.1	N	63.3	W
Luna 9	7.1	N	64.4	W
Surveyor 1	2.5	S	43.2	W
Surveyor 2	5.5	N	12	W
Luna 13	18.9	N	62	W
Surveyor 3	3.0	S	23.3	W
Surveyor 4	0.4	N	1.3	W
Surveyor 5	1.4	N	23.2	E
Surveyor 6	0.5	N	1.4	W
Surveyor 7	40.9	S	11.4	W
Luna 15	17	N	60	E
Apollo 11	1.1	N	23.8	E
Apollo 12	3.2	N	23.4	W
Luna 16	0.7	S	56.3	E
Luna 17	38.3	N	35	W
Apollo 14	3.7	S	17.5	W
Apollo 15	26.1	N	3.6	E
Luna 18	3.6	N	56.5	E
Luna 20	3.5	N	56.5	E
Apollo 16	8.6	S	15.5	E
Apollo 17	20.2	N	30.8	E
Luna 21	25.8	N	30.5	E
Luna 23	12.8	N	62.2	E
Luna 24	12.8	N	62.2	E

Apollo missions, facts and figures

Apollo 7	CDR:	Wally Schirra
	LMP:	Walt Cunningham
	CMP:	Donn Eisele
	Launched:	11 October 1968
	Returned:	22 October 1968
	Duration:	10 days 20.2 hours
	Mission:	The first test of the Apollo CSM in Earth orbit; the LM was not flown
Apollo 8	CDR:	Frank Borman
	LMP:	Bill Anders
	CMP:	Jim Lovell
	Launched:	21 December 1968
	Returned:	27 December 1968
	Duration:	6 days 3.0 hours
	Lunar orbit:	20.2 hours
	Mission:	Flying the Apollo CSM in lunar orbit; the LM was not flown
Apollo 9	CDR:	Jim McDivitt
	LMP:	Rusty Schweickart
	CMP:	Dave Scott
	Launched:	2 March 1969
	Returned:	13 March 1969
	Duration:	10 days 1.0 hours
	Mission:	Test of the complete Apollo spacecraft in Earth orbit
	EVA:	Schweickart made a 38-minute EVA to test the lunar spacesuit's PLSS
	LM:	Spider
	CSM:	Gumdrop
Apollo 10	CDR:	Tom Stafford
	LMP:	Gene Cernan
	CMP:	John Young
	Launched:	18 May 1969
	Returned:	26 May 1969
	Duration:	8 days 0 hours
	Lunar orbit:	2 days 13.7 hours
	Mission:	Test of the complete Apollo spacecraft in lunar orbit
	LM:	Snoopy
	CSM:	Charlie Brown
Apollo 11	CDR:	Neil Armstrong
	LMP:	Buzz Aldrin
	CMP:	Mike Collins
	Launched:	16 July 1969
	Returned:	24 July 1969
	Duration:	8 days 3.3 hours
	Landed on:	20 July 1969
	On surface:	21.6 hours
	Lunar orbit:	2 days 11.6 hours
	Lunar EVA:	2.5 hours
	Samples:	22 kg
	Mission:	First lunar landing achieved at Tranquility Base

	LM:	Eagle
	CSM:	Columbia

Apollo 12	CDR:	Pete Conrad
	LMP:	Al Bean
	CMP:	Dick Gordon
	Launched:	14 November 1969
	Returned:	24 November 1969
	Duration:	10 days 4.6 hours
	Landed:	19 November 1969
	On surface:	1 day 7.5 hours
	Lunar orbit:	3 days 17.0 hours
	Lunar EVA:	7.7 hours (3.9 + 3.8)
	Samples:	34.4 kg
	Mission:	Precision landing alongside Surveyor 3
	LM:	Intrepid
	CSM:	Yankee Clipper

Apollo 13	CDR:	Jim Lovell
	LMP:	Fred Haise
	CMP:	Jack Swigert
	Launched:	11 April 1970
	Returned:	17 April 1970
	Duration:	5 days 22.9 hours
	Mission:	Mission aborted
	LM:	Aquarius
	CSM:	Odyssey

Apollo 14	CDR:	Al Shepard
	LMP:	Ed Mitchell
	CMP:	Stu Roosa
	Launched:	31 January 1971
	Returned:	9 February 1971
	Duration:	9 days 0 hours
	Landed on:	5 February 1971
	On surface:	1 day 9.5 hours
	Lunar orbit:	2 days 18.7 hours
	Lunar EVA:	9.2 hours (4.7 + 4.5)
	Samples:	43.0 kg
	Mission:	Landing in Fra Mauro
	LM:	Antares
	CSM:	Kitty Hawk

Apollo 15	CDR:	Dave Scott
	LMP:	Jim Irwin
	CMP:	Al Worden
	Launched:	26 July 1971
	Returned:	7 August 1971
	Duration:	12 days 7.2 hours
	Landed on:	30 July 1971
	On surface:	2 days 18.9 hours
	Lunar orbit:	6 days 1.3 hours
	Lunar EVA:	19.1 hours (0.5 + 6.6 + 7.2 + 4.8)
	Samples:	76.8 kg
	Mission:	Landing at Hadley-Apennine

	EVA:	0.6 hour by Worden to retrieve SIM film
	LM:	Falcon
	CSM:	Endeavour
Apollo 16	CDR:	John Young
	LMP:	Charlie Duke
	CMP:	Ken Mattingly
	Launched:	16 April 1972
	Returned:	27 April 1972
	Duration:	11 days 1.9 hours
	Landed on:	20 April 1972
	On surface:	2 days 23.0 hours
	Lunar orbit:	5 days 5.9 hours
	Lunar EVA:	20.3 hours (7.2 + 7.4 + 5.7)
	Samples:	94.7 kg
	Mission:	Landing at Descartes-Cayley
	EVA:	1.4 hours by Mattingly to retrieve SIM film
	LM:	Orion
	CSM:	Casper
Apollo 17	CDR:	Gene Cernan
	LMP:	Jack Schmitt
	CMP:	Ron Evans
	Launched:	7 December 1972
	Returned:	19 December 1972
	Duration:	12 days 13.8 hours
	Landed on:	11 December 1972
	On surface:	3 days 3.0 hours
	Lunar orbit:	6 days 3.8 hours
	Lunar EVA:	22.1 hours (7.2 + 7.6 + 7.3)
	Samples:	110.5 kg
	Mission:	Landing at Taurus-Littrow
	EVA:	1.1 hours by Evans to retrieve SIM film
	LM:	Challenger
	CSM:	America

Apollo lunar samples

CHAPTER 3: Apollo 12

1. Sample 12007 (unbagged) 0.06 kg
2. Samples 12010 and 12015 (unbagged)
3. Sample 12016 (unbagged) 2 kg
4. Samples 12006 and 12020 (unbagged) 0.20 and 0.30 kg
5. Sample 12030 (bag 1D)
6. (bag 3D)
7. (bag 5D)
8. Sample 12055 (unbagged)
9. Samples 12035, 12052, 12055 and 12063 (unbagged)
10. Sample 12063
11. Sample 12064 (unbagged) 1.2 kg
12. Samples 12045 to 12047 (bag 15D) 0.40 kg in total

CHAPTER 4: Apollo 14

1. Samples 14041 to 14046 (bag 3N)
2. Samples 14047 and 14048 (bag 5N)
3. Samples 14049 to 14050 (bag 6N)
4. Samples 14051 and 14052 (bag 7N)
5. (bag 12)
6. (bag 10)
7. Samples 14082 and 14084 (bag 13N) 0.25 kg
8. Sample 14321 (unbagged) 9 kg
9. Samples 14053 and 14054 (bag 14N)
10. Samples 14055 and 14062 (bag 15N)
11. Samples 14066 and 14067 (bag 17N)
12. Sample 14310 (unbagged) 3.4 kg
13. Sample 14306 (bag 26N) 0.5 kg
14. Sample 14313 (bag 27N)
15 … 1.5 kg
16. Sample 14310
17. In sample 14053

CHAPTER 5: Apollo 15

1. Sample 15065 (bag 156) 1.58 kg
2. Sample 15075 (bag 157) 0.81 kg
3. Sample 15076 (bag 157) 0.40 kg
4. Sample 15085 (bag 158) 0.47 kg
5. Sample 15086 (bag 158) 0.22 kg
6. Sample 15205 (bag 161) 0.34 kg
7. Sample 15206 (bag 160) 0.09 kg
8. Sample 15016 (unbagged) 0.92 kg
9. Sample 15245 (bag 163) 0.12 kg
10. Sample 15295 (bag 188) 0.95 kg
11. Sample 15298 (unbagged) 1.73 kg
12. Sample 15299 (unbagged) 1.70 kg
13. Sample 15255 (bag 190) 0.24 kg

14. Sample 15256 (bag 190) 0.20 kg
15. Sample 15265 (bag 193) 0.31 kg
16. Sample 15266 (bag 193) 0.27 kg
17. Sample 15400 (bag 168)
18. Sample 15405 (bag 168) 0.51 kg
19. Sample 15418 (bag 194) 1.14 kg
20. Sample 15419 (bag 194) 0.02 kg
21. Sample 15425 (bag 195) 0.13 kg
22. Sample 15426 (bag 195) 0.22 kg
23. Sample 15415 (bag 196) 0.27 kg
24. Sample 15435 (bag 170) 0.40 kg
25. Sample 15455 (bag 198) 0.94 kg
26. Sample 15445 (bag 171) 0.29 kg
27. Sample 15310 (bag 172)
28. Sample 15459 (unbagged) 4.83 kg
29. Sample 15495 (bag 174) 0.91 kg
30. Sample 15475 (bag 203) 0.41 kg
31. Sample 15476 (bag 203) 0.27 kg
32. Sample 15499 (unbagged) 2.0 kg
33. Sample 15485 (bag 204) 0.10 kg
34. Sample 15486 (bag 204) 0.05 kg
35. Sample 15058 (unbagged) 2.67 kg
36. Sample 15059 (unbagged) 1.15 kg
37. Sample 15515 (bag 273) 0.14 kg
38. Sample 15505 (bag 255) 1.15 kg
39. Sample 15529 (bag 274) 1.53 kg
40. Sample 15535 (unbagged) 0.40 kg
41. Sample 15545 (bag 278) 0.75 kg
42. Sample 15595 (bag 281) 0.24 kg
43. Sample 15596 (bag 281) 0.23 kg
44. Sample 15597 (bag 281) 0.15 kg
45. Sample 15598 (bag 281) 0.14 kg
46. Sample 15555 (unbagged) 9.61 kg
47. Sample 15382 (rake) 3.2 grams
48. Sample 15386 (rake) 7.5 grams

CHAPTER 6: Apollo 16

1. Sample 60095 (bag 4), 0.05 kg
2. Sample 60035 (bag 351) 1.05 kg
3. Sample 60075 (bag 373) 0.18 kg
4. Sample 61155 (bag 371) 0.05 kg
5. Sample 61156 (bag 371) 0.06 kg
6. Sample 61175 (bag 364) 0.54 kg
7. Sample 61195 (bag 2) 0.59 kg
8. Sample 61135 (bag 362) 0.25 kg
9. Sample 61295 (bag 353) 0.18 kg
10. Sample 61015 (unbagged) 1.80 kg
11. Sample 61016 (unbagged) 11.73 kg
12. Sample 62235 (bag 5) 0.32 kg
13. Sample 62236 (bag 5) 0.06 kg
14. Sample 62255 (bag 7) 1.20 kg

15. Sample 62275 (bag 9) 0.44 kg
16. Sample 62295 (bag 10) 0.25 kg
17. Sample 60015 (unbagged) 5.6 kg
18. Sample 60025 (unbagged) 1.84 kg
19. Sample 64435 (bag 394) 1.08 kg
20. Sample 64455 (bag 397) 0.06 kg
21. Sample 64475 (bag 398) 1.03 kg
22. Sample 65075 (bag 403) 0.11 kg
23. Sample 65035 (bag 404) 0.45 kg
24. Sample 65315 (bag 405) 0.30 kg
25. Sample 65095 (bag 336) 0.56 kg
26. Sample 65055 (bag 337) 0.50 kg
27. Sample 65056 (bag 337) 0.08 kg
28. Sample 65015 (unbagged) 1.8 kg
29. Sample 66035 (bag 407) 0.21 kg
30. Sample 66055 (bag 408) 1.31 kg
31. Sample 66075 (bag 409) 0.35 kg
32. Sample 66095 (bag 410) 1.18 kg
33. Sample 68035 (bag 413) 0.02 kg
34. Sample 68115 (bag 340) 1.19 kg
35. Sample 68416 (bag 341) 0.18 kg
36. Sample 68815 (unbagged) 1.83 kg
37. Sample 69935 (bag 378) 0.13 kg
38. Sample 69955 (bag 380) 0.08 kg
39. Sample 60115 (bag 381) 0.14 kg
40. Sample 67035 (bag 373) 0.25 kg
41. Sample 67055 (bag 383) 0.22 kg
42. Sample 67015 (unbagged) 1.2 kg
43. Sample 67075 (bag 384) 0.22 kg
44. Sample 67095 (bag 385) 0.34 kg
45. Sample 67115 (bag 386) 0.24 kg
46. Sample 67435 (bag 415) 0.35 kg
47. Sample 67455 (bag 416) 0.94 kg
48. Sample 67475 (bag 418) 0.18 kg
49. Sample 67415 (bag 387) 0.18 kg
50. Sample 67935 (bag 389) 0.23 kg
51. Sample 67915 (unbagged) 2.5 kg
52. Sample 67975 (bag 393) 0.45 kg
53. Sample 63335 (bag 428) 0.07 kg
54. Sample 63355 (bag 429) 0.07 kg
55. Sample 60017 (unbagged) 2.1 kg

Chapter 7: Apollo 17

1. - (unbagged) put aside, but not returned to Earth
2. Sample 70135 (bag 10E) 0.45 kg
3. Sample 71035 (bag 476) 0.15 kg
4. Sample 71055 (bag 454) 0.67 kg
5. Sample 71135 (bag 477) 0.04 kg
6. Sample 71155 (bag 478) 0.03 kg
7. Sample 71175 (bag 479) 0.21 kg
8. Sample 71520-97 (bags 457 & 458) rake, in two bags
9. Sample 70035 (unbagged) 5.77 kg

10. Sample 72130-35 (bag 26E)
11. Sample 72155 (bag 28E) 0.24 kg
12. Sample 72160-64 (bag 29E)
13. Sample 72255 (bag 494) 0.46 kg, Sample 72215 (bag 514) 0.38 kg, Sample 52235 (bag 515) 0.06 kg
14. Sample 72275 (bag 495) 3.64 kg
15. Sample 72315 (bag 516) 0.13 kg
16. Sample 72335 (bag 517) 0.11 kg
17. Sample 72355 (bag 518) 0.37 kg
18. Sample 72375 (bag 519) 0.02 kg
19. Sample 72395 (bag 499) 0.54 kg
20. Sample 72530-59 (bag 501)
21. Sample 72415 (bag 503) 0.03 kg
22. Sample 72435 (bag 504) 0.16 kg
23. Sample 74115-19 (bag 41Y)
24. Sample 74220 (bag 509)
25. Sample 74240-49,85-87 (bag 510)
26. Sample 74260 (bag 511)
27. Sample 74255 (bag 512) 0.74 kg
28. Sample 74275 (bag 461) 1.49 kg
29. (bag 43Y)
30. Sample 75120-24 (bag 44Y)
31. Sample 75015 (bag 462) 1.00 kg, Sample 75035 (bag 463) 1.24 kg
32. Sample 75055 (bag 464) 0.95 kg
33. Sample 75060-66 (bag 465)
34. Sample 75080-89 (bag 467)
35. Sample 75075 (bag 466) 1.00 kg
36. Sample 70295 (bag 45Y) 0.36 kg
37. Sample 76130-37 (bag 47Y)
38. Sample 76220-24 (bag 534)
39. Sample 76240-46 (bag 312)
40. Sample 76015 (unbagged) 2.82 kg
41. Sample 76215 (bag 535) 0.65 kg
42. Sample 76235 (bag 556) 0.03 kg
43. Sample 76255 (bag 536) 0.41 kg
44. Sample 76275 (bag 537) 0.06 kg
45. Sample 76295 (bag 538) 0.26 kg
46. Sample 76320-24 (bag 557)
47. Sample 76315 (bag 539) 0.67 kg
48. Sample 76530,35-77 (bag 558)
49. Sample 76055 (unbagged) 6.41 kg
50. Sample 76335 (bag 560) 0.36 kg
51. Sample 76035 (bag 48Y) 0.38 kg
52. Sample 77510-19,25-26 (bag 540), Sample 77530-45 (bag 542)
53. Sample 77017 (bag 541) 1.73 kg
54. Sample 77215 (bag 543) 0.85 kg
55. Sample 77075 (bag 544) 0.17 kg
56. Sample 77115 (bag 561) 0.12 kg
57. Sample 77135 (bag 562) 0.34 kg
58. Sample 77035 (unbagged) 5.73 kg
59. Sample 78120-24 (bag 50Y)
60. Sample 78135 (bag 563) 0.13 kg

61. Sample 78220-24 (bag 545)
62. Sample 78235 (bag 564) 0.20 kg
63. Sample 78255 (bag 546) 0.05 kg
64. Sample 78530, 35-99 (bag 565)
65. Sample 78155 (bag 567) 0.40 kg
66. Sample 79135 (bag 480) 2.28 kg
67. Sample 79155 (bag 571) 0.32 kg
68. Sample 79175 (bag 481) 0.68 kg
69. Sample 79190 (bag 482) 0.37 kg
70. (bag 52Y) not returned to Earth
71. Sample 79220-28 (bag 483)
    Sample 79240-45 (bag 484)
    Sample 79260-65 (bag 485)
72. Sample 79215 (bag 486) 0.55 kg
73. Sample 70315 (bag 54Y) 0.15 kg
74. Sample 70215 (unbagged) 8.11 kg

Glossary

accretion The progressive assembly of a body by the incorporation by impact of smaller bodies. This is believed to be how the planets formed out of the Solar Nebula.

acid A substance which liberates hydrogen ions in solution. An acid reacts with a base to make a salt and water (only). A substance with acid properties (opposite to an alkali; a base) is said to be acidic.

age, absolute A tricky concept. Depending upon the context, it can mean the time since a rock unit formed; the time since a specific rock solidified from a molten state; the time that a rock has been exposed to cosmic rays; ....

agglutinate An impact-melt comprising a mixture of glass, mineral and rock fragments.

aggregate A rock that is a consolidation of fragments of other rocks.

albedo The reflectivity (expressed as a percentage) of a material; a dark material has a low value. On average, the Moon reflects 7 per cent of incident light. The darkest asteroids reflect only 1–3 per cent, so they are as dark as coal.

alkali A soluble hydroxide of a metal. A substance with alkali properties (opposite to those of an acid) is said to be alkaline. The light univalent metals (lithium, sodium and potassium) are called alkali metals; the bivalent metals (beryllium, magnesium and calcium) are alkali earth metals.

alumina Aluminium oxide (Al_2O_3)

aluminous melt group Impact-melt breccias containing a lot of aluminium.

amorphous Having no internal structural order; that is, glassy.

andesite A very fine-grained volcanic rock made largely of plagioclase feldspar with pyroxene as the minor constituent; it is the extrusive equivalent of diorite.

annealing The recrystallisation of minerals in the solid state after being melted.

anorthosite A rock comprising more than 90 per cent plagioclase feldspar. It is a slow-cooled plutonic rock with a conspicuous crystalline structure ('twinning'). When Gene Shoemaker studied the Surveyor 7 data from Tycho, he found that it matched anorthositic gabbro. The Apollo 11 regolith had tiny fragments which consisted of more than 70 per cent plagioclase (anorthositic gabbro). Apollo 15 was the first to sample anorthosite. It is the initial lunar crust that formed as the magma ocean cooled, and (being alkali poor) is calcic plagioclase feldspar. All the anorthositic samples from various lunar landing sites proved to be strikingly homogeneous. Anorthosites are usually referred to as ferroan after the iron-rich compositions of their olivine and pyroxene minority constituents, as opposed to the more magnesium-rich varieties of these mafic minerals.

ANT Anorthosite, Norite and Troctolite – the aluminous suite of endogenic igneous rock types in the ancient lunar crust; anorthosite is by far the predominant member of the group.

apatite A phosphate mineral.

Apennine Bench Formation The light-toned plain in the Imbrium Basin, in the gap between the Apennine Range and the Archimedes Plateau. It lined the basin floor almost immediately after the basin's formation, so it was initially believed to be impact-melt, but when sampled it proved to be a lava extrusion. Although it is within the basin, underlying the Mare Imbrium, it is not a mare-type lava, it is light because it is alumina-rich (or, rather, it is *not* enriched with dark mafic minerals). It is also KREEPy. In fact, it is the best example of a KREEPy basalt flow.

Apennine Range An arc of mountains forming the southeastern rim of the Imbrium Basin. They are a fragment of what once was a full ring. It peaks steeply on the inner side of the arc, but slopes much more gently on the outer side. The chain incorporates many radial fractures. The entire structure was produced essentially instantaneously by the giant impact that excavated the basin, and the mountains are, in effect, massive blocks of crust that were uplifted.

aphanitic A hard rock possessing a homogeneous ground-mass with a texture that is so fine that its crystals are not resolvable to the naked eye; very fine-grained.

Apollo Sample Number On arrival at the Lunar Receiving Laboratory, the lunar samples (both individual rocks and collections of fines) were assigned numbers. Unfortunately, the algorithm for giving the numbers varied from one mission to the next. When a sample was divided, its various parts were assigned subsidiary identifiers, which were appended to the primary identifier behind a comma.

armalcolite An mineral rich in titanium oxide that was found in the mare basalts. It is $(Fe,Mg)Ti_2O_5$. As it was first found in the Apollo 11 samples, it was named by combining the letters of the surnames of the crew (i.e. Armstrong Aldrin and Collins) with the descriptor 'ite'. It was subsequently identified on the Earth in large impact craters such as the Ries in Germany.

ash Tiny fragments of lava sprayed out of a volcanic vent. Because it has a high surface area to mass, it cools rapidly, falls as dust, and settles as a fine-grained blanket. On a slope, it will flow like a fluid.

'*Backroom*' The informal name for the Apollo Lunar Surface Science Support Room (in fact, a suite of rooms) at Mission Control in Houston.

basalt An aphanitic lava; the extruded form of gabbro. Lunar basalt was extruded at 1,400°C and had an extremely low viscosity. Its albedo is related to its chemical composition. The lighter variety is plagioclase feldspar enriched by mafic minerals (typically either pyroxene or olivine, which are mutually incompatible) in roughly equal proportions, but the darker type that flooded the basins is only about 30 per cent plagioclase feldspar. The very darkest lava is also enriched with titanium. Because the basin in-fill is called mare basalt, the lighter material is called non-mare basalt. Most of the mare basalt was extruded over the interval of 500 million years subsequent to the Great Bombardment. The non-mare basalt is believed to have more or less coincided with this bombardment. The non-mare basalt is more finely grained than the mare basalt and is virtually free of cavities, whereas the mare type is highly vesicular and often vuggy.

base A substance that reacts with acid to produce a salt and water (only). A substance with this property is said to be basic.

basin A massive depression left by the impact of an asteroid-sized body during the period of the Great Bombardment. By definition it is an impact crater wider than about 300 km in diameter, so large that it does not have a coherent wall; a ring of mountains (in many cases reduced to disjoint arcs) define the rim. The largest basins have concentric ranges. Although many basins have been flooded by basalt, the basin and the in-fill are distinct stratigraphic structures (this was not evident for many years; they were believed to be aspects of the same event). By a curious coincidence, therefore, both the deepest holes and the tallest mountains were made together and almost instantaneously. The Aitken Basin, just beyond the limb at the South Pole, is the largest basin. It is 2,500 km in diameter, and its peripheral mountains rise 9 km above the surrounding terrain. It has not been filled by mare material, however. The Orientale Basin (despite its name, by a quirk of convention it lies on the western limb) is the freshest basin.

bedrock The intact layer of rock immediately below the regolith. It is a regional geological unit in the overall stratigraphic structure.

bench An inflection on a slope (such as a mountain flank or a crater wall) which creates a relatively flat surface.

Binary Accretion Model A theory which accepts that the Earth and Moon formed together, by accretion, but claims they did so independently.

bombardment Given the formation of the lunar basins by a series of massive impacts soon after the Moon formed, it was argued that the early Solar System was such a dynamic environment that all rocky bodies were likewise battered by asteroidal (or planetesimal) impacts – this being called the Great Bombardment. Its timescale is not clear, however. Nor is it clear whether it was random or episodic. The theory of a cataclysm argued that all of the basins and most of the large craters formed during a single epoch, about 3.8–3.9 billion years ago. Others have argued that this marked simply the terminal phase of a prolonged bombardment.

boulder In terms of lunar field geology, a 'boulder' was defined as a rock that had at least one dimension in excess of a metre.

breccia A mechanically assembled rock. A regolith breccia is a clod of soil made by the shock of an impact compressing the regolith. A fragmental breccia is made up of angular fragments of many other shattered rocks, but with little or no melt. An impact-melt breccia is shattered rock welded together by impact melt. A granulitic breccia is a melt which has been recrystallised by high temperature and pressure, giving it a granulitic texture. A 'one-rock' breccia is a matrix in which the clasts are fragments of homogeneous material. If a breccia is shattered by a later impact and becomes bound up into another matrix, the result is a 'two-rock' breccia. And so on. In each generation, the consolidating matrix may be different and the clast mix may be varied, therefore complex rocks can result. Breccias are abundant on the Moon.

bulk sample A formal surface sampling procedure designed to collect all material within a given volume of the lunar regolith.

caldera If the roof of a magma chamber collapsed this can leave a large cavity in the surface.

Capture Model A theory that proposes the Earth and Moon were not only formed independently, but in different parts in the Solar System, and that the Moon was 'captured' as the result of an encounter with the Earth.

Cayley Formation The light-toned rolling plains in the highlands form a distinct morphological unit known as the Cayley Formation. It was initially interpreted as the result of volcanic processes filling the valleys, but Apollo 16 revealed it to be primarily fluidised brecciated material.

central peak The mountain complex in the centre of a crater where the material at the point of impact rebounded and solidified. As a result, this material is from far below the local bedrock.

cinder cone A mound of viscous lava or pyroclastic ash which builds up around a volcanic vent.

circular maria The maria within the lunar basins, as distinct from the mare lavas that flooded terrain outwith the basins.

clast An angular fragment caught up in a breccia matrix. Clasts can be rock (lithic) or individual minerals (non-lithic).

Collisional Ejection Model The theory which proposes that the Moon formed from material ejected from the Earth when a Mars-sized body struck it; once in orbit, the ejecta rapidly accreted to form the Moon. This theory has the advantage of explaining why the Moon's bulk density is similar to that of the Earth's mantle and crust.

complex crater A crater with a flat floor, a central peak and a terraced wall. Such craters are generally larger than about 25 km in diameter, and range up to near-basin size.

comprehensive sample A formal surface sampling procedure to collect a random selection of rocks within a given area in order to establish the characteristic types.

conglomerate A structural composite rock made of other rocks; it is similar to a breccia, but with rounded rather than angular fragments.

Copernican Era In terms of lunar stratigraphy, all features created since the Eratosthenian Era; in effect, the last 1 billion years.

core The central part of a planetary body that has undergone differentiation, with the iron-rich material sinking to form a metallic nucleus.

cosmic ray The rain of charged particles in space. The solar wind contributes ions of the lighter elements, but there is also a flux of relativistic nuclei of heavier elements from beyond the Solar System.

crater It was once thought that many of the circular depressions in the lunar surface were volcanic in origin, but it turned out that they are virtually all the result of impacts. In effect, there is a distribution of sizes running all the way from the smallest zap-pit up to the largest basin. A primary crater is the result of a high-energy impact by space debris – typically at a speed of 20 km/second. A secondary is caused by the fall of ejecta (which must be travelling at less than the lunar escape velocity of 2.4 km/second, so is of lower energy). Many craters up to 25 km across are actually secondaries from basin formation. Until recently, craters with dark rims (halos) were thought to be pyroclastic vents, but it turned out that they are impacts that excavated dark material from beneath a light-toned overburden.

crust The outer part of a planetary body that has undergone differentiation, with a low-density silicate-rich material forming a crystalline envelope.

crystal As a silicate-rich lava cools, its minerals precipitate from solution to create a homogeneous rock. The atoms link up to create a regular lattice, the shape of which depends on the elements involved. Certain elements can fit into a specific shape, and are called 'compatibles'; all others are called 'incompatibles'. There can be elements in common, but as soon as a crystal starts to take on a particular form the incompatibles are excluded. Because different crystals form at different temperatures, the magma chemistry evolves and becomes progressively enriched by a succession of different incompatibles. If gas is released into the solution and becomes trapped, the resulting rock will have cavities. If the rock cools rapidly, the cavities will be spherical and have smooth walls, but if it cools slowly some of the metals in the gas may condense to form crystals on the wall of the cavity, partially filling it and making its shape irregular. A spherical cavity is referred to as a vesicle and an irregular one as a vug. If the molten magma crystallises too rapidly for a lattice to form, the result is a homogenous mass of glass (obsidian is the glassy form of rhyolite, for example).

dark mantle Some patches of the lunar surface are considerably darker than others. Apollo 17 sampled one and found it to be pyroclastic. It had been hoped that the fire fountains that produced it were 'recent', but they were ancient and the dark glass had been thoroughly mixed into the regolith.

date Given an assumption of primordial abundance, the ratios of radioactive decay products can measure the age of a rock. Elements with half-lives of billions of years are accurate to about 100 million years.

Descartes Formation A patch of domical hills that are light in tone and display a furrowed facia. They are located south of Descartes Crater and west of the Kant Plateau, and form a distinct morphological unit. It was once interpreted as silicic volcanism, but turned out not to be so.

diabase rock This is an igneous rock in which the interstices between the feldspar crystals are filled with grains of pyroxene; dolerite.

diorite The intrusive form of andesite.

dome A volcanic mound; similar to a cone but asymmetrical and with an irregular slope.

dunite An igneous rock composed almost entirely of olivine.

ecliptic The plane in which the Earth/Moon system orbits around the Sun.

effusion The flow of lava from a volcanic vent.

ejecta The debris thrown out by the formation of an impact crater. If the material splashes out and mantles the immediate environs of the crater, then it is called the ejecta blanket.

Eratosthenian Era In terms of lunar stratigraphy, features created between the end of the Imbrian Era and the start of the Copernican Era; i.e. the interval from 3.2 billion years to 1 billion years ago.

extrusion Lava that has flooded onto the surface.

facia In terms of lunar surface morphology, any distinctive surface texture that can be used to delineate a rock unit on a geological map.

fault scarp A cliff formed by vertical crustal movement along a fault.

feldspar See: plagioclase feldspar

fillet A bank of dust that has collected alongside the base of a rock.

fire fountain The spray of fine droplets of molten rock that is explosively vented when a volatile-rich lava nears the surface.

Fission Model The model in which the Earth and Moon were once a single body. As it condensed from the Solar Nebula, the conservation of angular momentum required the body to rotate faster, this led first to the formation of an equatorial bulge and then, at a critical point, to that bulge splitting off to form a satellite.

Fra Mauro Formation Patches of hummocky terrain peripheral to the Imbrium Basin form a distinct morphological unit made of crustal rock that was displaced by that impact.

gabbro A general term for plutonic rocks made of plagioclase and clino-pyroxene that crystallised sufficiently slowly to form granular crystals; its extruded form is basalt.

gardening When a small meteorite strikes the lunar surface at high speed, it either vaporises or melts both itself and the target material. The material immediately beneath the point of contact is compressed sufficient to create regolith breccia. Below this, the shock merely serves to shatter coherent structure. Large impacts excavate bedrock which is then broken down into ever-smaller pieces by the relentless rain of meteoroids.

geological unit A stratigraphically distinct body of rock with a single source and time of deposition.

geophone A seismometer designed to serve as part of a network of sensors to investigate the structure of the subsurface.

glass If a high-viscosity melt solidifies into a state with no internal order, the result is glass rather than a crystal. Glass can also be made by an extreme pressure-pulse, such as occurs at the point of a meteoritic impact.

graben A linear trench formed when a strip of land drops down between a pair of faults opened by extensional forces.

grain The individual crystals in a rock, the texture of which varies from coarsely to finely grained.

granite A silicic gabbro with a granular texture; its extruded form is rhyolite.

granulitic rock A rock that recrystallised in its solid state (i.e. without remelting) by high temperature and pressure.

highlands On the Moon, the light rock that forms most of the southern hemisphere of the Nearside and almost all of the Farside.

igneous Of, or relating to, material formed from a liquid state.

ilmenite An iron–titanium oxide mineral; $FeTiO_3$. It is a major constituent of the mare basalts, and its presence explained why the maria are so dark. The fact that the proportions of this mineral were different at each mare site visited explained why the various maria had different albedos.

Imbrian Era The impact that made the Imbrium Basin 'sculpted' a wide variety of features on much of the Nearside, so when the stratigraphers set out to infer the relative ages of the lunar surface features they came to regard this impact as the "zero point" on the chronology, with other detail being classified as pre-Imbrian or post-Imbrian. The post-Imbrian was later divided into the Eratosthenian and Copernican

Eras. The Apollo samples dated the Imbrium impact at 3.84 billion years ago, and the Imbrian Era was defined as running from this event through to 3.2 billion years ago. The pre-Imbrian has been divided in order to accommodate the Nectarian Era.

Imbrium System In stratigraphic terms, the lunar surface features that were created during the Imbrian Era.

impact flux The number of impacts mapped as a function of time.

impact melt That portion of the target rock that is melted by the shockwave of an impact passing through it. It is the zone between the rock that is vaporised and the basement that is merely fractured. Although it coats the cavity of the crater, in the case of large impacts the melt can pool and leave a surface similar to a lava flow. It was once thought that the maria were vast sheets of impact melt, but this is not the case.

'instant rock' A casual term for regolith breccia, prompted by the fact that the rock is created "in an instant" by a high-energy impact shockwave compressing the regolith.

isostasy The forces which govern the level at which materials of different densities settle in the crust of a planetary body.

KREEP A material that is relatively enriched in potassium, phosphorus and rare earth elements. It is believed that KREEP accumulated in the deep mantle and was excavated by the double-whammy of the Procellarum and Imbrium impacts. It would have settled while the Moon's surface was a magma ocean, because it tends not to participate in crystallisation and would therefore be absent from the scum of low-density material that formed the initial crust. Radioactive elements would have helped to maintain this material molten. When it solidified as the final act of the fractionation of the magma ocean it formed a KREEPy gabbro. As 'KREEP' refers to the chemical composition of a rock, not to its minerals, the term is really an adjective. The scale of KREEPiness is compared to the amount in the Fra Mauro Formation; the levels being low, medium or high-KFM (i.e. 'KREEP at Fra Mauro'). Orbital remote-sensing established that the radioactive elements uranium, potassium and thorium were concentrated in Procellarum in precisely the areas that were most KREEPy. LKFM is an impact-melt that has a composition richer in iron and magnesium than typical for the anorthositic crust, and is reprocessed crustal material.

lath A mineral with a lenticular shape.

lava The extruded form of magma.

lava channel The landform in which a low-viscosity lava extrusion flows downhill from a volcanic vent. On Earth the rate is modest, the flow is laminar, and the lava exploits existing terrain. However, on the Moon the rate was much higher, the flow was turbulent, and the lava excavated the terrain to make a sinuous rille much larger than its terrestrial equivalent. The longest channel on the Nearside is Schroter's Valley on the Aristarchus Plateau, but Hadley Rille, which Apollo 15 visited, is only slightly smaller.

libration Any angular motion that fails to complete a full circle. Because the Moon's orbit around Earth is elliptical and its rotation is tidally locked, at certain times it is possible to 'peek' around the edge to view some Farside features; over time 59 per cent of the surface is visible from Earth.

limb The edge of the Moon's disk, viewed either from Earth or from a spacecraft.

lithosphere The solid outer shell of a differentiated planet, including the crust and the uppermost rigid layer of the mantle.

lobate scarp The small scalloped cliff that forms at the end of a lava flow or the collapsed overriding front of a thrust fault.

mafic A mineral enriched in iron and/or magnesium at the expense of silicon and aluminium. Being denser, these minerals (such as olivine and pyroxene) formed deep in the lunar magma ocean and subsequently rose through faults in the basin floors and were extruded as the mare basalts.

magma ocean As the Moon accreted, its surface was molten to a depth of several hundred kilometres. Over a period of several hundred million years, the buoyant plagioclase feldspar differentiated from the silicate-rich mantle and crystallised as the primitive crust.

magnesian suite A family of plutonic rocks that is distinguished from the ferroan anorthosites by a high fraction of magnesium-rich silicates. It includes norite, dunite and troctolite. Unlike the homogeneous anorthositic crust, the magnesian suite is chemically diverse, indicating that the mantle reservoirs had distinctive characteristics.

mantle That section of a planetary body that is sandwiched between the crust and the core. It is the main source of magma for volcanic activity.

maria The dark lavas that cover about 30 per cent of the Nearside of the Moon and 2 per cent of the Farside; 16 per cent overall. The absolute ages measured from samples 'fixed' the relative ages from stratigraphic analysis and revealed that the maria were still forming during the Eratosthenian Era, but not in the Copernican Era.

mascon Analysis of the Doppler variation of the signals from the Lunar Orbiters led to the discovery that the Moon's gravitational field is uneven. The variations in gravity were interpreted as mass concentrations (mascons) associated with the circular maria, notably Serenitatis, Nectaris, Crisium, Imbrium and Orientale. The older, partially obliterated basins, many of which are not lava-filled, do not show mascons. This distinction indicated that the lunar crust had evolved during the period of basin formation. Early on, when the crust was thin and pliable, this was able to react to isostatic forces resulting from the removal of crustal rock in the act of basin formation, with the basin floor rising to compensate. But later, as the mantle cooled, the crust thickened and was able to resist isostatic restoration. Consequently, the later basins did not adjust, and when dense silicate-rich lavas were extruded from fractures in their floors this created gravitational anomalies. Because mascons perturb low orbits, a spacecraft must manoeuvre to maintain a specific orbit.

massif An isolated massive block within a group of mountains. A lunar massif is a block of crust that was tilted and uplifted by the shock of a major impact.

matrix The host material (ground-mass) of a breccia that binds together the clasts.

maturation The gradual accumulation of agglutinates in the regolith, darkening it and suppressing the spectral features of its mineral composition.

megaregolith The brecciated portion of the Moon's anorthositic crust, extending to a depth of several tens of kilometres.

metamorphic rock A rock formed by reprocessing existing rock by the application of heat sufficient to induce mineralogical, structural and chemical changes but without actually remelting the minerals. On the Moon, the only reprocessed rock is impact-melt.

mineral The naturally occurring substances with crystalline structures that are the raw material of rocks. A mineral that contains water in its structure is said to be hydrated; these are common in the Earth's crust but conspicuously absent on the Moon.

month, sidereal The time that the Moon takes to make one orbit around the Earth; 27.3 days. Since the Moon's axial spin is 'tidally locked' to the Earth, this is also the time that the Moon takes to rotate on its axis.

month, synodic The time that the Moon takes to run a cycle of phases from 'full' to 'full'. At 29.5 days, this is longer than the axial rotation because the Earth/Moon system is travelling around the Sun, which requires the Moon to complete more than one orbit of the Earth to re-establish a specific alignment with respect to the Sun.

Nectarian Era That period of lunar history between the formation of the Nectaris Basin and the Imbrian Era; i.e. between 3.92 and 3.84 billion years ago. Events prior to this are referred to simply as pre-Nectarian.

norite A coarsely grained plutonic rock in which the plagioclase feldspar combines with pyroxene as the mafic component.

olivine A mafic silicate; $(Mg,Fe)_2SiO_4$. It is not as common as pyroxene in lunar basalt, and can be absent.

outcrop A naturally occurring exposure of bedrock.

plagioclase feldspar This group of minerals form 60 per cent of the outer 15 km of Earth's crust. In German, 'feldspar' means 'field crystal'. Plagioclase feldspar is a subset of the feldspar group; $(Ca,Na)(Al,Si)_4O_8$. Much terrestrial plagioclase is sodium-based, but the Moon is deficient in sodium and its plagioclase is calcic. Feldspathic laths make a light-coloured rock. Anorthosite is extremely pure plagioclase and is white; norite is plagioclase enriched by pyroxene; troctolite is plagioclase enriched by olivine.

plain On the Moon, a landscape that has been rendered more or less flat either by a volcanic extrusion or by a splash of ejecta.

plate tectonics This theory was devised in the late 1960s to explain the large-scale structure of Earth's crust in terms of lithospheric 'plates' that jostle one another.

pluton A body of molten rock that crystallises as it rises through the crust. Because it cools slowly without disruptive flow, if favours crystal-growth and produces a granular rock which is called gabbro.

plutonic rock A coarsely grained igneous rock that solidifies underground; strictly, at *great* depth, because a rock that solidifies just beneath the surface is referred to as hypabyssal.

phenocryst A conspicuous mineral crystal in a porphyritic rock.

porphyry An extremely finely grained basaltic matrix containing phenocrysts. To old-timers, a porphyritic rock is called porphrite.

pristine A petrological term for an igneous rock which solidified from endogenic magma.

province A collection of geographically distributed but morphologically related geological units.

pyroclastic Literally 'fire-shattered fragment'. A volatile-rich basaltic material that is explosively ejected from a volcanic vent. An ash flow can behave like a fluid, and flow down hill as a 'hot avalanche' (a nuée ardent) prior to coming to rest.

pyroxene A mafic silicate; $(Ca,Fe,Mg)_2Si_2O_6$. It is the most common mafic mineral in lunar basalt.

quartz A silica mineral. Although common in the Earth's crust, it is extremely scarce on the Moon.

radiogenic heating The heating within a planetary body (possibly to melting point) as the result of the decay of radioactive elements.

rare earth elements A group of elements that are rare in Earth's crust: lanthanum, cerium, neodymium, praseodymium, promethium, samarium, europium, gladolinium, terbium, erbium, thulium, dysprosium, holmium, ytterbium and lutetium.

ray The light-toned material that is deposited in a radial pattern around a newly formed crater. It is bright because it is finely pulverised rock and the fractured crystals are very reflective.

refractory elements Those elements that boil out of solution only at high temperatures. Being depleted in volatiles, the lunar crust is dominated by refractories such as aluminium, calcium, magnesium and titanium.

regolith The proper name for a 'soil' composed solely of rock fragments, forming an unconsolidated mass of debris on top of bedrock. As a result of the gardening process, the lunar regolith has a seriate distribution of fragment sizes.

rhyolite A light-coloured silica-rich viscous basalt; the extruded form of granite.

rock An aggregate of minerals. In terms of lunar field geology, a rock was defined as any unitary fragment having a mass exceeding 50 grams. In order to assist in communicating by radio, samples were described (in order of increasing size) as 'pebble-', 'walnut-', 'grapefruit-', 'football-', 'boulder-' and 'house-sized'. The term 'block' was used to describe any rock larger than fist-sized.

scoria A vesicular cindery dark lava that texturally resembles pumice but is darker and denser; such a lava is said to be scoriaceous.

sculpture Any surface feature caused by the fall of ejecta during the formation of a basin.

shatter cone A striated pattern imposed on a finely grained rock by intense shock. This characteristic was first observed in rock shattered by nuclear tests, then in lunar rock shattered by impact; only later was it seen in terrestrial landforms that proved to be impact craters.

shockwave A transient increase in pressure, characterised by the phrase 'megabars for microseconds'.

siderophile elements Literally, a group of 'iron loving' elements: sodium, zinc, germanium, arsenic, selenium, bromine, silver, cadmium, indium, antimony, tellurium, rhenium, osmium, iridium, gold, thallium, nickel, bismuth and tin. They join with iron to make compounds, and tend to 'follow' iron in igneous processes. Siderophile compounds are more common on the Moon than in the terrestrial crust, but this is due to accumulation of meteoritic debris (one type of meteorite is predominately composed of iron and nickel).

silica Crystalline silicon dioxide; SiO_2. It has a high melting point and forms a variety of silicates in rock.

silicate Silica can join with calcium, aluminium, iron, magnesium, sodium and potassium to make a variety of minerals featuring the silicon tetrahedron (SiO_4).

solar wind The plasma that flows out from the Sun. It is mostly ionised hydrogen (free protons and electrons) but also contains heavier nuclei.

spall The ejection of surficial material as a result of focusing a seismic shockwave. In addition to applying to individual rocks, the 'chaotic terrain' antipodal to the major lunar basins is believed to be an example of spalling on a global scale.

spinel A magnesium-iron-chromium-aluminium oxide mineral with black crystals. It is a minority constituent in the lunar basalts.

stratigraphy The study of a planetary surface in terms of the layering of the rock units and their surface exposures.

superposition A common-sense law of stratigraphy which asserts that newer rock units overlay older ones.

swale A shallow valley.

tectonics The process of deformation of a planetary crust to create surface relief.

terminator The line of longitude corresponding to sunrise or sunset on the lunar surface. As the Moon orbits Earth, it turns on its axis and the terminators track at a corresponding rate.

thermal history The rate at which heat has been produced and lost by the Moon, mapped as a function of time. It is inferred from analysis of rock samples.

transient lunar phenomena A general term for a variety of phenomena reported over the years and interpreted as proof of contemporary volcanism (typically in the form of gaseous emissions), but very difficult to confirm as such.

trap A volcanic vent that issues lava of extremely low viscosity which spreads far and wide. The best terrestrial examples are the Siberian, Deccan and Columbia River lava fields. The lunar maria are trap extrusions.

troctolite A coarsely grained plutonic rock in which the plagioclase feldspar combines with olivine as the mafic component.

tuff A loosely consolidated rock consisting of 'welded' pyroclastic.

twinning If the crystals in a rock share a common alignment, this is often apparent to the unaided eye. In plagioclase feldspar the laths resemble the spines of books on a shelf.

upland fill The material that rendered more or less flat the low-lying terrain in the lunar highlands, forming plains. Two forms were envisaged: a low-viscosity lava lower in mafic silicates than the dark maria but, being slightly enriched in silica, sufficiently viscous to create the Cayley Formation; and a silicic lava of greater viscosity that produced the domical hills of the Descartes Formation. However, both proved to be basin ejecta.

USGS The United States Geological Survey, informally known as 'The Survey'.

vent A crustal aperture through which volcanic material (lava, ash, gas) is erupted.

venting The release of gas from rock; depending on the pressure, it can range form a gentle 'leak' to a violent 'jet'.

vesicle As magma approaches the surface, its pressure rapidly diminishes and the volatiles in the melt boil out of solution. If the gas is unable to escape, it makes spherical cavities in the rock. If the magma is under directional stress as it solidifies (i.e. it is flowing) then the vesicles may be distorted.

vesicle pipe A tubular cavity formed by gas venting under pressure through a rock while it is plastic.

viscosity A measure of the resistance of a fluid to flow. A really runny fluid has a low viscosity, and a really sticky fluid has a high viscosity.

vitrophyre An igneous rock whose matrix is glassy.

volatile elements Those elements that boil out of solution at low temperatures. The Moon is strikingly deficient in volatiles such as sodium, sulphur and chlorine.

volcanic rock Igneous rock that solidifies after being extruded onto the surface.

volcano A vent through which lava extrudes onto the surface. The term is usually reserved for the extrusion of lava of sufficient viscosity to construct a mound (if the magma is of sufficiently low viscosity to flow far and wide, it is referred to as a trap).

vug If the gas in a vesicle cools sufficiently slowly for the volatile minerals to recrystallise, this will partially fill the cavity, making it irregular.

wrinkle ridge When a mare plain sinks isostatically, this applies a compressional force that causes the material to form patterns of ridges.

xenolith A fragment of rock from deep within the crust that is torn off the wall of a magma feed pipe and carried to the surface. They are very rare, and much prized for the insight they yield into the lower crust. The term means 'strange rock'.

zap-pit Tiny pits made in the surface of a rock that is exposed to micrometeoroid rain.

Chronological bibliography

The most comprehensive source detailing the activities of the Apollo astronauts on the lunar surface is the *Apollo Lunar Surface Journal* edited by Eric Jones, which is on the Internet. For background reading on lunar geology, I can thoroughly recommend *To A Rocky Moon* by Don Wilhelms (1993) and *Once and Future Moon* by Paul Spudis (1996). For a documented review of the hard science, consult *The Lunar Sourcebook* (1991) which was edited by Grant Heiken, David Vaniman and Bevan French.

The Moon's Face: A Study of the Origin of its Surface
  Features
Gilbert, G.K.
Bulletin of the Philosophical Society of Washington, 1893

The Origin of Continents and Oceans
Wegener, Alfred
Methuen, 1924

The Meteoritic Origin of Lunar Craters
Baldwin, Ralph
Popular Astronomy, 1942

The Meteoritic Origin of Lunar Structures
Baldwin, Ralph
Popular Astronomy, 1943

The Moon's Lack of Folded Ranges
Chamberlin, R.T.
Journal of Geology, 1945

The Story of the Moon
Fisher, Clyde
Doubleday, 1945

Origin of the Moon and its Topography
Daly, R.A.
Proceedings of the American Philosophical Society, 1946

The Meteoritic Impact Origin of the Moon's Surface
  Features
Dietz, R.S.
Journal of Geology, 1946

The Face of the Moon
Baldwin, Ralph
University of Chicago Press, 1949

The Exploration of Space
Clarke, Arthur
Temple Press, 1951

Across The Space Frontier
von Braun, Wernher
Viking, 1952 (collected articles from Collier's Magazine)

The Planets: Their Origin and Development
Urey, Harold,
Yale University Press, 1952

On the Origin of the Lunar Surface Features
Kuiper, Gerard
Proceedings of the National Academy of Sciences, 1954

The Lunar Surface
Gold, Thomas
Monthly Notices of the Royal Astronomical Society, 1955

The Origin and Significance of the Moon's Surface
Urey, Harold
Vistas in Astronomy, 1956

Satellite!
Bergaust, Erik and Beller, William
Lutterworth, 1957

Satellites and Spaceflight
Burgess, Eric
Chapman and Hall, 1957

The Strange World of the Moon
Firsoff, V.A.
Basic Books, 1959

The Exploration of the Moon
Kuiper, Gerard
Vista in Astronautics, 1959

Photographic Lunar Atlas
Kuiper, Gerard *et al.*
University of Chicago, 1960

Photometry of the Moon
Minnaert, M.
In *Planets and Satellites* by Kuiper, Gerard and
  Middlehurst, Barbara (Editors)
University of Chicago Press, 1961

Project Mercury: Countdown for Space
National Geographic, May 1961

The Moon
Wilkins, H.P. and Moore, Patrick
Faber & Faber, 1961

*An Atlas of the Moon's Far Side: The Lunik-III
  Reconnaissance*
Barabashov, N.P. *et al.*
Sky Publishing, 1961

Recent Radiometric Studies of the Planets and the
  Moon
Sinton, William
In *Planets and Satellites* by Kuiper, Gerard and
  Middlehurst, Barbara (Editors)
University of Chicago Press, 1961

The Structure of the Moon's Surface
Fielder, Gilbert
Pergamon, 1961

On the Possible Presence of Ice on the Moon
Watson, K., Murray, B.C. and Brown, H.
J. Geophys. Res., vol. 66, pp. 1598–1600, 1961

America on the Moon
Holmes, Jay
L.B. Lippincott, 1962

Interpretation of Lunar Craters
Shoemaker, Gene
in *Physics and Astronomy of the Moon*, Kopal, Z.
  (Editor)
Academic Press, 1962

Stratigraphic Basis for a Lunar Time Scale
Shoemaker, Gene and Hackman, Robert
IAU Symposium, Leningrad, 5–11 December 1960; and
  in *The Moon*, Kopal, Z. and Mikhailov, Z.K.
  (Editors)
Academic Press, 1962

Origin and History of the Moon
Urey, Harold
In *Physics and Astronomy of the Moon*, Kopal, Z.
  (Editor)
Academic Press, 1962

Interplanetary Correlation of Geologic Time
Shoemaker, Gene, Hackman, Robert and
  Eggleton, Richard
Advances in Astronautical Sciences, 1962

Exploration of the Moon's Surface
Shoemaker, Gene
American Scientist, 1962

Survey of the Moon
Moore, Patrick
Eyre & Spottiswoode, 1962

*America's Race for the Moon: The New York Times
  Story of Project Apollo*
Sullivan, Walter (Editor)
Random House, 1962

The Measure of the Moon
Baldwin, Ralph
University of Chicago Press, 1963

The Lunar Surface: Introduction
Wright, Frederick, Wright, F.H. and Wright, Helen
In *The Moon, Meteorites, and Comets*, Kuiper, Gerard
  and Middlehurst, Barbara (Editors)
University of Chicago Press, 1963

The Moon: Target for Apollo
Chester, Michael and McClinton, David
Putnam, 1963

Pride and Power: The Rationale of the Space Program
Van Dyke, Vernon
University of Illinois Press, 1964

The Sky At Night
Moore, Patrick
Eyre & Spottiswoode, 1964

Ranger VI Photographs of the Moon
NASA SP-61, 1964

The Geology of the Moon
Shoemaker, Gene
Scientific American, 1964

A Fundamental Survey of the Moon
Baldwin, Ralph
McGraw-Hill, 1965

Principles of Physical Geology
Holmes, Arthur
Nelson, 1965

The Apollo Program
Lutman, C.C.
Air University Review, May–June 1965

Lunar Geology
Fielder, Gilbert
Lutterworth, 1965

Zond-3 Photographs of the Moon's Far Side
Lipskiy, Y.N
Sky & Telescope, 1965

Eyes on the Moon
Smith, G.M., Vrebalovich, J. and Willingham, O.E.
Astronautics & Aeronautics, March 1966

Astronomy, Space, and the Moon
Tifft, William
Astronautics & Aeronautics, September 1966

What Luna 9 Told Us About The Moon
Lipskiy, Y.N
Sky & Telescope, 1966

*The Moon Capture Theory of Hoerbiger After Fifty Five
  Years*
Sykes, Egerton
Markham House, 1966

First Men to the Moon
von Braun, Wernher
Holt, Rinehart & Winston, 1966

Ranger IX Photographs of the Moon
NASA SP-112, 1966

Assault On The Moon
Burgess, Eric
Hodder & Stoughton, 1966

Apollo Accident Report
Aviation Week & Space Technology, February 1967

Collecting and Processing Samples of the Moon
McLane, James
Astronautics & Aeronautics, August 1967

11 Days Aboard Apollo 7
Life Magazine, 6 December 1968

Apollo 8: Guide to the High Adventure
Newsweek, 23 December 1968

Exploration of the Moon by Spacecraft
Kopal, Zdenek
Oliver & Boyd, 1968

To the Moon: A Distillation of the Great Writings from Ancient Legend to Space Exploration
Wright, Hamilton, Wright, Helen and Rapport, Samuel (Editors)
Meredith, 1968

America in Space: The First Decade
Anderton, David
NASA EP-48, 1968

Apollo 8: Man around the Moon
US Government Printing Office, 1968

Mascons: Lunar Mass Concentrations
Muller, Paul and Sjogren, William
Science, 1968

The Encyclopedia of Space
Allward, Maurice (Editor)
Hamlyn, 1968

The Sky At Night: 2
Moore, Patrick
Eyre & Spottiswoode, 1968

Gemini: A Personal Account of Man's Venture into Space
Grissom, Gus
Macmillan, 1968

Surveyor 7 Mission Report
Gault, D.E. *et al.*
Jet Propulsion Laboratory TR-32-1264, 1968

Lunar Atlas
Alter, Dinsmore (Editor)
Dover, 1968

How An Idea No One Wanted Grew Up To Be the LEM
Life Magazine, 14 March 1969

Apollo 9 Album
Life Magazine, 28 March 1969

The Moon: Man's First Goal In Space
National Geographic, February 1969

First Men Around the Moon
Parker, P.J.
Spaceflight, March 1969

Man's Vision of the Lunar Voyage: Suddenly Real
Life Magazine, 30 May 1969

Apollo 8: A Most Fantastic Voyage
National Geographic Magazine, May 1969

The Unexpected Payoff of Project Apollo
Alexander, T.
Fortune, July 1969

Off to the Moon
Life Magazine, 4 July 1969

Apollo's Great Leap for the Moon
Life Magazine, 25 July 1969

Special Report: Apollo 11 Lunar Landing
Aviation Week & Space Technology, 28 July 1969

Apollo 9 Tests Lunar Module
Parker, P.J.
Spaceflight, July 1969

Apollo 10: The Last Rehearsal
Parker, P.J.
Spaceflight, August 1969

Down to the Moon ... and the Giant Step
Life Magazine, 8 August 1969

Man on the Moon: An Assessment
Rabinowitch, Eugene and Lewis, Richard (Editors)
Special Issue of Bulletin of the Atomic Scientists, September 1969

Man on the Moon
Parker, P.J.
Spaceflight, September/October 1969

The Exploration of the Moon
Hess, W. *et al.*
Scientific American, October 1969

First Explorers on the Moon: The Incredible Story of Apollo 11
National Geographic Magazine, December 1969

To the Moon and Back
Life Magazine, Special Edition, 1969

Surveyor Program Results
NASA SP-184, 1969

Apollo 11: Preliminary Science Report
NASA SP-214, 1969

The Post-Apollo Space Program: Directions for the Future
NASA Space Task Group
US Government Printing Office, 1969

Apollo 11: On the Moon
The New York Times, 1969.

Code Name Spider: Flight of Apollo 9
NASA EP-68, 1969

Apollo Lunar Surface Experiments Package
Space World, November 1969

The Moon
Kopal, Zdenek
D. Reidel, 1969

Mission Report: Apollo 10
NASA EP-70, 1969.

The Lunar Controversy
McCall, G.J.H.
Journal of the British Astronomical Association, 1969

History of Rocketry and Space Travel
von Braun, Wernher and Ordway, Frederick
Thomas Y. Crowell, 1969

Man on the Moon
Mansfield, John
Stein & Day, 1969

Man on the Moon
Fairley, Peter
Arthur Barker Limited, 1969.

U.S. on the Moon
Newman, Joseph
US News and World Report Inc., 1969

Invasion of the Moon, 1969: The Story of Apollo 11
Ryan, Peter
Penguin, 1969

Observations of the Lunar Regolith
Shoemaker, Gene *et al.*
Journal of Geophysical Research, 1969

Man on the Moon: The Impact on Science, Technology, and International Cooperation
Rabinowitch, Eugene and Lewis, Richard (Editors)
Basic Books, 1969

We Reach the Moon: The New York Times Story of Man's Greatest Adventure
Wilford, John Noble
Bantam Books, 1969

Analysis of Apollo 8: Photography and Visual Observations
NASA SP-201, 1969

Conquest of the Moon
Sutton, Felix
Grosset & Dunlap, 1969

Appointment on the Moon: The Inside Story of America's Space Adventure
Lewis, Richard
Viking, 1969

Project Apollo: The Way to the Moon
Booker, P., Frewer, G.C. and Pardoe, G.
American Elsevier Pub. Co., 1969

Footprints on the Moon
Barbour, John
The Associated Press, 1969

Lunar Orbiter Photographic Data: Data Users' Note
Beeler, Mary and Michlovitz, K.
NSSDC, Goddard Space Flight Center, 1969

Apollo on the Moon
Cooper, Henry
Dial Press, 1969

America's Next Decades in Space: A Report for the Space Task Group
NASA, 1969

Soviet Space Exploration - The First Decade
Shelton, William
Arthur Barker, 1969

Journey to Tranquillity: The History of Man's Assault on the Moon
Young, Hugo, Silcock, Bryan and Dunn, Peter
Johathan Cape, 1969

Five-Day Mission Plan to Investigate The Geology of the Marius Hills Region of the Moon
Elston, D.P and Willingham, C.R.
US Geological Survey, 1969

Preliminary Lunar Exploration Plan of the Marius Hills Region of the Moon
Karlstrom, T.N.V., McCauley, J.F. and Swann, G.A.
US Geological Survey, 1969

The Triumph of Apollo 12
Parker, P.J.
Spaceflight, February/March 1970

The Next Decade in Space: A Report of the Space Science and Technology Panel of the President's Science Advisory Committee
President's Science Advisory Committee, March 1970

The Joyous Triumph of Apollo 13
Life Magazine, 24 April 1970

Four Days of Peril Between Earth and Moon: Apollo 13, Ill-Fated Odyssey
Time Magazine, 27 April 1970

Apollo 13: "Houston, We've Got a Problem"
NASA EP-76, 1970

Apollo 12 Lunar Samples: Trace Element Analysis of a Core and the Uniformity of the Regolith
Ganapathy, R. *et al.*
Science, 30 October 1970

Seismic Data from Man-Made Impacts on the Moon
Latham, G. *et al.*
Science, 6 November 1970

Petrogenesis of Apollo 11 Basalts and Implications for Lunar Origin
Ringwood, A.E.
Journal of Geophysical Research, 10 November 1970

Moon Rocks
Cooper, Henry
Dial Press, 1970

The Sky At Night: 3
Moore, Patrick
BBC, 1970

Decision to go to the Moon
Logsdon, John
The MIT Press, 1970

The Invasion of the Moon, 1957–1970
Ryan, Peter
Penguin, 1971

The Old Moon & the New
Firsoff, V.A
A.S. Barnes, 1970

A Search for Carbon and its Compounds in Lunar Samples from Mare Tranquillitatis
Kvenvolden, K.A. and Ponnamperuma, C. (Editors)
NASA SP-257, 1970

Proceedings of the First Lunar Science Conference
Levinson, A.A. (Editor)
Pergamon Press, 1970 (in 3 volumes)

Geology of the Moon: A Stratigraphic View
Mutch, Thomas
Princeton University Press, 1970

Apollo 12: Preliminary Science Report
NASA SP-235, 1970

The Lunar Rocks
Mason, Brian and Melson, William
Wiley Interscience, 1970

Moon Flight Atlas
Moore, Patrick
Rand McNally, 1970

Guide to Lunar Orbiter Photographs
Hansen, T.P
NASA SP-242, 1970

Apollo 12: A New Vista for Lunar Science
NASA EP-74, 1970

The Moon as Viewed by Lunar Orbiter
Kosofsky, L.J. and El-Baz, F.
NASA SP-200, 1970

First on the Moon
Armstrong, Neil Collins, Michael and Aldrin Edwin
Michael Joseph, 1970

*Summary of Lunar Stratigraphy: Telescopic
    Observations*
Wilhelms, Don
US Geological Survey, 1970

Preliminary Examination of the Apollo 12 Lunar Samples
Davies, J.E
Spaceflight, January 1971

Lunar Geology in the Light of the Apollo Programme
O'Hara, M.J.
Endeavour, January 1971

White Tracks on the Moon
Life Magazine, 26 February 1971

Noble Gas Abundances in Lunar Material: Cosmic Ray
    Spallation Products and Radiation Ages from the
    Sea of Tranquility and the Ocean of Storms
Bogard, D.D
Journal of Geophysical Research, 10 April 1971

Apollo in Lunar Orbit
Esenwein, George, Roberson, Floyd and Winterhalter,
    David
Astronautics & Aeronautics, April 1971

Apollo 14: A Visit to Fra Mauro
Baker, David
Spaceflight, May/June/October 1971

The Climb up Cone Crater
Hall, Alice
National Geographic Magazine, July 1971

From the Good Earth to the Sea of Rains
Time Magazine, 9 August 1971

The Magnetism of the Moon
Dyal, P. and Parkin, C.W.
Scientific American, August 1971

Preliminary Examination of Lunar Samples from
    Apollo 14
Science, 20 August 1971

The Lunar Rocks
Mason, Brian
Scientific American, October 1971

Expedition to Hadley–Apennine
Baker, David
Spaceflight, October/November/December 1971

Lunar Photos Reveal New Details
Gregory, William
Aviation Week & Space Technology, 20 December 1971

Apollo 14: Science at Fra Mauro
Froehlich, Walter
NASA, 1971

Understanding the Earth
Gass, I., Smith, P.J. and Wilson, R. (Editors)
Open University, 1971

Proceedings of the Second Lunar Science Conference
Levinson, A.A. (Editor)
The MIT Press, 1971 (in 3 volumes)

*Analysis of Apollo 10: Photography and Visual
    Observations*
NASA SP-232, 1971

Apollo 14: Photographic Data Package
NSSDC, Goddard Space Flight Center, 1971

Apollo 14: Science at Fra Mauro
Froehlich, Walter
NASA EP-91, 1971

Moon Rocks and Minerals
Levinson, A.A. and Taylor, S.
Pergamon Press, 1971

Lunar Hadley Rille: Consideration of its Origin
Greeley, Ron
Science, 1971

The Russian Space Bluff
Vladimirov, Leonid
Tom Stacy Ltd, 1971

*On the Moon with Apollo 15: A Guidebook to Hadley
    Rille and the Apennine Mountains*
Simmons, Gene
NASA, 1971

Apollo 15 at Hadley Base
NASA EP-94, 1971

Apollo 14: Preliminary Science Report
NASA SP-272, 1971

Lunar Orbiter Photographic Atlas of the Moon
Bowker, David and Hughes, J. Kenrick
NASA SP-206, 1971

*Preliminary Description of Apollo 15 Sample
    Environments*
Swann, G.A.
US Geological Survey, 1971

Apollo 11 Mission Report
NASA SP-238, 1971

What Made Apollo a Success?
NASA SP-287, 1971

Atlas and Gazetteer of the Near Side of the Moon
Gutschewski, G., Kinsler, D.C. and Whitaker, E.
NASA SP-241, 1971

Project Ranger: A Chronology. Pasadena
Hall, R. Cargill
JPL/HR-2, 1971

Lunar Photographs from Apollos 8, 10, and 11
Musgrove, Robert
NASA SP-246, 1971

Moon: Possible Nature of the Body that Produced the
    Imbrian Basin, from the Composition of Apollo 14
    Samples
Ganapathy, R. *et al.*
Science, 7 January 1972

Lunar Gravity via Apollo 14 Doppler Radio Tracking
Sjogren, W.L. *et al.*
Science, 14 January 1972

Geologic Setting of the Apollo 15 Samples
Science, 28 January 1972

The Apollo 15 Lunar Samples: A Preliminary
    Description
Science, 28 January 1972

To the Mountains of the Moon
National Geographic Magazine, February 1972

Lunar Research: No Agreement on Evolutionary
    Models
Hammond, A.L
Science, 25 February 1972

The Third Lunar Science Conference
Page, T.L.
Sky & Telescope, March/April 1972

The Moon and Sixpence of Science
Wasserburg, G.J.
Astronautics & Aeronautics, April 1972

Heat Flow and Convection Demonstration Experiments
    Aboard Apollo 14
Grodzka, P.G. and Bannister, T.C.
Science, 5 May 1972

Apollo 16 Explores Lunar Highlands
Aviation Week & Space Technology, 15 May 1972

Lunar Crust: Structure and Composition
Toksoz, N.M. *et al.*
Science, 2 June 1972

Photo Geodesy from Apollo
Light, D.L.
Photogrammetric Engineering, June 1972

Apollo 16 Geochemical X-Ray Fluorescence
    Experiment: Preliminary Report
Adler, I. *et al.*
Science, 21 July 1972

Mission to Descartes
Baker, David
Spaceflight, July/August 1972

*Analysis of Surveyor 3 Material and Photographs
    Returned by Apollo 12*
NASA SP-284, 1972

*On the Moon with Apollo 16: A Guidebook to the
    Descartes Region*
Simmons, Gene
NASA, 1972

*Preliminary Geologic Investigation of the Apollo 16
    Landing Site*
US Geological Survey, 1972

Apollo 15: Preliminary Science Report
NASA SP-289, 1972

Apollo 16: Preliminary Science Report
NASA SP-315, 1972

*Compositions of Major and Minor Minerals in Five
    Apollo 12 Crystalline Rocks*
French, Bevan *et al.*
NASA, 1972

One Last Fiery Hurrah for Apollo
Life Magazine, 19 December 1972

Moons and Planets
Hartmann, William
Wadsworth, 1972

*Documentation and Environment of the Apollo 16
    Samples: A Preliminary Report*
Muehlberger, W.R. *et al.*
US Geological Survey, 1972

*Preliminary Report on the Geology and Field Petrology at
    the Apollo 17 Landing Site*
Muehlberger, W.R. *et al.*
US Geological Survey, 1972

Preliminary Geologic Investigation of the Apollo 15
    Landing Site
Swann, G.A. *et al.*
in *Apollo 15 Preliminary Science Report*
NASA SP-289, 1972

*On the Moon with Apollo 17: A Guidebook to Taurus-
    Littrow*
Simmons, Gene
NASA, 1972

Petrology and Stratigraphy of the Fra Mauro Formation at the Apollo 14 Site
Wilshire, H.G. and Jackson, E.D
NASA, 1972

Apollo 16 at Descartes
Froehlich, Walter
NASA EP-97, 1972

The Last Apollo
Baker, David.
Spaceflight, February/March/April 1973

Documentation and Environment of the Apollo 17 Samples: A Preliminary Report
Muehlberger, W.R. *et al.*
US Geological Survey, 1973

Electron Microprobe Analyses of Minerals from Apollo 15 Mare Basalt Rake Samples
Dowty, Eric *et al.*
Department of Geology & Institute of Meteoritics, University of New Mexico, 1973

Apollo 17 at Taurus Littrow
Anderton, David
ANASA EP-102, 1973

Apollo 17: The Final Flight
National Geographic, September 1973

Thirteen: The Flight that Failed
Cooper, Henry
Dial Press, 1973

Apollo 17: Preliminary Science Report
NASA SP-330, 1973

The National Aeronautics and Space Administration
Hirsch, Richard and Trento, Joseph
Praeger, 1973

Pictorial Guide to the Moon
Alter, Dinsmore (Editor)
Thomas Y. Crowell Co., 1973

Apollo 17 Index: Mapping Camera and Panoramic Camera Photographs
Johnson Space Center, 1973

Photogeology of Dark Material at the Taurus-Littrow Region of the Moon
Lucchitta, B.K.
Proceedings of the Fourth Lunar and Planetary Science Conference, 1973

To Rule the Night
Irwin, Jim and Emerson, William
Hodder & Stoughton, 1973

Vitrification Darkening in the Lunar Highlands and Identification of the Descartes Material
Adams, J.B. and McCord, T.B.
Proceedings of the Fourth Lunar Science Conference, 1973

Volcanic Production Rates: Comparison of Oceanic Ridges, Islands and the Columbia River Plateau Basalts
Baksi, A.K. and Watkins, N.D.
Science, 1973

Turbulent Lava Flow and the Formation of Lunar Sinuous Rilles
Hulme, G.
Modern Geology, 1973

Apollo 16 Exploration of Descartes: A Geological Summary
Science, 1973

Geological Exploration of Taurus-Littrow: Apollo 17 Landing Site
Science, 1973

The Moon in the Post-Apollo Era
Kopal, Zdenek
D. Reidel, 1974

Mapping of the Moon: Past and Present
Kopal, Zdenek and Carder, R.W.
D. Reidel, 1974

Return To Earth
Aldrin, Buzz and Warga, Wayne
Bantam, 1974

Apollo Scientific Experiments Data Handbook
Lauderdale, W.W. and Eichelman, W.F. (Editors)
Lyndon B. Johnson Space Center, NASA TM X-58131, 1974

Orange Glass: Evidence for Regional Deposits of Pyroclastic Origin on the Moon
Adams, J.B., Pieters, C. and McCord, T.B.
Proceedings of the Fifth Lunar Science Conference, 1974

Apollo
Chappell, Russell
NASA, 1974

Lunar Basin Formation and Highland Stratigraphy
Howard, K., Wilhelms, D. and Scott, D.H.
Reviews of Geophysics and Space Physics, 1974

The Voyages of Apollo: The Exploration of the Moon
Lewis, Richard
Quadrangle/New York Times Books, 1974

Carrying The Fire: An Astronaut's Journeys
Michael Collins
Farrar, Straus & Giroux, 1974

Lunar Magnetism
Fuller, M.
Reviews of Geophysics and Space Physics, 1974

Man's Conquest of Space
National Geographic Society, 1974

Lunar Dark Mantle Deposits: Possible Clues To Early
    Mare Deposits
Head, J.W.
*Proceedings of the Fifth Lunar and Planetary Science
    Conference*, 1974

Magnetism and the Interior of the Moon
Dyal, P., Parkin, C.W. and Dailey, W.D.
Reviews of Geophysics and Space Physics, 1974

Hello Earth
Worden, Al
Nash of Los Angeles, 1974

Thickness of Mare Material in the Tranquillitatis and
    Nectaris Basins
DeHon, R.A.
Proceedings of the Fifth Lunar Science Conference,
    1974

Orange Material in the Sulpicius Gallus Formation of
    the Southwestern Edge of Mare Serenitatis
Lucchitta, B.K. and Schmitt, H.H.
*Proceedings of the Fifth Lunar and Planetary Science
    Conference*, 1974

Multiringed Basins: Orientale and Associated Features
Moore, Henry *et al.*
Proceedings of the Fifth Lunar Science Conference,
    1974

*Lunar Science: A Post-Apollo View – Scientific Results
    and Insights from the Lunar Samples*
Taylor, S.R.
Pergamon Press, 1975

The Moon
Wood, J.A
Scientific American, 1975

Apollo 17 Landing Site Geology
Lyndon B. Johnson Space Center, 1975

Apollo Expeditions to the Moon
Cortright, Edgar (Editor)
NASA SP-350, 1975

The Future of the U.S. Space Program
Levine, Arthur
Praeger Publishers, 1975

Apollo Program Summary Report
Lyndon B. Johnson Space Center, 1975

On High-Alumina Mare Basalts
Ridley, W.I.
*Proceedings of the Sixth Lunar and Planetary Science
    Conference*, 1975

The Moon After Apollo
El-Baz, Farouk
Icarus, 1975

Lunar Mineralogy
Frondel, Clifford
Wiley, 1975

Lunar Volcanism in Space and Time
Head, J.W.
Reviews of Geophysics and Space Physics, May 1976

What's New on the Moon?
French, Bevan
NASA EP-131, 1976

Volcanoes
Francis, Peter
Penguin, 1976

*Moon Morphology: Interpretations Based on Lunar
    Orbiter Photography*
Schultz, P.H.
University of Texas, 1976

*Lunar-Sample Processing in the Lunar Receiving
    Laboratory, High-Vacuum Complex*
White, David
Lyndon B. Johnson Space Center
NASA Technical Note D-8298, 1976

A Map of the Earth's Moon
National Geographic, 1976

Geological Structure of the Eastern Mare Basins
DeHon, R.A. and Waskom, J.D.
Proceedings of the Seventh Lunar Science Conference,
    1976

Volcanic Features of the Nearside Equatorial Lunar
    Maria
Guest, J.E. and Murray, J.B.
Journal of the Geological Society of London, 1976

*The Moon Book: Exploring the Mysteries of the Lunar
    World*
French, Bevan (Editor)
Penguin, 1977

*Geology of the Apollo 14 Landing Site in the Fra Mauro
    Highlands*
US Geological Survey, 1977

Chemistry of Apollo 12 Mare Basalts: Magma Types
    and Fractionation Processes
Rhodes, J.M. *et al.*
Proceedings of the Eighth Lunar Science Conference,
    1977

Apollo 17: The Most Productive Lunar Expedition
NASA Mission Report MR-12, 1977

*Geology of the Apollo 14 Landing Site in the Fra Mauro
    Highlands*
Swann, G.A. *et al.*
US Geological Survey, 1977

*Destination Moon: A History of the Lunar Orbiter
    Program*
Byers, Bruce
NASA TM X-3478, 1977

Lunar Impact: A History of Project Ranger
Hall, R. Cargill
NASA SP-4210, 1977

Geology on the Moon
Guest, J.E. and Greeley, R.
Wykeham, 1977

Cratering in the Solar System
Hartmann, William
Scientific American, January 1977

The Deep Seismic Structure of the Moon
Goins, N.R., Dainty, A. and Toksoz, M.N.
Proceedings of the Eighth Lunar and Planetary Science Conference, 1977

The Ries Crater of Southern Germany: A Model for Large Basins on Planetary Surfaces
Chao, Ed
Geologisches Jahrbuch, 1977

Sun, Moon, and Sanding Stones
Wood, J.E.
Oxford University Press, 1978

Apollo Over the Moon: A View from Orbit
Masursky, H., Colton, G.W. and El-Baz, F.
NASA SP-362, 1978

Moonport: A History of Apollo Launch Facilities and Operations
Benson, C. and Faherty, W.B.
NASA SP-4204, 1978

Thickness of Mare Flow Fronts
Gifford, A.W. and El-Baz, F.
Proceedings of the Ninth Lunar and Planetary Science Conference, 1978

Mare Volcanism in the Herigonius Region of the Moon
Greeley, Ron and Spudis, Paul
Proceedings of the Ninth Lunar and Planetary Science Conference, 1978

Lunar KREEP Volcanism
Hawke, B.R and Head, J.W.
Proceedings of the Ninth Lunar and Planetary Science Conference, 1978

Imbrian-Age Highland Volcanism on the Moon: The Gruithuisen and Mairan Domes
Head, J.W. and McCord, T.B.
Science, 1978

Mare Crisium: The View From Luna 24
Head, J.W. *et al.*
Pergamon, 1978

An Apollo Perspective
Logsdon, John M. *et al.*
Astronautics & Aeronautics, December 1979

The Solar System
Wood, J.A.
Prentice Hall, 1979

Apollo: Ten Years Since Tranquility Base
Hallion, Richard P. and Crouch, Tom (Editors)
Smithsonian Institution Press, 1979

Chariots for Apollo: A History of Manned Lunar Spacecraft
Brooks, Courtney, Grimwood, James and Swenson, Loyd
NASA SP-4205, 1979

The Lunar Interior
Goins, N.R., Toksoz, N. and Dainty, A.M.
Proceedings of the Tenth Lunar and Planetary Science Conference, 1979

Thickness of the Western Mare Basalts
DeHon, R.A.
Proceedings of the Tenth Lunar and Planetary Science Conference, 1979

Multispectral Mapping of the Apennine Region: Identification and Distribution of Regional Pyroclastics
Hawke, B.R. *et al.*
Proceedings of the Tenth Lunar and Planetary Science Conference, 1979

Alphonsus-Style Dark Halo Craters
Head, J.W. and Wilson, L.
Proceedings of the Tenth Lunar and Planetary Science Conference, 1979

The Origin of the Earth And Moon
Ringwood, A.E.
Springer-Verlag, 1979

ALSEP Termination Report
Bates, J.R. *et al.*
NASA RP-1036, 1979

Vertical Movement in Basins: Relation to Mare Emplacement, Basin Tectonics and Lunar Thermal History
Solomon, S.C. and Head, J.W.
Journal of Geophysical Research, 1979

Space Rocks: Getting Our Hands on the Universe
French, Bevan
Air & Space, Fall 1980

Stratigraphy of Part of the Lunar Near Side
Wilhelms, Don
NASA, 1980

The Geology and Petrology of the Apollo 11 Landing Site
Beaty, D.W. and Albee, A.L.
Proceedings of the Eleventh Lunar and Planetary Science Conference, 1980

Lunar Remote Sensing and Measurements
Moore, H.J. *et al.*
US Geological Survey, 1980

Stratigraphy of Part of the Lunar Near Side
Wilhelms, Don
US Geological Society, 1980

Composition and Structure of the Deep Lunar Interior
Delano, J.W. and Taylor, S.R.
Proceedings of the Eleventh Lunar and Planetary Science Conference, 1980

An Iron Core in the Moon Generating an Early Magnetic Field
Runcorn, S.K.
Proceedings of the Eleventh Lunar and Planetary Science Conference, 1980

Rocks of the Early Lunar Crust
Odette, James
Proceedings of the Eleventh Lunar and Planetary Science Conference, 1980

Earthlike Planets: Surfaces of Mercury, Venus, Earth, Moon, Mars
Murray, Bruce, Malin, Michael and Greeley, Ronald
W.H. Freeman, 1981

Red Star in Orbit: The Inside Story of the Soviet Space Programme
Oberg, James
Random House, 1981

Geology of the Apollo 16 Area, Central Lunar Highlands
Ulrich, George, Hodges, C.A. and Muehlberger, W.R. (Editors)
US Geological Survey, 1981.

Geologic Investigation of the Taurus-Littrow Valley
Wolfe, E.W. *et al.*
US Geological Society, 1981

Harold Clayton Urey–In Memoriam
Sagan, Carl
Icarus, 1981

Characterisation and Distribution of Pyroclastic Units in the Rima Bode Region of the Moon
Gaddis, L.R. *et al.*
Proceedings of the Twelfth Lunar and Planetary Science Conference, 1981

The Geological Investigation of the Taurus-Littrow Valley, Apollo 17 Landing Site
Wolfe, Edward *et al.*
US Geological Survey, 1981

The New Solar System
Beatty, Kelly, O'Leary, Brian and Chaikin, Andrew (Editors)
Book Club Associates, 1981

The History of Manned Space Flight
Baker, David
New Cavendish Books, 1981

The Moon
Moore, Patrick
Mitchell Beazeley, 1981

The Moon: Our Sister Planet
Cadogan, Peter
Cambridge University Press, 1981

Lunar Dark Haloed Impact Craters: Origins And Implications For Early Mare Volcanism
Bell, A.F. and Hawke, B.R.
Journal of Geophysical Research, 1981

Remote Sensing Studies Of Lunar Dark Halo Craters
Hawke, B.R and Bell, J.F.
Proceedings of the Twelfth Lunar and Planetary Science Conference, 1981

Nickel for Your Thoughts: Urey and the Origin of the Moon
Brush, Stephen
Science, 3 September 1982

Planetary Science: A Lunar Perspective
Taylor, S.R.
Lunar and Planetary Institute, 1982

Photographic Atlas of the Planets
Briggs, G.A. and Taylor, F.W.
Cambridge University Press, 1982

The Lunar Surface Sounder: Stratigraphy and Structural Evolution of Southern Mare Serenitatis
Sharpton, V.L. and Head, J.W.
Journal of Geophysical Research, 1982

The Lunar Regolith: Chemistry, Mineralogy and Petrology
Papike, J.J. *et al.*
Reviews of Geophysics and Space Physics, 1982

All We Did Was Fly To The Moon
Lattimer, Dick (Editor)
The Whispering Eagle Press, 1983

Beginning and End of Lunar Mare Volcanism
Schultz, P.H. and Spudis, P.
Nature, 1983

The Geology of the Terrestrial Planets
Carr, M.H. (Editor)
NASA SP-469, 1984

Apollo 16 Site Geology and Impact Melts: Implications for the Geologic History of The Lunar Highlands
Spudis, Paul
Proceedings of the Fifteenth Lunar and Planetary Science Conference, 1985

The Moon
Wilhelms, Don
In *The Geology Of The Terrestrial Planets* by Carr, M.H. (Editor)
NASA SP-469, 1984

Far Travelers: The Exploring Machines
Nicks, Oran
NASA SP-480, 1985

Composition and Evolution of the Lunar Crust in the Descartes Highlands
Stoffler, D. *et al.*
Journal of Geophysical Research, 1985

Remote Sensing of Lunar Pyroclastic Mantling Deposits
Gaddis, L.R and Pieters, C.M.
Icarus, 1985

Catalog of Apollo 15 Rocks
Ryder, Graham
Lyndon B. Johnson Space Center, 1985

Geology and Petrology of the Apollo 15 Landing Site:
    Past, Present and Future Understanding
Spudis, Paul and Ryder, Graham
Eos Transactions, AGU, 1985

*The University of Arizona's Lunar and Planetary
    Laboratory, its Founding and Early Years*
Whitaker, Ewen
Lunar and Planetary Laboratory, 1985

The Making of the Earth
Fifield, Richard
New Scientist, 1985

Space Travel
von Braun, Wernher
Harper & Row, 1985

Meteorites, the Moon and the History of Geology
Marvin, Ursula
Journal of Geological Education, May 1986

Some Observations on the Geology of the Apollo 15
    Landing Site
Swann, G.A.
in *Workshop on the Geology and Petrology of the Apollo
    15 Landing Site* by Ryder Graham and Spudis, Paul
    (Editors)
Lunar and Planetary Institute, 1986

Origin of the Moon
Hartmann, W., Phillips, R.G. and Taylor, G.J. (Editors)
Lunar and Planetary Institute, 1986

Early History of Selenogony
Brush, Stephen
In *Origin of the Moon* by Hartmann, W., Phillips, R.G.
    and Taylor, G.J. (Editors)
Lunar and Planetary Institute, 1986

Status and Future of Lunar Geoscience
NASA SP-484, 1986

Solar System Log
Wilson, Andrew
Jane's, 1986

The Geologic History of the Moon
Wilhelms, Don, McCauley John and Trask, Newell
US Geological Survey, 1987

The Origin of the Moon
Taylor, S.R
American Scientist, September 1987

The Formation of Hadley Rille and Implications for the
    Geology of the Apollo 15 Region
Spudis, Paul and Greeley, Ron
Lunar Planetary Science, 1987

Planetary Landscapes
Greeley, Ron
Allen & Unwin, 1987

The Birth of the Earth
Fisher, David
Columbia University Press, 1987

Heroes in Space: From Gagarin to Challenger
Bond, Peter
Blackwell, 1987

A History of Modern Selenogony: Theoretical Origins
    of the Moon from Capture to Crash 1955–1984
Brush, Stephen
Space Science Reviews, 1988

Liftoff: The Story of America's Adventure in Space
Collins, Michael
Grove Press, 1988

For All Mankind
Hurt, Harry
Atlantic Monthly Press, 1988

Countdown
Borman, Frank and Serling, Robert
Morrow, 1988

Race Into Space: The Soviet Space Programme
Harbey, Brian
Ellis Horwood, 1988

Earthquakes
Bolt, Bruce
Freeman, 1988

Geological Structures and Moving Plates
Park, R.G.
Backie, 1988

Earth
Lambert, David
Cambridge University Press, 1988

Drifting Continents and Shifting Theories
Le Grand, H.E.
Cambridge University Press

A Trip to the Moon
Weaver, Kenneth
Air and Space, June/July 1989

A Smooth Spot in Tranquility
Wilhelms, Don
Air and Space, June/July 1989

The Impact of Lunar Laser Ranging on Gravitational
    Theory
Mulholland, Derral
Astronomy Quarterly, 1989

*"One Small Step" – The Apollo Missions, the Astronauts,
    the Aftermath: A Twenty Year Perspective*
Furniss, Tim
G.T. Foulis & Co., 1989

*Moon Trip: A Personal Account of the Apollo Program
and Its Science*
King, Elbert
University of Houston, 1989

The First Lunar Landing: 20th Anniversary
Armstrong, Neil, Aldrin, Edwin and Collins, Michael
NASA EP-73, 1989

*Journey into Space: The First Three Decades of Space
Exploration*
Murray, Bruce
W.W. Norton & Co., 1989

Men from Earth
Aldrin, Buzz and McConnell, Malcolm
Bantam, 1989

Volcanoes
Decker, Robert and Decker, Barbara
Freeman, 1989

Apollo, the Race to the Moon
Murray, Charles and Cox, Catherine
Simon & Schuster, 1989

*Where No Man Has Gone Before: A History of Apollo
Lunar Exploration Missions*
Compton, David
NASA SP-4214, 1989

Apollo 11 Moon Landing
Shayler, David
Ian Allen Ltd, 1989

The Origin of the Solar System: The Capture Theory
Dormand, John and Woolfson, Michael
Ellis Horwood, 1989

Cosmic Catastrophes
Chapman, Clark and Morrison, David
Plenum, 1989

Mining the Moon
Burt, D.M.
American Scientist, November 1989

Theories of the Origin of the Solar System 1956–1985
Brush, Stephen
Reviews of Modern Physics, January 1990

The Solar System
Encrenaz, T. and Bibring, J-P.
Springer-Verlag, 1990

Apollo in its Historical Context
Logsdon, John *et al.*
George Washington University Space Policy Institute,
1990

Exploring Space: Voyages in the Solar System and Beyond
Burrows, William
Random House, 1990

How Did Impact Processes on Earth and Moon Become
Respectable in Geologic Thought?
Elston, Wolfgang
Earth Sciences History, 1990

Global Tectonics
Kearey, Philip and Vine Fred
Blackwell, 1990

The Restless Earth
Gregory, K.J.
Guiness Publishing, 1990

Observatories on the Moon
Burns, J.O., Duric, N. Taylor G.J. and Johnson, S.W.
Scientific American, March 1990

Atlas of the Moon
Rukl, Antonin
Hamlyn, 1990

Exploring the Planets
Hamblin, Kenneth and Christiansen, Eric
Macmillan, 1990

Moonwalker
Duke, Charlie
Oliver-Nelson Books Inc, 1990

Missions to the Planets
Moore, Patrick
Cassell, 1990

The Lunar Sourcebook: A User's Guide To The Moon
Heiken, Grant, Vaniman, David and French, Bevan
(Editors)
Cambridge University Press, 1991

*First Among Equals: The Selection of NASA Space
Science Experiments*
Naugle, John
NASA SP-4215, 1991

Evolution of the Moon: Apollo Model
Schmitt, H.H.
American Mineralogist, 1991

NASA Engineers and the Age of Apollo
Fries, Sylvia
NASA SP-4104, 1992

Space Exploration
Davies, J.K.
Chambers, 1992

Project Apollo in Retrospect
Gregory, William
In *Blueprints for Space: Science Fiction to Science Fact*
by Ordway, Frederick and Liebermann, Randy
Smithsonian Institution Press, 1992

An Argument for Human Exploration of the Moon and
Mars
Spudis, Paul
American Scientist, May 1992

*Angle of Attack: Harrison Storms and the Race to the
Moon*
Gray, Mike
W.W. Norton & Co., 1992

To a Rocky Moon: A Geologist's History of Lunar Exploration
Wilhelms, Don
University of Arizona Press, 1993

Manned Space Flight
Bockstiegel, Karl-Heinz (Editor)
Carl Heymanns Verlag, 1993

Outpost on Apollo's Moon
Burgess, Eric
Columbia University Press, 1993

Lunar Impact Basins: New Data for the Western Limb and Far Side from the Galileo Flyby
Head, J.W. *et al.*
Journal of Geophysical Research, 1993

Galileo Imaging Observations of Lunar Maria and Related Deposits
Greeley, Ron *et al.*
Journal of Geophysical Research, 1993

Galileo Multispectral Imaging of the North Pole and Eastern Limb Regions of the Moon
Belton, M.J.S. *et al.*
Science, 1994

NASA: A History of the U.S. Civil Space Program
Launius, Roger
Krieger, 1994

Moonshot: The Inside Story of America's Race to the Moon
Shepard, Alan and Slayton, Deke
Virgin, 1994

Remembering Apollo
Special Issue of Discover Magazine, July 1994

Clementine Maps the Moon
Bruning, David
Astronomy, July 1994

The Moon Voyagers
Chaikin, Andrew
Astronomy, July 1994

Lost Moon: The Perilous Voyage of Apollo 13
Lovell, Jim and Kluger, Jeffrey
Houghton Mifflin, 1994

A Man on the Moon
Chaikin, Andrew
Viking, 1994

The Scientific Legacy of Apollo
Taylor, G.J.
Scientific American, July 1994

Apollo's Gift: The Moon
Ryder, Graham
Astronomy, July 1994

DEKE!
Slayton, Deke and Cassutt, Michael
Forge, 1994

Pionering Apollo – James E. Webb of NASA
Lambright, W.H.
Johns Hopkins University Press, 1995

History of the Lunar Surface
Hartmann, William
University of Arizona Press, 1996

Volcanoes of the Solar System
Frankel, Charles
Cambridge University Press, 1996

Once and Future Moon
Spudis, Paul
Smithsonian Institute Press, 1996

Planetary Volcanism
Cattermole, Peter
Wiley-Praxis, 1996

The Clementine Bistatic Radar Experiment: Evidence for Ice on the Moon
Nozette, S., Lichtenberg, C.L., Spudis, P.D., Bonner, R., Ort, W., Malaret, E., Robinson, M. and Shoemaker, E.M.
Science, vol. 274, pp. 1495–1498, 1996

Moon Missions
Mellberg, W.F.
Plymouth Press, 1997

Radar Mapping of the Lunar Poles: A search for Ice Deposits
Stacy, N.J.S., Campbell, D.B. and Ford, P.G.
Science, vol. 276, pp. 1527–1530, 1997

Deconstructing the Moon
Jayawardhana, Ray
Astronomy, September 1998

Lunar Prospector Results
Science, 4 September 1998

NASA's New Moon
Foust, Jeffrey
Sky & Telescope, September 1998

Apollo
Bean, Alan and Chaikin, Andrew
Greenwich Workshop Press, 1998

Fluxes of Fast and Epithermal Neutrons from Lunar Prospector: Evidence for Water Ice at the Lunar Poles
Feldman, W.C., Maurice, S., Binder, A.B., Barraclough, B.L., Elphic, RC. and Lawrence, D.J.
Science, vol. 281, pp. 1496–1500, 1998

Premature Lunar Assumptions
Schmitt, H.H. and Kulcinski, G.L.
Space News, vol. 9, pp. 23–24, 1998

The Last Man on the Moon
Cernan, Eugene and Davis Don
St Martins Press, 1999

The Race
Schefter, James
Century Books

Reanalysis of Clementine Bistatic Radar of the Lunar
    South Pole
Simpson, R.A. and Tyler, G.L.
J. Geophys. Res., vol. 104, pp. 3845–3862, 1999

Topography of the Lunar Poles from Radar
    Interferometry: A Survey of Cold Trap Locations
Margot, J.-L., Campbell, D.B., Jurgens, R.F. and Slade,
    M.A.
Science, vol. 284, pp. 1658–1660, 1999

No Water Ice Detected from Lunar Prospector
    Impact
NASA Press release 99–119, 13 October 1999

Polar Hydrogen Deposits on the Moon
Feldman, W.C., Lawrence, D.J., Elphic, R.C.,
    Barraclough, B.L., Maurice, S., Genetay, I. and
    Binder, A. B.
J. Geophys. Res. Planets, vol. 105, pp. 4175–4196,
    2000

Solar-Wind Hydrogen at the Lunar Poles
Schmitt, H.H., Kulcinski, G.L., Santarius, J.F., Ding,
    J., Malecki, M.J. and Zalewski, M.J.
Proc. Space 2000 Conf., pp. 653–660, Albuquerque,
    New Mexico, 27 February to 2 March 2000

The Peaks of Eternal Light on the Lunar South Pole
Kruijff, M.
*Fourth International Conference on Exploration and
    Utilization of the Moon*, ESA/ESTEC, SP-462,
    September 2000

The Moon – A Biography
Whitehouse, David
Headline, 2001

Tracking Apollo to the Moon
Lindsay, Hamish
Springer, 2001

Evidence for Water Ice Near the Lunar Poles
Feldman, W.C., Maurice, S., Lawrence, D.J., Little,
    R.C., Lawson, S.L., Gasnault, O., Wiens, R.C.,
    Barraclough, B.L., Elphic, R.C., Prettyman, T.H.,
    Steinberg, J.T. and Binder, A.B.
J. Geophys. Res., vol. 106, pp. 23232–23252, 2001

Space Weathering Effects on Lunar Cold Traps
Crider, D.H. and Vondrak, R.R.
Proc. Lunar and Planet. Sci. Conf., 1922, 12–16 March
    2001

Apollo – The Epic Journey to the Moon
Reynolds, D.W.
Tehabi Books, 2002

Apollo – The Lost and Forgotten Missions
Shayler, D.J.
Springer-Praxis, 2002

Modelling the Stability of Volatile Deposits in Lunar
    Cold Traps
Crider, D.H. and Vondrak, R.R.
*Workshop on Moon beyond 2002: Next steps in lunar
    science and exploration*, Taos, New Mexico, 3006,
    12–14 September 2002

Apollo EECOM – Journey of a Lifetime
Leibergot, Sy and Harland, D.M.
Apogee Books, 2003

The Moonlandings – An Eyewitness Account
Turnill, Reginald
Cambridge University Press, 2003

Radar Imaging of the Lunar Poles: Long-Wavelength
    Measurements Reveal a Paucity of Ice in the Moon's
    Polar Craters
Campbell, B.A., Campbell, D.B., Chandler, J.F., Hine,
    A.A., Nolan, M.C. and Perillat, P.J.
Nature, vol. 246, pp. 137–138, 2003

Space Weathering of Ice Layers in Lunar Cold
    Traps
Crider, D.H. and Vondrak, R.R.
Adv. Space Res. vol. 31, pp. 2293–2298, 2003

Two Sides of the Moon
Scott, David and Leonov, Alexei
Simon & Schuster, 2004

*Lunar Exploration – Human Pioneers and Robotic
    Surveyors*
Ulivi, Paolo and Harland, D.M.
Springer-Praxis, 2004

Apollo
Murray, Charles and Cox, C.B.
South Mountain Books, 2004

The Robotic Lunar Exploration Program: An
    Introduction to the Goals, Approach and
    Architecture
Watzin, J.G., Burt, J. and Tooley, C.
American Institute of Aeronautics and Astronautics,
    *First Space Exploration Conference: Continuing the
    Voyage of Discovery*, Orlando, Florida, 30 January
    to 1 February 2005.

Constant Illumination at the Lunar North Pole
Bussey, D.B.J., Fristad, K.E., Schenk, P.M., Robinson,
    M.S. and Spudis, P.D.
Nature, vol. 434, p. 842, 2005

Apollo – The Definitive Sourcebook
Orloff, R.W. and Harland, D.M.
Springer-Praxis, 2006

Surveyor Lunar Exploration Program
Apogee Books, 2006

Return to the Moon
Schmitt, H.H.
Copernicus Books, 2006

No Evidence for Thick Deposits of Ice at the Lunar
  South Pole
Campbell, D.B., Campbell, B.A., Carter, L.M., Margot,
  J.-L. and Stacy, N.J.S.
Nature, vol. 443, pp. 835–837, 2006

The First Men on the Moon – The Story of Apollo 11
Harland, D.M.
Springer-Praxis, 2007

Soviet and Russian Lunar Exploration
Harvey, Brian
Springer-Praxis, 2007

Praxis Manned Spaceflight Log: 1961–2006
Furniss, Tim, Shayler, D.J. and Shayler, M.D.
Springer-Praxis, 2007

Lunar and Planetary Rovers
Young, Anthony
Springer-Praxis, 2007

On the Moon – The Apollo Journals
Heiken, Grant and Jones, Eric
Springer-Praxis, 2007

How Apollo Flew to the Moon
Woods, W.D.
Springer-Praxis, 2007

For a more comprehensive bibliography of the Apollo programme, in all its aspects, consult:

Launius, R.D. and Huntley, J.D. (compilers) (1994) 'An annotated bibliography of the Apollo program', *Monographs in Aerospace History*, No. 2. NASA History Office.

This monograph is available on the Internet at:

http://www.hq.nasa.gov/office/pao/History/Apollobib/cover.html

Index

Marshall Space Flight Center (MSFC), 107, 132
Masursky, Hal, 42
Mattingly, Ken, 69, 219, 224, 267; *see Apollo 16*
McCandless, Bruce, 24, 27, 29, 33
McCauley, Jack, 193
McDivitt, Jim, 106
McGill, George, 343
Meteor Crater (Earth), 69
Milton, Dan, 193
Mitchell, Ed, 100–102, 122, 202, 214, 234, 245; *see
        Apollo 14*
Moon,
    accretion, 18
    Alphonsus Crater, 5–7, 269
    anorthosite, 37, 134, 136, 143, 150, 152, 186, 202,
        214, 219, 254, 264, 295, 327, 330, 347–348
    anorthositic gabbro, 18, 37, 38, 191, 197, 299, 320,
        322
    Apennine Bench Formation, 98, 113, 190–191, 360
    Apennine Range, 97–98, 100, 105, 113, 130, 132, 183,
        186, 190–191, 198, 214, 262, 272
    Apollonius Crater, 347
    Archimedes Crater, 98, 190
    armalcolite, 36
    Aristarchus Crater, 5
    Aristarchus Plateau, 360
    ashflow tuff, 193, 208
    Aristillus Crater, 113–114, 132, 147, 186, 191
    Autolycus Crater, 1, 113–114, 132, 147, 186, 191
    basalt, 2, 14, 33, 36, 38, 54, 59, 64, 93, 94, 121, 127,
        130, 138, 143, 147, 152, 156, 173, 179, 180, 189,
        190–191, 245, 254, 263, 264, 276, 280, 318, 336,
        341, 347–349
    base surge, 186, 191, 295
    basins, 36, 38, 67, 186, 193, 219, 251, 257, 262, 265,
        267, 269, 295
    breccias, 85, 93–94, 121–122, 127, 136, 138, 143, 147,
        150, 152–153, 186, 197, 199, 202, 207–208,
        213–214, 219, 223, 226, 229, 231, 234–235, 240,
        245–246, 248, 251, 254, 257, 262, 264, 266–267,
        274, 295, 297, 299, 305, 318, 320, 322, 326–327,
        330, 332
    Carpathian Mountains, 10
    Cayley Crater, 193
    Cayley Formation, 193–195, 197, 199, 202, 204, 207,
        213, 219, 223, 226, 234, 240, 246, 251, 257, 262,
        264, 266, 267, 272, 354, 360
    Censorinus Crater, 67
    Central Highlands, 5, 193, 195, 196, 263
    'circular maria', 18, 360
    'cold' Moon, 2, 18, 33, 64, 192, 193, 262
    Copernican Era, 117, 152, 193, 267
    Copernicus Crater, 3, 5, 10, 43, 54, 64, 269, 343
    core, 79, 191, 354
    crust, 18, 38, 65, 66, 69, 95, 134, 150, 186, 191,
        263–264, 269, 301, 354
    dark-halo craters, 5, 269, 270, 272–273, 292, 299, 305,
        307, 310–311
    dark mantle, 71, 97, 168, 270, 272–273, 276, 280, 290,

        292, 294, 310–311, 332, 343, 345
    Davy Rille, 43, 97, 269
    Descartes Crater, 193
    Descartes Formation, 193–194, 195, 214, 219, 223,
        226, 234, 245, 246, 257, 266, 267
    dunite, 301, 322
    Eratosthenes Crater, 10
    europium, 38
    far side, 1, 8, 10, 12, 95, 269, 361
    feldspathic rock, 18, 37, 64, 94, 138, 143, 150, 186,
        208, 213–214, 264, 348
    fire fountains, 147, 186, 272, 310–311, 341
    Flamsteed Ring, 12–13, 42, 343
    fluidised fragmental breccia, 257, 272, 360
    formation, xvii, 18, 36
    Fra Mauro Crater, 42, 43, 67, 68
    Fra Mauro Formation, 67–69, 71, 73, 75, 77, 82, 85,
        93, 94–95, 97, 114, 186, 193, 202, 245, 251, 262,
        263, 266, 267, 270, 354, 360
    fragmental breccias, 93, 262, 326
    fumarolic alteration, 307, 310
    gabbro, 18, 150, 156, 190, 264, 276, 278, 280, 285,
        290, 292, 307, 311, 318, 320, 322, 326, 336
    gardening, 4–5, 36, 127, 152, 186, 276, 280, 290,
        310–311, 318, 327, 336, 341, 355, 361
    gas emissions, 95, 219, 356, 360
    Gassendi Crater, 269
    glass, 37–38, 54, 136, 147, 150, 169, 186, 202, 208,
        219, 226, 231, 235, 245, 251, 254, 274, 292, 301,
        310, 320, 327, 330, 332, 336, 341
    'glows', 5
    gravitational field, 12, 18
    Hadley Rille, 97–99 101–102, 111–129, 169, 173–179,
        183, 188–191
    'heat engine', 95, 191–192, 193, 269, 272, 355
    Hercules Promontory, 349
    Hipparcus Crater, 42, 67
    'hot' Moon, 2, 18, 69, 192, 354
    Humorum Basin, 95, 269
    ilmenite, 64, 285, 292, 318
    Imbrian Era, 186
    Imbrium Basin, 14, 67, 69, 73, 93, 97, 98, 134, 147,
        183, 186, 189–191, 193, 262, 265, 269–270, 272,
        349, 360
    Imbrium impact, 93
    impact melt, 14, 33, 64, 190
    impact-melt breccias, 94, 262, 264, 320, 326
    Julius Caeser Crater, 3, 16
    LKFM melt, 266
    Kant Plateau, 194–196
    Kepler Crater, 5, 11
    KREEP, 64–65, 94–95, 127, 147, 190, 266, 359, 360
    Lansberg Crater, 14
    Le Monnier Crater, 349–350
    Lichenberg Crater, 344
    'light plains', 38, 193, 194
    Littrow Crater, 71, 270–271, 345, 350
    mafic silicates, 18, 37, 121, 150, 156, 190, 193, 264,
        320, 355
    magnetic field, 79

Printing: Mercedes-Druck, Berlin
Binding: Stein+Lehmann, Berlin